FOOD PRESERVATION PROCESS DESIGN

Food Science and Technology International Series

A complete list of books in this series appears at the end of this volume.

Food Preservation Process Design

Dennis R. Heldman
Heldman Associates
Mason, Ohio

ELSEVIER

AMSTERDAM • BOSTON • HEIDELBERG • LONDON
NEW YORK • OXFORD • PARIS • SAN DIEGO
SAN FRANCISCO • SINGAPORE • SYDNEY • TOKYO
Academic Press is an imprint of Elsevier

Academic Press is an imprint of Elsevier
30 Corporate Drive, Suite 400, Burlington, MA 01803, USA
525 B Street, Suite 1800, San Diego, California 92101-4495, USA
84 Theobald's Road, London WC1X 8RR, UK

Library of Congress Cataloging-in-Publication Data
Heldman, Dennis R.
Food preservation process design / Dennis R. Heldman.
p. cm.
Includes bibliographical references and index.
ISBN 978-0-12-372486-1 (hardback)
1. Food—Preservation. I. Title.
TP371.2.H46 2011
664′.028—dc22 2010046577

British Library Cataloguing-in-Publication Data
A catalogue record for this book is available from the British Library.

For information on all Academic Press publications
visit our website at www.elsevierdirect.com

Printed in the United States of America

11 12 13 9 8 7 6 5 4 3 2 1

Contents

Preface vii

1 **Introduction** 1
History of preservation processes 6
The quantitative approach 9
Bibliography 17

2 **Kinetic Models for Food Systems** 19
Rate equations and rate constants 20
First-order model 25
Multiple-order models 27
Agent intensity models 30
Thermal process models 37
Uniform parameters 43
List of symbols 46
Bibliography 47

3 **Kinetics of Inactivation of Microbial Populations** 49
Characteristics of microbial survivor curves 50
Kinetic parameters for microbial populations 54
Applications of kinetic parameters 59
Definition of microbial inactivation 76
Kinetic parameters for alternative preservation technologies 77
List of symbols 77
Bibliography 78

4 **Kinetics of Food Quality Attribute Retention** 87
Characteristics of quality retention kinetics 88
Kinetic parameters for product quality retention 91
Applications of kinetic parameters for quality attributes . . 95
Impacts of preservation processes on quality attributes . . . 105
List of symbols 106
Bibliography 107

5	**Physical Transport Models**	**111**
	Physical properties	**112**
	Heating and cooling in containers	**125**
	Ohmic heating	**138**
	Microwave heating	**139**
	Ultra-high pressure applications	**141**
	List of symbols	**143**
	Bibliography	**144**
6	**Process Design Models**	**147**
	The process design parameter	**148**
	General approaches to preservation process design	**150**
	Process design targets	**166**
	Integrated impacts of preservation processes	**171**
	Design of a microwave process	**182**
	Design of an ohmic heating process	**193**
	Design of ultra-high pressure processes	**197**
	Design of pulsed-electric-field processes	**202**
	Design of combined processes	**209**
	List of symbols	**212**
	Bibliography	**214**
7	**Process Validation and Evaluation**	**217**
	Process validation for microbial inactivation	**219**
	Alternative approaches to validation	**229**
	Process validation for alternative process technologies	**237**
	List of symbols	**243**
	Bibliography	**243**
8	**Optimization of Preservation Processes**	**245**
	The HTST concept	**246**
	Applications to nonliquid foods	**249**
	List of symbols	**263**
	Bibliography	**265**
9	**Designing Processes in the Future**	**267**
	Assembly of kinetic parameters	**268**
	Transport models	**271**
	Process models	**272**
	Opportunities for evolving process technologies	**273**
	Bibliography	**276**
	Appendix	**279**
	Index	**341**

Preface

The preservation processes for foods have evolved over several centuries, but recent attention to nonthermal technologies suggests the initiation of a new direction in food preservation. This book documents the quantitative approaches to preservation process design and prepares food science professionals for the food preservation challenges of the future—such as evaluating emerging preservation technologies and selecting appropriate food preservation technologies.

The text focuses on the three primary elements of food preservation process design:

1. Kinetic models for changes in food components, including microbial populations—the background, statistics, and applications of kinetic models used to describe changes in components of food during a preservation process.
2. Transport models for food systems—the primary transport models needed to describe the changes in physical characteristics within a food structure during a preservation process.
3. Process design models—the integration of kinetic and transport models, as needed predict the process time required to accomplish the desired objectives of the preservation process.

The concepts presented build on the strong, successful history of thermal processing of foods, using examples from these preservation processes. Significant attention has been given to the fate of food quality attributes during the preservation process and the concepts for optimizing the process parameters to maximize the retention of food quality.

Food Preservation Process Design is an ideal text for a capstone or senior design course at the fourth year of the undergraduate program in food science. The information in the book also provides the basis for a graduate-level course on preservation processes. The examples, tabular data, and the computational approaches are designed to stimulate individual or team efforts in process design. In addition, the content should be an excellent reference for food industry professionals involved in preservation process design.

The first chapter provides historical background on food preservation processes, with an emphasis on quantitative aspects. Attention has been given to positive outcomes from successful food preservation technologies as a basis for evaluating alternative process technologies. The introduction to the book emphasizes the challenges associated with experimental verification of preservation processes, and the opportunities for optimizing the processes to maximize retention of product quality attributes.

Chapter 2 presents the background on kinetic models currently used for food preservation process design. The evolution from reaction rate kinetics is reviewed, and examples are used to illustrate the evaluation of the appropriate kinetic parameters for first- and multiple-order models. The relationships of the typical kinetic parameters to the traditional parameters from thermal processing are presented, along with a justification for a more uniform set of parameters for the future.

Typical kinetic parameters for inactivation of microbial populations are presented in Chapter 3. Some of the best available kinetic parameters for both vegetative pathogens and pathogenic spores are presented in tabular form, along with background on the conditions of measurement. These parameters include examples for alternative process technologies. The variability associated with kinetic parameters, as well as the influence of product composition on the magnitude of the parameters, has been considered with examples illustrating the use of the kinetic parameters in process design.

Chapter 4 covers the kinetic parameters for typical food product quality attributes. Most of the available parameters are for nutrient and color changes as a function of temperature. Examples illustrate the use of kinetic models to predict the retention of quality attributes during a preservation process and provide the basis for optimizing the retention of quality.

The fundamental aspects of transport models are presented in Chapter 5, as background for food preservation process design.

The prediction models for physical properties based on product composition have been provided along with typical transport models for thermal energy exchange. Emphasis has been placed on models for prediction of temperature within the food product structure during typical preservation processes and on the unique relationships occurring during the application of alternative process technologies.

In Chapter 6, the emphasis is on process design and the integration of appropriate kinetic and transport models. The process design parameter for food preservation is established, with specific attention to microbiological safety, as well as product spoilage. The impact of product structure on uniform application of the process, as well as the influence on process design, is illustrated. The subsequent impact of the process on product quality attributes is illustrated through the use of examples.

The validation of the preservation process is the subject of Chapter 7. The challenges associated with process validation when attempting to confirm probabilities of survivors is illustrated through examples. The appropriate use of surrogate microorganisms, chemical tracers, and other approaches to measuring the impact of the process being evaluated is discussed, with some of the unique concerns and requirements for alternative technologies considered.

The process design approach presented in this book provides the ideal opportunity for optimization of preservation processes, as demonstrated in Chapter 8. The unique relationship of the magnitudes of kinetic parameters for microbial populations as compared to product quality attributes provides the basis for maximizing quality retention, while achieving the microbial safety and product shelf-life. The extension of these concepts to alternative preservation technologies is also explored.

The final chapter of the book is a brief look at the future of food preservation process design, with an emphasis on the need for more and improved kinetic parameters for both microbial populations and quality attributes. Some of the challenges associated with alternative preservation technologies are also discussed.

In closing, I would like to acknowledge the feedback and encouragement from many colleagues as the content of this book evolved. These colleagues include many students enrolled in courses where several of the concepts covered in this volume were presented and tested. The comments from all have been valuable in finalizing the concepts shared throughout these pages.

Dennis R. Heldman

Introduction

1

1.1 History of preservation processes 6
1.2 The quantitative approach 9
Bibliography. 17

People have been preserving foods for centuries! Of course, the processes used for preservation have evolved at different points in history, but the evaluation and design of processes have become quantitative as more scientific research on the processes has been completed. The overall purpose of this book is to illustrate the applications of the most recent research for quantitative evaluation and description of preservation processes. These illustrations should strengthen the quantitative basis of current preservation process design and provide the background to identify information needed to enhance quantitative design of processes in the future.

The primary focus of food preservation has been on controlling microbial populations, with a specific emphasis on pathogenic microorganisms. According to Potter and Hotchkiss (1995), the primary preservation technologies for foods include the following:

Heat: The use of thermal energy to increase the temperature of a food is the most recognized and widely used agent for food preservation. Elevated temperatures cause a decline in microbial populations and extend the shelf life of the product by eliminating microorganisms causing food spoilage and food-borne illness in humans. Many shelf-stable foods are available to consumers as a result of thermal processing. These processes have been

Food Preservation Process Design
ISBN: 978-0-12-372486-1

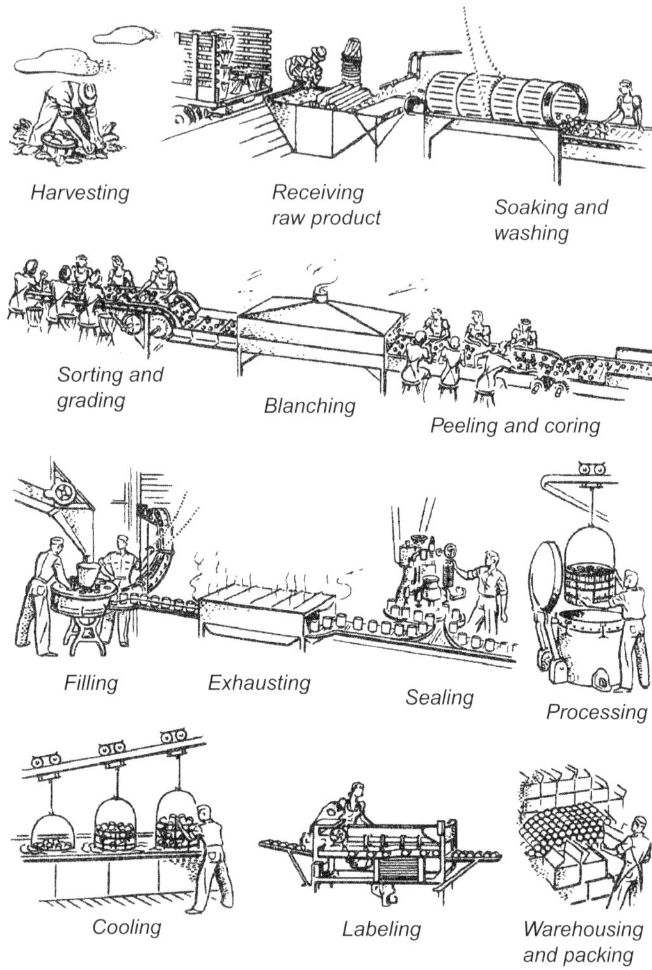

Figure 1.1 Typical steps in the heat preservation process (from Jackson & Shinn, 1979).

described in a quantitative manner for many years and provide a fundamental basis or structure for describing other preservation processes (Figure 1.1).

Refrigeration: The use of reduced temperatures to extend food product shelf life has a long history. Ice has been used for centuries to reduce the temperature of foods and prevent spoilage. In general, the reduction of a food product temperature does not reduce the microbial population but prevents microbial growth

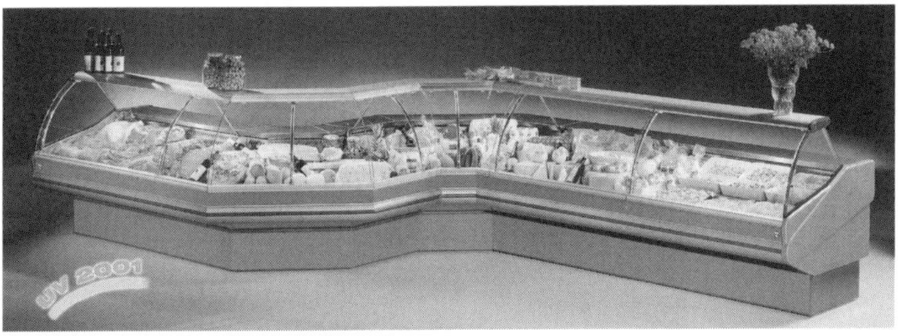

Figure 1.2 A refrigerated storage cabinet for food products (Nuline Refrigeration www.nulinerefrigeration.com.au/5.html).

Figure 1.3 An array of packaged dry foods (www.gdargaud.net/Antarctica/WinterDCe.html).

and the associated deterioration of other food quality attributes (Figure 1.2).

Dehydration: Drying foods may have been one of the earliest forms of preservation. Exposure of many foods to thermal energy from the sun causes water to evaporate from the product. Sufficient reductions of moisture content inhibit the growth of microorganisms, and the product spoilage associated with microbial growth (Figure 1.3).

Acidity: Adjustments in the pH of a food is a popular preservation step for many products. This type of preservation occurs in different ways in different foods, ranging from naturally low pH (high acid) foods to fermentation processes where growth of selected microorganisms causes an adjustment in the pH of the product, and the inhibition of growth of pathogens and spoilage microorganisms. Often, the pH of the food is used in combination with other processes, such as thermal, to accomplish preservation.

Water activity: Many food components (natural or added) influence the growth of microbial populations in products. Elevated concentrations of sugars and salts cause microbial cell dehydration, which is the diffusion of water from the cell, leading to inhibition of growth or complete inactivation. These same impacts occur in dry and intermediate moisture foods. The magnitude of product water activity has become an indictor used in control of food deterioration, including spoilage due to microbial growth.

Smoke: A traditional method of preservation for meat and meat products involves the use of smoke to control microbial growth. The impact of the method is due to the influence of smoke components on microorganisms, a mild temperature increase for an extended period of time along with a reduced moisture content of the food, at least near the product surface.

Atmospheric composition: The shelf life of many food products has been extended by reducing or eliminating the concentration of oxygen in the atmosphere or gas in direct contact with the product. This approach has been effective for products with deterioration caused by aerobic spoilage microorganisms. Several packaging systems have been developed using these concepts. However, there are obvious concerns and limitations to this approach when anaerobic pathogens or spoilage microorganisms are present in the product.

Additives: Many chemicals inhibit the growth of microbial populations or inactivate microorganisms, and a few of these additives have been approved for use in foods at low levels, as preservatives. Most of the additives are specific for certain spoilage microorganisms and for specific product applications.

Radiation: Various wavelengths within the electromagnetic spectrum are effective for inactivation of microorganisms, and many have been evaluated as preservation processes for food products. Only a limited number of products preserved by radiation have been made available to consumers, due to the negative perception of the technology.

Alternatives: During the past 15 years, several alternative technologies have evolved for evaluation as preservation processes for food products. These technologies include ultra-high pressure, microwave or ohmic heating, and pulsed electric fields.

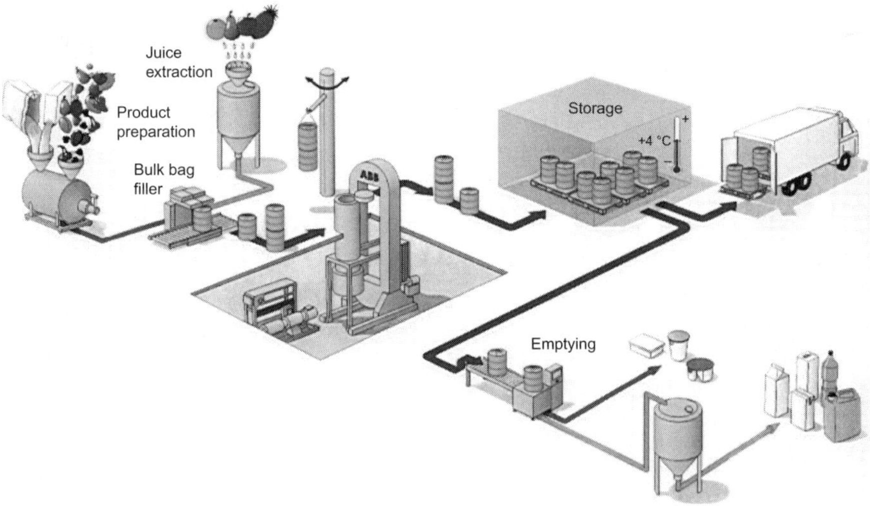

Figure 1.4 Use of high pressure for food preservation (from Singh & Yousef, 2005).

Sufficient information on the influence of these processes on microbial populations in foods must be assembled to allow quantitative evaluation of the processes (Figure 1.4).

All of the preceding approaches to preservation of foods have contributed to the safety and stability of foods available to consumers by controlling or eliminating microbial populations in foods. Many of the technologies are used in combination with another technology, and do not accomplish the desired result independently. Only heat (or the thermal process) and radiation have been demonstrated to cause a reduction in a target microbial population and have been quantified in a consistently predictable manner. Due to negative consumer perceptions about radiation, it is unlikely that consumers will accept food products from radiation preservation in the near future. Due to this situation, radiation has not been included as a preservation technology for analysis in this book. Heat has been used in combination with many of the other preservation technologies mentioned. In addition, other technologies used in combination with heat influence the effectiveness of thermal processes. Recent developments with ultra-high pressure and pulsed electric fields suggest that these technologies are similar to thermal and may be used in combination with other technologies.

In summary, this book focuses on the quantitative evaluation of preservation processes for food products. The process design concepts build on the long and successful history of thermal process design but extend the analysis to combination processes and to nonthermal technologies, such as ultra-high pressure and pulsed electric fields. In addition, the analysis covers concepts needed to estimate the impact of a process on food components, including nutrients and other product-quality attributes. Finally, the book explores opportunities to optimize preservation processes to achieve process efficiency and product quality retention.

1.1 History of preservation processes

Although the history of food preservation dates back many centuries to the use of thermal radiation from the sun to create dry foods, the work of Nicholas Appert is recognized as the first successful controlled process. Appert (1810) developed a system for sealing food in glass bottles and used thermal energy to increase the temperature of the product to levels exceeding 100°C. His work was stimulated by a prize offered by the French Directory in 1795, in response to the need to provide sufficient and safe foods to Napoleon's troops. By 1809, Appert had succeeded in preserving certain foods by immersing glass containers, containing the food, in boiling water. He was awarded the prize from the French government. Appert's accomplishments are recognized as the beginning of thermal processing (commercial sterilization) to create shelf-stable foods (Figure 1.5).

Nearly 50 years passed before a fellow Frenchman, Louis Pasteur, discovered that the origin of food spoilage was the growth of microorganisms. Today, Pasteur is recognized for another thermal preservation process: Pasteurization.

Many developments and occurrences have contributed to the evolution of preservation process design. Following the breakthrough discoveries of Appert and Pasteur, developments were very slow for nearly 100 years. The pioneers of thermal processing research and application in the United States were Prescott and Underwood (1897). These researchers completed the important research on microbiology of canned foods. These developments were accompanied by new methods for manufacturing metal

Figure 1.5 Portrait of Nicholas Appert (Appert, Nicholas. 1810. *L'art de conserver*. Chez Patri et Cie).

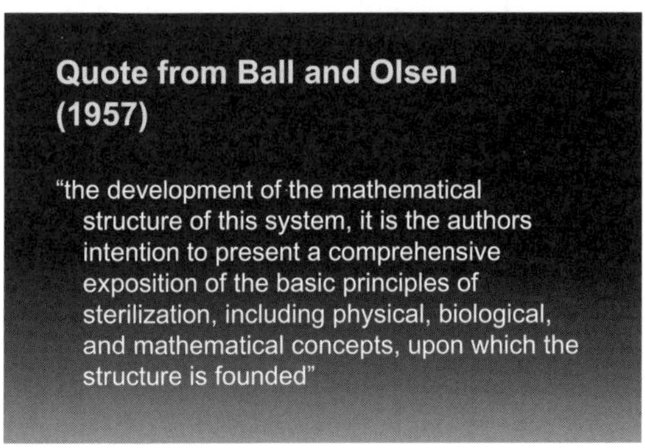

Figure 1.6 Quote from Ball and Olson (1957).

cans, specifically for food applications. In the 1920s, research by Bigelow (1922) and Ball (1923) began to provide the basis for quantification of the process and introduced opportunities for predictive process design. The book *Sterilization in Food Technology* by Ball and Olson (1957) provided exhaustive documentation of these developments. Among the contributions during this time was the application of the thermocouple to the measurement of temperatures during experimental processing of foods. Shortly thereafter, Stumbo (1965) published *Thermobacteriology in Food Processing* and provided an additional view on the need for predicting process times for thermal processing of shelf-stable foods. The consistent trend toward quantification of the preservation process during this time period is best emphasized by the quote from the preface of Ball and Olson (1957) in Figure 1.6.

In the early 1900s, a series of food poisoning outbreaks and deaths due to *Cl. Botulinum* toxin in canned foods prompted the research and contributions of Bigelow (1920) and Ball (1927). These outbreaks led to the establishment of the National Canners Association (NCA) (later National Food Processors Association [NFPA], now Grocery Manufacturers Association [GMA]). The NCA established laboratories in 1913 to assist the food canning industry in responding to food safety challenges. During the following century, significant discoveries were published on quantifying the process and the impact of thermal processing on nutrients in foods. These discoveries were followed by a new focus on research to improve the precision of process design for thermal processes. These same food safety concerns also resulted in the Food, Drug and Cosmetic Act of 1938, and the specific regulations for thermally processed low-acid foods packaged in hermetically sealed containers, as published in the Code of Federal Regulations (CFR 21.113). These historical developments emphasize the importance of these processes, and the motivation for continuing to refine process design for all preservation processes for foods (Figure 1.7).

Starting in the middle of the twentieth century, a series of refinements to the prediction methods for preservation processes were initiated. The work of C. Olin Ball and C. R. Stumbo stimulated most of these refinements. A few of the key contributions to these

Figure 1.7 The Code of Federal Regulations.

improvements and extensions include Pflug and Esselen (1963); Pflug, Blaisdell, and Kopelman (1965); Teixeira, Dixon, Zahradnik, and Zinsmeinster (1969); and Manson and Cullen (1974). These published works have created a structure to be used in the evaluation and optimization of processes for all types of preservation processes. Additional references to publications from these and other researchers will be introduced throughout this book.

1.2 The quantitative approach

The approach to process design presented in the book has three significant components. When viewed in general terms, the components of the approach include the following:

Kinetics of reactions: During preservation, the process impacts all components of the food. The impacts of the process are evident in many ways, depending on the food component being considered. Much of the published literature describes the impact of elevated temperatures on the decline in microbial population as a function of time during the process. More recently, the influence of thermal processes on the concentration or intensity of other food components (quality attributes) have been measured and published. For most situations, first-order models and the appropriate rate constants have been used to describe the impact of the process on the food component. These same models and parameters should be used for all preservation processes so that the effectiveness of different technologies can be compared, and combinations of technologies can be evaluated. The published literature also provides quantitative data on the influence of agent intensity on the rate constants. These relationships have been described by models normally associated with the kinetics of chemical reactions, and the constants associated with these kinetic models (Figure 1.8).

Physical transport models: The application of most preservation processes can be detected by measurement of one or more physical parameters within the product. A physical transport model can be used to describe these parameters. These types of models become the methods for predicting the intensity of the physical parameter at any location within the product structure.

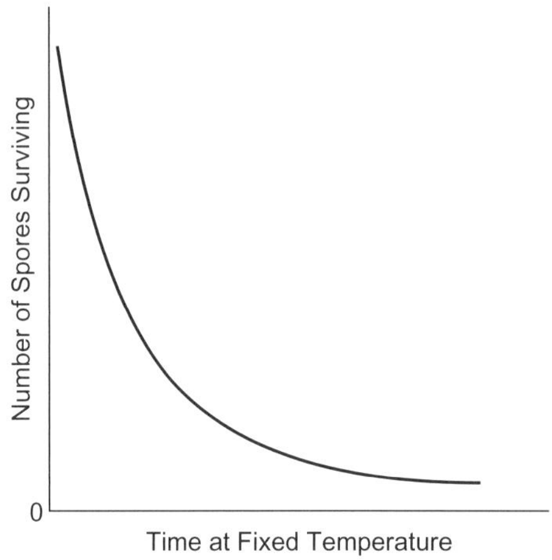

Figure 1.8 A survivor curve for microbial spores.

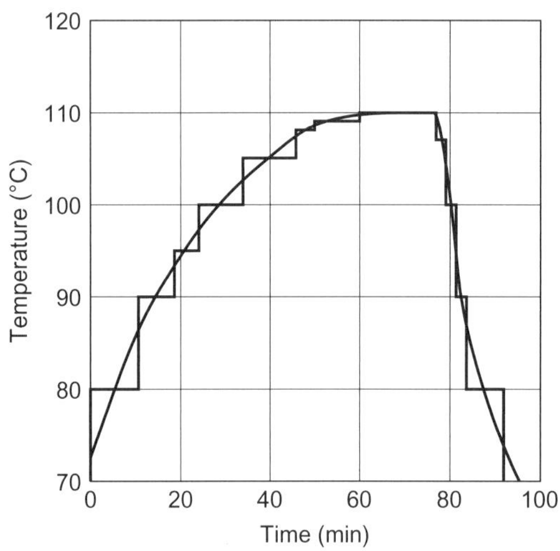

Figure 1.9 Typical heating curve for food in a can (from Earle, 1983).

Obviously, the model and the complexity of the description will vary with physical characteristics of the product, ranging from low-viscosity liquids to homogeneous solids. The published literature provides insight on the transport of thermal energy within foods and the prediction of temperature distribution histories. Similar models are available for predicting the intensity of other physical parameters during food preservation (Figure 1.9).

Preservation process design: The design models for preservation processes involve the integration of the appropriate kinetic

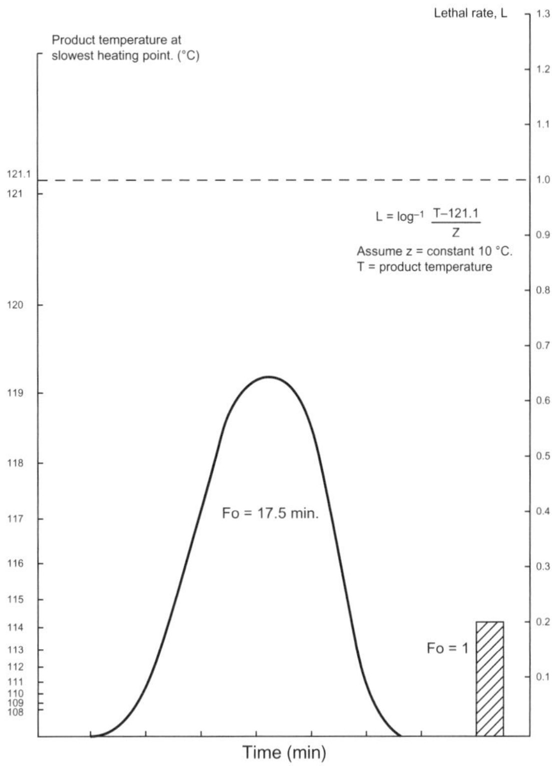

Figure 1.10 A typical lethal rate curve (FAO Corporate Document Repository www.fao.org/docrep/003/t0007e/t0007e02.htm).

model with the appropriate physical transport model. Usually, the output from the successful completion of this integration step is the concentration or intensity of a specific food product component. The traditional use of this integration step has been in thermal process design, where the output is the population of microorganisms surviving the preservation process. The primary focus of the work of C. Olin Ball (Ball & Olson, 1957) and C. R. Stumbo (1965) was to provide the tools needed to accomplish this integration step when applied to the reduction of the microbial populations (usually pathogens) during thermal processing in the manufacture of shelf-stable food products. A limited number of attempts have been made to demonstrate the power of this process design step for predicting changes in other components of the food during the process or for applying other preservation technologies (Figure 1.10).

1.2.1 Experimental validation of processes

A key factor in the design of any process is validation. For processes associated with the preservation of foods, the validation must be accomplished through a series of experimental measurements. Because the systems and the products pose many challenges, the experimental procedures require careful consideration to ensure that the process predicted by the design is an acceptable description of the process. Detailed descriptions of experimental procedures used in validation is not the focus of this book, but we will outline the concepts and present key references for detailed information. The following issues need to be addressed during validation:

Food components or attributes: The kinetic parameters used in process design models are based on experimental measurements of defined food components or attributes, and they are usually measured under controlled laboratory conditions. In many cases, these measurements are completed with the component or attribute carried in a substrate different from the food being considered in the process design. For example, kinetic constants for the microbial survivor curve are often measured using pure cultures of a specific microorganism in a buffer solution. The impacts of the product structure and other product components on the parameters must be validated in an experiment involving the actual product.

Intensity of agent: The outputs from any process simulation depend on the intensity of the agent being used for preservation. This intensity must be monitored throughout the actual preservation process. These measurements require the use of appropriate measurement techniques at the appropriate locations within or near the product structure or container to ensure that the process can be validated. For example, the continuous monitoring of temperature at appropriate locations during the process is critical to ensuring a validation of the thermal process design.

Measurement precision: The most critical measurement for the process validation is the magnitude of the primary product component or attribute during the process and specifically at the completion of the process. The design of most processes is based on a specific target magnitude of a given component or attribute. The purpose of the process may be to reduce the magnitude of the component or attribute to insignificant levels. One of the most challenging examples is the detection of surviving pathogens or spoilage microorganisms after completing a preservation process. Process models are capable of predicting probabilities of survivors, but the measurements must provide appropriate validation of these probabilities. Often, trace amounts of key food components may be impacted by the preservation process and may be equally challenging to measure and monitor (Figure 1.11).

Pathogenic microorganisms: Preservation processes for foods are established to eliminate the threat of food-borne disease. Validating a preservation process for pathogens presents many challenges. When a process is being validated under commercial manufacturing conditions, pathogens cannot be introduced into the environment. These situations are usually accommodated by using surrogate microorganisms, that is, nonpathogenic organisms that respond to the preservation process in the same manner

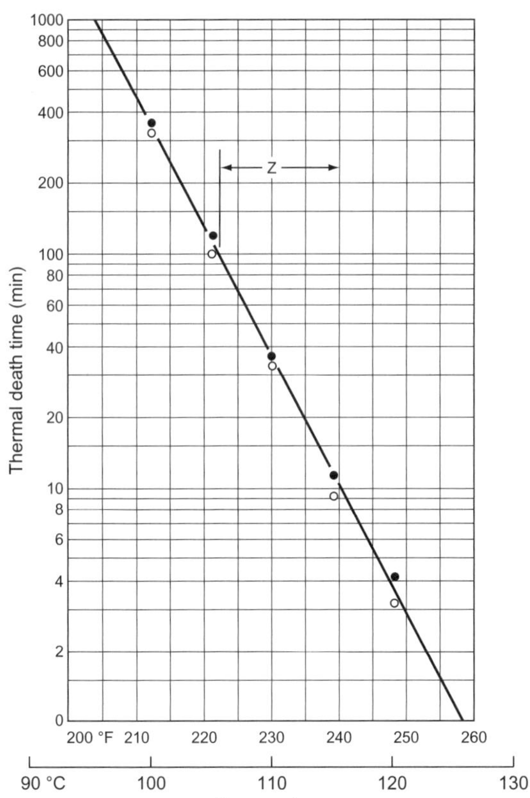

Figure 1.11 A thermal death time curve for microbial spores (from Earle, 1983).

as the actual pathogen. Even the handling of pathogenic microorganisms under controlled laboratory environments requires extreme care and understanding.

Process scale-up: The scale up of any process is one of the more challenging steps in validation. This step is even more critical when considering preservation processes. The basic information used in process design is obtained in laboratory-scale experiments. In most cases, the results of these experiments are for specific microbial populations or quality attributes and are expressed in terms of kinetic parameters. Because the experiments have been designed for specific microorganisms or quality attributes, the influence of the process on other product components or attributes may not be evident. These impacts are usually evaluated during pilot-scale experiments, as an important step to commercial-scale operations. The magnitude of validation experiments at pilot-scale, or commercial-scale, presents a significant challenge. For a statistically valid evaluation, the numbers of product containers to be included in the validation may become prohibitive. This places more importance on all steps associated with preservation process design.

1.2.2 Successful food preservation processes

One of the goals of food preservation processes is to ensure that food-borne illness among consumers is nonexistent or minimized. Over the past 70 to 80 years, the outbreaks of food-borne illness from shelf-stable foods have been infrequent, and the food industry in the United States has maintained an exceptional record. Much of this success has been based on the development of preservation processes using thermal energy to increase the temperature of food products for appropriate time periods to eliminate the hazards associated with pathogenic microorganisms. The shelf-stable products available to consumers are a natural extension of food preparation occurring in the kitchen of the consumer. Consumers have accepted the quality attributes of these products, and the success of thermal preservation is due to the similarity of the preservation process and the typical food preparation by the consumer.

The safety record for other food preservation processes used to manufacture shelf-stable foods has been impressive as well. The use of dehydration has created a variety of new and different dry foods. In a similar manner, adjustments in water activity have provided consumers with new and safe foods. Other approaches have been used for independent preservation of foods. As indicated earlier, the use of radiation preservation for shelf-stable foods has been demonstrated as technically successful but has had limited impact in the marketplace due to lack of acceptance by consumers.

1.2.3 Emerging preservation processes

Over the past 50 years, many alternatives to thermal processes for food preservation have been proposed and evaluated. Much of the motivation for the continuing investigation of alternative preservation technologies has been the reduction in the negative impacts of the thermal process on quality attributes of food products. During the 1950s, significant efforts were devoted to developing irradiation as an alternative to thermal processing. Unfortunately, applications of this alternative technology, where improvements in quality were demonstrated, were never realized due to the lack of consumer acceptance. A variety of modest developments based on applications of the irradiation technology continue to be pursued.

During the past 20 years, there has been renewed interest in evaluation of an array of alternative technologies for food preservation. The focus of these investigations has been on several technologies identified by the Food and Drug Administration (FDA), and evaluated by an IFT/FDA Task Force (2001). Some of these alternatives depend on the impact of temperature and time to cause reductions in microbial populations. The mechanisms for inactivation of microorganisms by other technologies are not thermal, although several cause a product temperature rise during application of a process. According to published information, sufficient information has been assembled for the successful process design of several of these alternative technologies. These technologies will continue to receive consideration as preservation processes for food products. A review by Sun (2005) suggests that high pressure, pulsed electric fields, radiofrequency, high-intensity pulsed light, ultrasound, and irradiation are the most promising nonthermal processes. This review emphasizes the importance of using combinations of two or more technologies and the concept of hurdle technologies. In addition, Sun (2005) concluded that ultra-high pressure and pulsed electric fields are technologies with significant promise for use in food preservation.

In this book, we explore the design of preservation processes for several of the emerging or alternative technologies. When sufficient input parameters for process design models are available, they are illustrated and documented. The goal is to demonstrate the process design for several of the emerging technologies, including traditional thermal processing technologies, and to present valid comparisons. These comparisons include evaluations of process efficiency and effectiveness, as well as the impacts on product quality attributes. The status of each of the potential preservation technologies was reviewed in the 2001 IFT/FDA Task Force Report and will be evaluated in the various chapters of this book.

1.2.4 Food product quality considerations

As previously suggested, a continuing motivation for investigation of alternative preservation processes has been to reduce the impact of the process on quality attributes of the food product. Over the past 25 years, there has been significant growth in the research literature for the parameters needed to evaluate processes. Most of the research has focused on thermal processes and on the kinetic parameters required to evaluate the influence of thermal processes on food quality attributes.

Several of the early investigations on kinetic parameters for food quality attributes compared the parameters to those for microbial populations. These investigations revealed that the use of higher temperature for short time periods improved the retention of product quality attributes and still maintained the desired microbial safety or product shelf life. The results of these studies stimulated initiatives on aseptic processing and packaging, as well as other variations on traditional thermal processes, in an effort to reduce the impact of the thermal process on product quality attributes (Figure 1.12).

The availability of kinetic parameters for food quality attributes has provided the basis for many new process design opportunities. The early publications of Teixeira et al. (1969) have demonstrated that even complex thermal processes can be optimized. These examples will be used to illustrate the steps involved in using kinetic parameters of both microbial populations and quality attributes, followed by integration with physical process mechanisms, to optimize all types of preservation processes. The goal of these analyses is to present a quantitative approach to optimize preservation processes, independent of the type of process used for preservation.

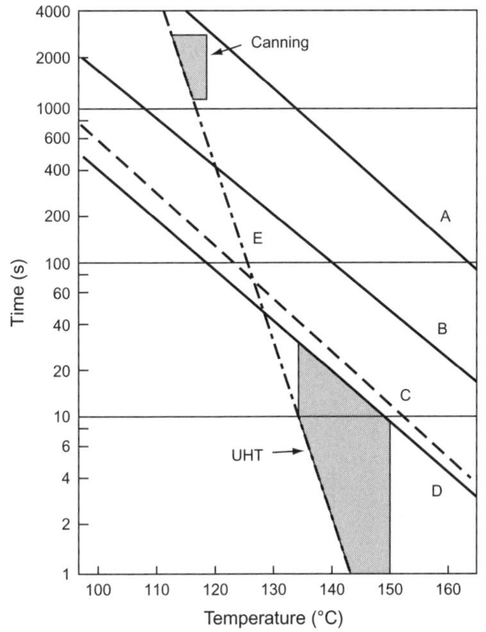

Figure 1.12 Quality retention improvements by using higher temperatures for shorter times (from Fellows, 1988; after Killeit, 1986).

Bibliography

Appert, N. (1810). *L'art de conserver*. Paris: Chez Patri et Cie.

Ball, C. O. (1923). Thermal process time for canned foods. Part 1. *Bulletin National Research Council*, *7*(37), 76.

Ball, C. O. (1927). Theory and practice in processing. *The Canner*, *64*(5), 27.

Ball, C. O. (1936). Apparatus for a method of canning. U.S. Patent 2,020,303.

Ball, C. O., & Olson, F. C. W. (1957). *Sterilization in food technology*. New York: McGraw-Hill.

Bigelow, W. D. (1922). A contribution to our knowledge of processing. *The Canner*, *54*(8), 33.

Bigelow, W. D., Bohart, G. S., Richardson, A. C., & Ball, C. O. (1920). Heat penetration in processing canned foods. *National Canners Association Bulletin*, 16L.

Earle, R. L. (1983). *Unit operations in food processing* (2nd ed.). Oxford: Pergamon Press.

Fellows, P. (2000). *Food processing technology: principles and practice* (2nd ed.). New York: Woodhead Publishing; CRC Press.

IFT/FDA. (2001). Kinetics of microbial inactivation for alternative food processing technologies. *Journal Food Science* (Special Supplement), 1–108.

Jackson, J. M., & Shinn, B. M. (1979). *Fundamentals of food canning technology*. Westport, CT: AVI Pub. Co, Inc.

Killeit, U. (1986, March–April). The stability of vitamins. *Food Europe*, 21–24.

Manson, J. E., & Cullen, J. F. (1974). Thermal process simulation for aseptic processing of foods containing discrete particulate matter. *Journal of Food Science*, *39*, 1084.

Pflug, I. J., & Esselen, W. B. (1963). Food processing by heat sterilization. Chapter 36. In J. L. Heid & M. A. Joslyn (Eds.), *Food Processing Operations*, (Vol. 2, pp. 410–479). Westport, CT: The AVI Pub Co., Inc.

Pflug, I. J., Blaisdell, J. L., & Kopelman, I. (1965). Developing temperature-time curves for objects that can be approximated by a sphere, infinite plate or infinite cylinder. *ASHRAE Transactions*, *71*(1), 238–248.

Potter, N. N., & Hotchkiss, J. H. (1995). *Food science* (5th ed.) (p. 608). New York: Chapman & Hall.

Prescott, S. C., & Underwood, W. L. (1897). Microorganisms and sterilization processes in the canning industry. *Technology Quarterly*, *10*(1).

Singh, R. P., & Yousef, A. E. (2001). Technical elements of new and emerging food technologies. *FAO Report*. http://www.fao.org/Ag/ags/Agsi/Nonthermal/nonthermal_1.htm.

Stumbo, C. R. (1965). *Thermobacteriology in food processing*. New York: Academic Press.

Sun, D.-W. (Ed.), (2005). *Emerging technologies for food processing*. San Diego, CA: Elsevier Academic Press.

Teixeira, A. A., Dixon, J. R., Zahradnik, J. W., & Zinsmeinster, G. E. (1969). Computer optimization of nutrient retention in thermally processing of conduction-heating foods. *Food Technology*, *23*, 137.

Kinetic Models for Food Systems

2

2.1 Rate equations and rate constants 20
2.2 First-order model 25
2.3 Multiple-order models 27
2.4 Agent intensity models 30
2.5 Thermal process models 37
2.6 Uniform parameters 43
List of symbols 46
Bibliography 47

A food is a dynamic system. Each component of a food is changing continuously, beginning with changes that occur at the time of the harvest or assembly of raw food materials. These changes continue during handling, processing, and distribution of the product, and the changes are influenced by ingredients incorporated into the product during formulation of the final food product. The changes occur at different rates, depending on the exposure of the product to external environments and the intensity of environmental factors during the chain of events between harvest or assembly and the time of consumption.

The changes occurring within the food system have a variety of impacts on the food product, including changes in the microbiological population and modification of some product quality attributes. For many food products, the microbiological safety depends on intentional reduction in the population of microbial pathogens during a preservation process. Changes in quality attributes of the product also occur at different rates, depending on conditions during a process or within a storage environment. The shelf life of a

Food Preservation Process Design
ISBN: 978-0-12-372486-1

food product depends on rates of reactions occurring within the product during storage and distribution.

Kinetic models provide a structural framework for quantitatively describing the changes occurring in a food system. The measurement or availability of appropriate kinetic constants for these changes enable us to estimate the magnitudes of change in a given food component during a process, storage, or other event prior to food consumption. Most kinetic models account for the intensity of external agents (temperature, pressure, etc.) impacting the food components as a function of time. In summary, kinetic models are a significant component of any preservation process design model.

Several authors have documented the development and application of kinetic models for food applications. Van Boekel (1996) has provided a general overview of kinetic models in food science and importance of statistics. Villota and Hawkes (2007) present a review of reaction kinetics in food systems and provide comprehensive information on kinetic constants for food constituents under a variety of conditions. The kinetic models associated with inactivation of microbial populations in foods have been reviewed by the IFT/FDA Panel (2001).

2.1 Rate equations and rate constants

The origins of kinetic models used to describe changes in food systems are found in physical chemistry. The models were developed to describe changes during chemical reactions. These models have become a significant component in the analysis of chemical reactions, and the results of the analyses provide insights about the mechanisms involved in the reactions. In these analyses, the first test is an evaluation of the rate equation and the dependence of the reaction rate on the concentration of reactants involved in the reaction.

When considering chemical reactions, the analysis usually begins with monitoring the number of moles of each reactant involved in the reaction. These basic expressions have been used to predict concentrations of food components or microbial populations in the product. For a reaction with two reactants (A and B) involved in a reaction to create two products (C and D):

$$A + B \rightarrow C + D \tag{2.1}$$

the rates associated with the reaction become

$$\text{Rate of reaction} = d[C]/dt = d[D]/dt = -d[A]/dt = -d[B]/dt \tag{2.2}$$

In most applications, the rate of reaction is determined by measuring the rate of disappearance of one or more of the reactants, although measuring the rate of appearances of a reaction product may be appropriate as well. The reaction rate constant (k) is incorporated in the following manner:

$$\text{Rate of reaction} = k[A] = k[B] \qquad \text{for “first order”}$$

$$\text{Rate of reaction} = k[A]^2 = k[A]\,[B] \qquad \text{for “second order”}$$

In general, the rate equation becomes

$$-d[A]/dt = k[A]^n \tag{2.3}$$

where the parameter (n) is the order of the reaction.

When considering chemical reactions occurring within foods, two factors that need to be considered are the potential dependence of the reaction rate on

- Initial concentration
- Time

Either or both of these factors could influence the order of reaction (n). As emphasized by Van Boekel (1996), these factors should be evaluated to avoid misinterpretation of results from experiments conducted to measure rate constants.

An evaluation of the influence of initial concentration can be accomplished by expressing Eq. (2.3) as

$$\text{Ln}\,\{-d[A_o]/dt\} = \text{Ln}\,k + n_c\ \text{Ln}\,\{[A_o]\} \tag{2.4}$$

where n_c is the order of reaction when considering initial concentration of reactants.

When Eq. (2.3) is expressed as

$$\text{Ln}\,\{-d[A]/dt\} = \text{Ln}\,k + n_t\ \text{Ln}\,\{[A]\} \tag{2.5}$$

then n_t is the order of the reaction when considering the change in reaction rates at different times during the reaction.

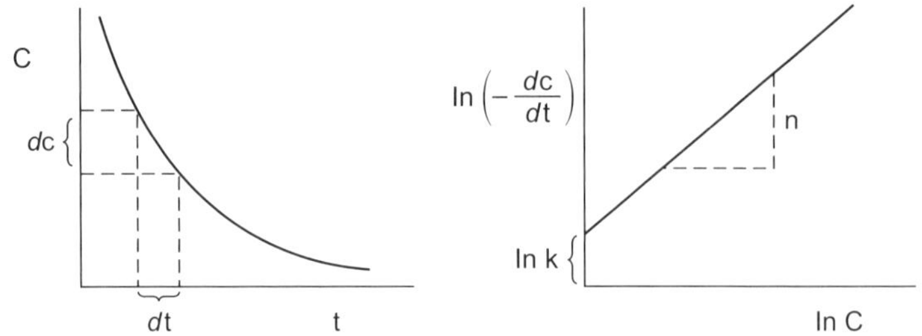

where : C = concentration
k = rate constant
n = reaction order
t = time

Figure 2.1 A graphical description of reaction order (from Villota & Hawkes, 2007).

A graphical representation of Eq. (2.5) is presented in Figure 2.1. The difference in the parameters used to describe order of reaction provides insight into the reaction mechanisms involved in the reaction. When $n_t > n_c$, an inhibitor has likely evolved during the course of the reaction. When $n_t < n_c$, the reaction is likely to be autocatalytic.

Example 2.1

The concentration of the primary reactant in a deterioration reaction occurring in a food product during storage is changing in the following manner.

Time	Concentration
0 days	15,000 units
1	1500
2	540
3	420
4	300
8	150
12	95
15	84
17	68
35	20
51	8
70	2

Determine the order of reaction based on rates as a function of reaction time.

Given:
The concentrations of the primary reactant have been measured over a range of times.

Approach:
Equation (2.5) will be used to estimate the order of reaction by plotting the natural log of reaction rate versus the natural log of concentration.

Solution:
1. The rates of reaction are estimated by assuming linearity between concentration measurements, with the rate given at the midpoint in time, as indicated in the following table:

Time	Rate	Concentration, C	Ln (Rate)	Ln (C)
0.5 days	13,500/day	8250 units	9.5	9.02
1.5	960	1020	6.87	6.93
2.5	120	480	4.79	6.17
3.5	120	360	4.79	5.89
6.0	37.5	225	3.62	5.42
10.0	13.75	122.5	2.62	4.81
13.5	3.67	89.5	1.30	4.49
16.0	8.0	76	2.08	4.33
26.0	2.67	44	0.98	3.78
43.0	0.75	14	−0.29	2.64
60.5	0.32	5	−1.15	1.61

2. A plot of Ln (Reaction Rate) versus Ln (Concentration) is presented in Figure 2.2.

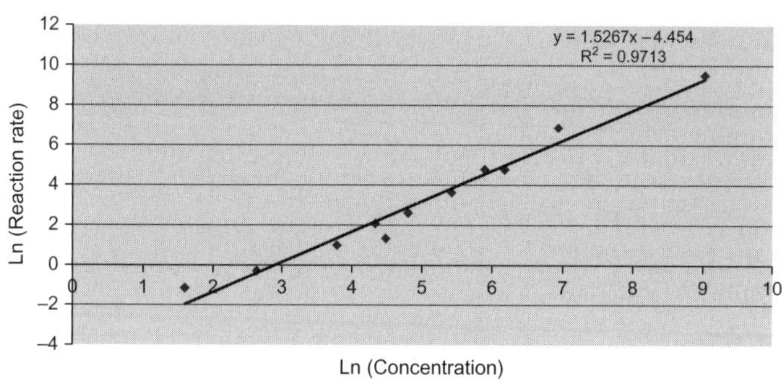

Figure 2.2 A plot of Reaction Rate versus Concentration for Example 2.1.

(*Continued*)

Example 2.1 (Continued)

3. The results of the analysis from Figure 2.2 indicate that the slope (n) of the relationship is 1.53. The reaction presented in this example is not first-order.
4. The intercept on the vertical axis is −4.45. This observation indicates that Ln k = −4.45, and $k = 1.17 \times 10^{-2}$/day. A more direct approach to estimating rate constants will be presented later in this chapter.

Although examples involving food and related systems are limited, some changes may be described by a zero-order model. Deterioration reactions, such as auto-oxidation and nonenzymatic browning, are best described by the linear relationship between concentration and time as illustrated in Figure 2.3. The expression for the zero-order reaction is

$$-d[A]/dt = k_o \tag{2.6}$$

After integration, the expression becomes

$$A_o - A = k_o\, t \tag{2.7}$$

where the concentration (A) at any time (t) is a function of the initial concentration (A_0) and the zero-order rate constant (k_0).

Rate constant are estimated from experimental data by determining the slope of the relationship between concentration and time.

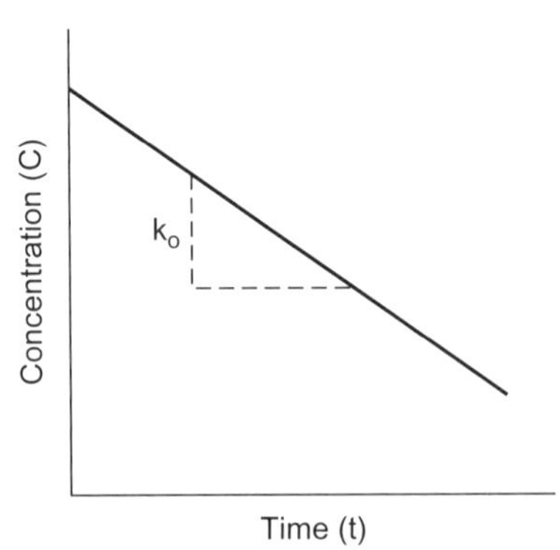

Figure 2.3 An illustration of a zero-order relationship and rate constant (from Villota & Hawkes, 2007).

2.2 First-order model

The general model for a first-order reaction is

$$-d[A]/dt = k[A] \tag{2.8}$$

where k is the first-order rate constant. This model describes the change in the reactant (A) as a function of time (t) for many situations involving food products. This is the most popular model for describing changes in food systems, even in situations where the model may not provide a good description of the experimental data. By rearrangement

$$-\int d[A]/[A] = k\int dt \tag{2.9}$$

and integration using appropriate limits, the solution to the differential equation becomes

$$\mathrm{Ln}\,(A/A_o) = -kt \tag{2.10}$$

or

$$[A] = [A_o]\,\exp\,(-kt) \tag{2.11}$$

This solution has been used to predict the concentration of the primary reactant as a function of initial concentration and time, given the magnitude of the first-order rate constant.

Example 2.2

An analysis of total ascorbic acid in a model food system with a water activity of 0.1 and at 20°C provided the following profile of degradation as a function of time:

Time (Weeks)	Concentration (Ratio)
0	1.0
1	0.951
4	0.897
8	0.903
12	0.688

(*Continued*)

Example 2.2 (Continued)

Time (Weeks)	Concentration (Ratio)
16	0.599
24	0.472
32	0.357
48	0.204
64	0.158
70	0.097

The concentration ratio is the ascorbic acid concentration at any time as compared to the initial concentration. Estimate the rate constant for this first-order reaction.

Given:
Experimental data for ascorbic acid concentrations as a function of time have been measured.

Approach:
By using a first-order model as presented as Eq. (2.10), a plot of the natural logarithm of the concentration ratio versus time provides a relationship with the slope equal to the first-order rate constant (k).

Solution:
1. The plot of the logarithm of concentration ratio versus time is as follows: The slope of the relationship in Figure 2.4 is -0.0321 per week. This value is the magnitude of the first-order rate constant, so

$$k = 3.185 \times 10^{-6}/\text{min}$$

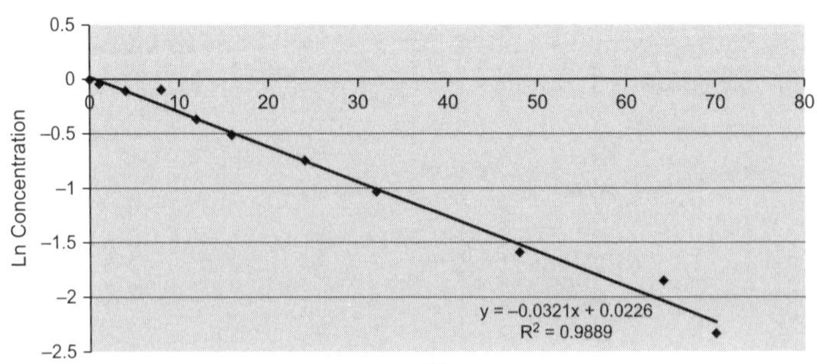

Figure 2.4 A plot of the logarithm of concentration ratio versus time for data presented in Example 2.2.

The previous example is typical of many situations with food products. As illustrated in the example, the rate constant (k) is based on measurements of the primary reactant in the food product. In this situation, the rate constant is a true first-order rate constant and is independent of the initial concentration of ascorbic acid. In most formulated food products, the results are influenced by break-down products of the reaction and are best described by pseudofirst-order kinetics. These observations emphasize that we must take care when using a rate constant to predict changes in a food product, when the rate constant has been measured in a substrate other than the food being evaluated.

2.3 Multiple-order models

As suggested by Eq. (2.3), kinetic models for multiple-order reactions may have applications for food products. For a second-order reaction, the expression becomes

$$-\int d[A]/[A]^2 = k\int dt \qquad (2.12)$$

The solution, after integration, becomes

$$1/[A] - 1/[A]_o = k_2 t \qquad (2.13)$$

where [A] is the initial concentration and "k_2" is the second-order rate constant.

It is evident from Eq. (2.13) that the units of a second-order rate constant must include concentration units. The magnitudes of second-order rate constants are determined by plotting the inverse of the concentration versus time as illustrated in Figure 2.5.

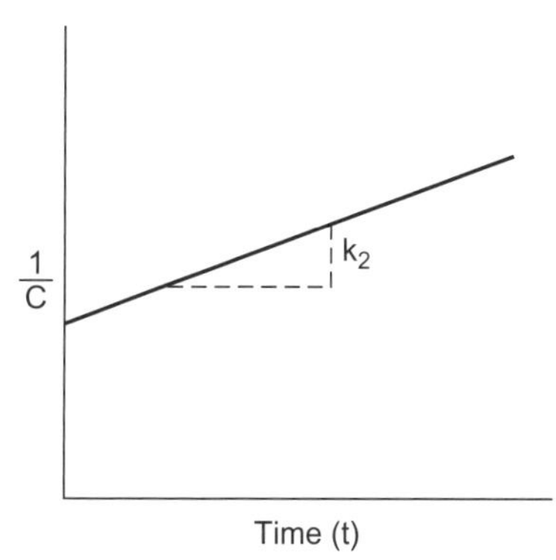

Figure 2.5 The second-order reaction relationship (from Villota & Hawkes, 2007).

Example 2.3

The concentration of a primary reactant has been measured as a function of time to obtain the following results:

Time	Concentration
0 min	7500 units
5	300
10	100
15	80
20	60
35	40
50	25
75	15

Evaluate the concentration-time date using a second-order model, and determine the second-order rate constant.

Given:
Experimental values of the primary reactant have measured at 5-minute intervals over a period of 75 minutes.

Approach:
The relationship of concentration and time are plotted as the inverse of concentration versus time to evaluate the slope. The magnitude of the reaction rate constant is determined from the slope of the relationship.

Solution:

1. The inverse of the concentration at each time interval is computed to obtain the following:

Time	Inverse concentration
0	1.33×10^{-4}
5	3.33×10^{-3}
10	1.00×10^{-2}
15	1.25×10^{-2}
20	1.67×10^{-2}
35	2.50×10^{-2}
50	4.00×10^{-2}
75	6.67×10^{-2}

2. A plot of the experimental values provides the relationship: The slope of the relationship obtained from the analysis and

presented in Figure 2.6 is 9×10^{-4}, so the second-order rate constant is

$$k_2 = 9 \times 10^{-4}/\text{min unit}$$

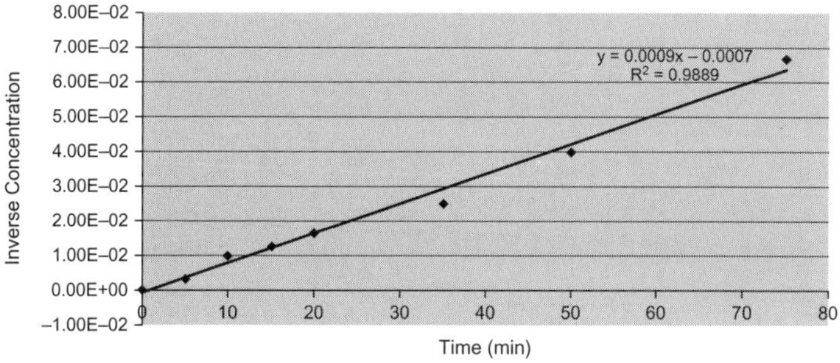

Figure 2.6 A plot of inverse concentration versus time for data from Example 2.3.

Several other kinetic models have applications to reactions occurring in food systems. The unique characteristic of many of these models is that the concentration of the reactant decreases or increases toward an equilibrium value as the reaction time increases. A typical model proposed for enzyme reactions in food systems is the Michaelis-Menten equation:

$$\mathbf{E} + \mathbf{S} \leftrightarrow \mathbf{ES} \leftrightarrow \mathbf{E} + \mathbf{P} \quad (2.14)$$

where **E** = enzyme concentration, **S** = substrate concentration, and **P** = reaction product concentration.

This model includes four rate constants; two forward rate constants for each part of the reaction, and two reverse rate constants. The Michaelis-Menten model describes the relationship illustrated in Figure 2.7, where rate of reaction (V) reaches an equilibrium based on the availability of substrate. Based on the plot, two constants are determined: the maximum reaction rate (V_{max}), and the Michaelis-Menten constant (K_M).

Villota and Hawkes (2007) have provided several other examples of reactions in food systems. In many situations, these reactions proceed toward an equilibrium concentration for one or more of the reactants.

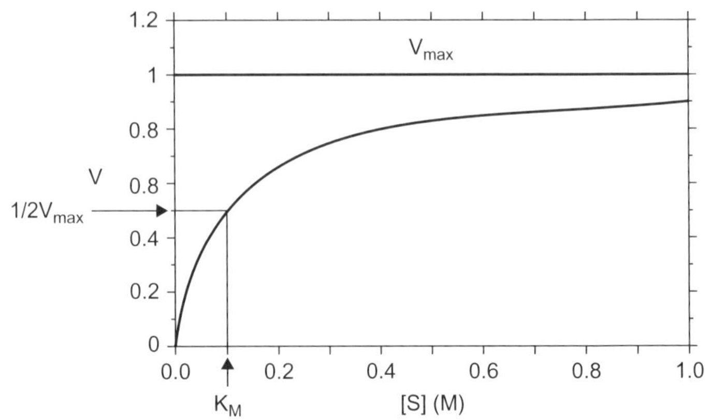

Figure 2.7 A Michaelis-Menten plot (from Villota & Hawkes, 2007).

2.4 Agent intensity models

As previously indicated, the magnitudes of rate constants are established by measuring concentrations as a function of time under controlled conditions. Although reaction rates may depend on several parameters, reactions in food products are very sensitive to temperature. The Arrhenius equation (1889) describes the influence of the intensity (or magnitude) of temperature on the reaction rate constant in a food product as

$$k = k^{o} \exp[-E_A/R\mathbf{T}] \tag{2.15}$$

where k^o = a pre-exponential factor, R = the gas constant, **T** = absolute temperature, and E_A = the activation energy.

The Arrhenius equation has been used to correlate the reaction rate constants in food systems over typical temperature ranges associated with preservation processes and storage of food products. These investigations have resulted in the quantification of activation energies for reactions in food products, as illustrated in Table 2.1. It should be evident that the ranges of activation energies associated with various reactions occurring in food products provide insight on the relative impact of temperature on these changes occurring in foods. Based on discussions by Villota and Hawkes (2007), it is

Table 2.1 Typical activation energy ranges for reactions in food products

Reaction	Activation energy (kJ/mole)
Enzyme reactions	0–34
Chlorophyll degradation	20–115
Ascorbic acid retention	20–165
Anthocyanins	30–125
α-Tocopherol	40–55
trans-Retinol	40–120
Betalains	40–120
Nonenzymatic browning	40–165
Hydrolysis of disaccharides	40–65
Lipid oxidation	40–105
Spore destruction	250–335
Vegetative cell inactivation	210–625
Protein denaturation	335–500

Source: Adapted from Villota and Hawkes (2007).

unlikely that the magnitude of these activation energies for food systems can be interpreted in terms of collision or transition state theories. The reactions occurring in most food system are far too complex, with multiple reactants or reactions involved. Significant care must be imposed when using the Arrhenius equation to predict rate constants beyond the range of temperatures used in the original measurements of rate constants for establishing the activation energy (E_A). In complex systems, there are no assurances that the influence of temperature, as indicated by the activation energy, will remain the same outside the range of measurement.

Expressing Eq. (2.15) as

$$\mathrm{Ln}\, k = -E_A/R\,\mathbf{T} + \mathrm{Ln}\, k^o \tag{2.16}$$

shows that the activation energy (E_A) can be quantified by a plot of Ln k versus the inverse of absolute temperature. Although the pre-exponential factor (k^o) can be evaluated from the analysis, no significance has been attached to the magnitude of this constant for a reaction occurring in a food product.

Example 2.4

The rate constants for retention of chlorophyll in broccoli juice have been measured over a range of process temperatures, as follows:

Temperature (°C)	Rate constants (1/min)
80	0.0085
90	0.0149
100	0.0229
110	0.0489
120	0.0943

Estimate the Activation Energy Constant to describe the influence of temperature on chlorophyll retention in broccoli juice from the data provided.

Given:
Rate constants for chlorophyll retention are provided at five temperatures.

Approach:
The Activation Energy Constant is determined from the slope of the relationship of the natural logarithm of the rate constants versus the inverse of absolute temperature.

Solution:

1. The rate constants and corresponding temperatures are arranged in the following manner:

Ln k	1/T
−4.868	2.83×10^{-3}
−4.206	2.76×10^{-3}
−3.777	2.68×10^{-3}
−3.018	2.61×10^{-3}
2.361	2.54×10^{-3}

2. A plot of Ln k versus the inverse of absolute temperature would appear as follows: From the relationship in Figure 2.8, a slope of -8.47637×10^3 and an intercept of 19.105 are obtained. Using

$$\text{Slope} = E_A/R = -8.4764 \times 10^3 \text{ K}$$

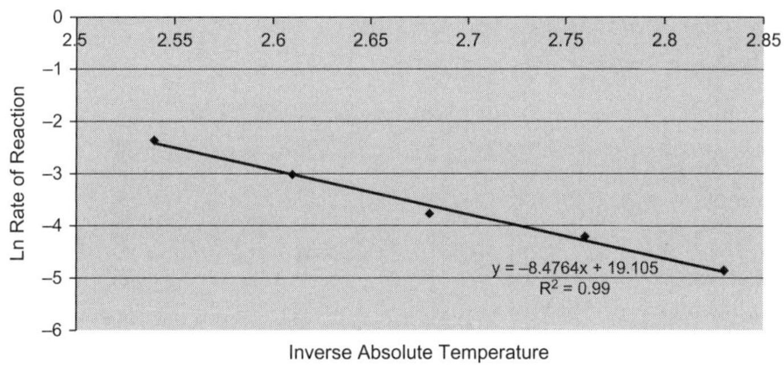

Figure 2.8 An Arrhenius plot of data from Example 2.4.

$$E_A = (-8.47637 \times 10^3\ K)(8.3144\ J/mole\ K)$$
$$= 70.476\ kJ/mole$$

and from the intercept

$$k^o = 1.98 \times 10^8/min$$

The Activation Energy Constants (E_A) describe the impact of temperature on reactions occurring in food products. Villota and Hawkes (2007) have assembled the magnitudes of these constants and many corresponding rate constants. These reactions include vitamin degradation in many different food systems, and pigment degradation or development in several foods. The same types of analysis and similar models have been used to characterize the reductions in microbial populations during thermal processes and similar changes in product texture during processes and product storage.

Pressure has emerged as a potential preservation technology, and the intensity of pressure may influence the rate of change in food product attributes. These changes include the rates of inactivation for microbial populations and the retention of food product quality attributes. The influence of pressure on reaction rate constants has been described by expressions derived from the Eyring relationship. The fundamental basis for these expressions is the relationship between the rates of reaction and the formation rates of an activated complex or quasi-equilibrium condition. The basic expression, at a constant temperature, is

$$[d \ Ln \ k/dP]_T = -\Delta V/R \ \mathbf{T} \tag{2.17}$$

where P = pressure, R = gas constant, k = reaction rate constant, **T** = absolute temperature, and ΔV is the Activation Volume (m^3/mole).

After integration, the expression becomes

$$Ln \ k = Ln \ k^o - \Delta V \ P/R \ \mathbf{T} \tag{2.18}$$

or

$$k = k^o \exp[-\Delta V \ P/R \ \mathbf{T}] \tag{2.19}$$

In theory, the Activation Volume (ΔV) represents the change in volume of the reactants as compared to the volume of the activated complex. When the magnitude is negative, the reaction rate increases with increasing pressure.

In practice, the magnitude of the Activation Volume is determined from measuring rate constants (k) over a range of pressures. Using the following form of Eq. (2.18),

$$Ln \ k = Ln \ k_{ref} - (\Delta V/R \ \mathbf{T}) [P - P_{ref}] \tag{2.20}$$

and rate constants (k) measured over a range of pressures (P), the reference rate constant (k_{ref}) and the Activation Volume (ΔV) are determined. The results from the experimental measurement of rate constants are plotted as the natural logarithm of the rate constant (Ln k) versus pressure differences ($P - P_{ref}$), and the slope of the linear relationship is (ΔV/R **T**).

Research leading to the publication of Activation Volume Constants (ΔV) for reactions occurring in a food system during application of pressure is limited at this time. The limited information includes the magnitudes of Activation Volume Constants for inactivation of microbial populations and enzyme systems and provides some insights on the kinetics of reactions occurring during preservation of foods using ultra-high pressures.

Example 2.5

The following inactivation rate constants for tomato polygalacturonase were measured at various pressures and 25°C by Fashin, van Loey Indrawati, Ludikhuyze, and Hendrickx (2002):

Pressure (MPa)	Rate constants (1/min)
350	0.099
375	0.210
400	0.451
425	1.010
450	1.757
500	3.126

Determine the Activation Volume Constant (ΔV) to describe the relationship between the enzyme inactivation rates and pressure.

Given:
Rate constants for inactivation of tomato polygalacturonase are provided at six different pressures.

Approach:
Use Eq. (2.20) and a plot of Ln k versus ($P - P_{ref}$) to estimate the Activation Volume Constant from the slope of the relationship.

Solution:
1. The data can be presented in the following manner by using 350 MPa as the reference pressure:

Pressure difference	Ln k
25	−1.56
50	−0.796
75	0.00995
100	0.5636
150	1.1398

2. A plot of Ln k versus pressure difference is as follows: The slope of the relationship in Figure 2.9 is 0.0222. Based on Eq. (2.20), the expression for the slope is

$$\Delta V/R\mathbf{T} = 0.0222/\text{MPa}$$

(*Continued*)

Example 2.5 (Continued)

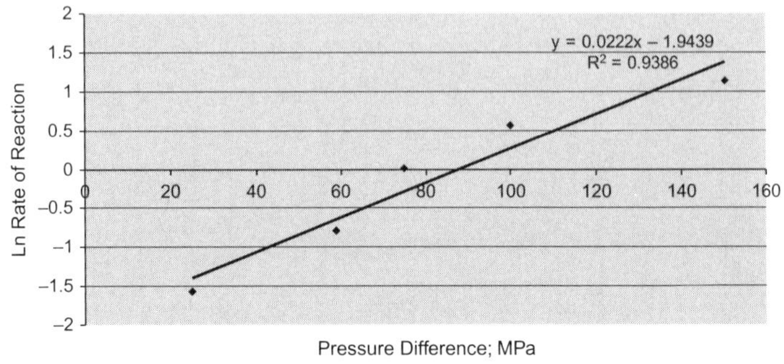

Figure 2.9 A plot of data from Example 2.5, used to estimate the Activation Volume Constant (ΔV).

and

$$\Delta V = -(0.0222/\text{MPa})(8.31441\ \text{m}^3\ \text{Pa/mole K})(25 + 273\ \text{K})$$

$$\Delta V = -55.01 \times 10^{-6}\ \text{m}^3/\text{mole} = -55.01\ \text{cm}^3/\text{mole}$$

Note: The Activation Volume Constant is negative as indicated by Eq. (2.20).

The magnitude of the Activation Volume Constant (ΔV) is expected to a function of temperature. The results presented by Fachin et al. (2002) indicated that the Activation Volume Constant for inactivation of tomato polygalacturonase increased with temperature over the range from 5 to 40°C. The increase in magnitude was not consistent over the entire temperature range. The change in magnitude was small, between 5 and 20°C, but increased significantly between 20 and 40°C. Apparently the impact of pressure on enzyme inactivation is more dramatic at higher temperatures. These types of observations can be very useful in process design and in selection of the optimum process parameters. Future investigations to explore the influence of pressure on food product components should include the influence of temperature.

2.5 Thermal process models

The published literature on food preservation processes includes significant amounts of kinetic data on the survival of microbial populations during thermal processes. Ball and Olson (1957) and Stumbo (1965) provided an in-depth description of the approaches for measurement of these types of data. Teixeira (2007) provided the use of kinetic parameters in thermal process design. During the original measurements of microbial survivor numbers as a function of time at elevated constant temperatures, researchers recognized that survivors decreased with time in a logarithmic manner. These types of data were analyzed by plotting the log of survivors versus time to create a linear relationship. The results of the analysis was the definition of the Decimal Reduction Time (D): the time needed for a 90% reduction in the microbial population or the time required for a one log-cycle reduction in the population of microorganisms. For these types of analyses, the equation for the survivor curve becomes

$$N/N_o = 10^{-t/D} \tag{2.21}$$

where N_o = initial population, N = population at any later time, t = time, and D = decimal reduction time.

Note that the typical survivor curve equation is based on log10. Because first-order reaction kinetics are described by the relationship between the natural logarithm of concentration as a function of time, the relationship between the Decimal Reduction Time (*D*) and a first-order reaction rate constant (k) becomes

$$k = \frac{2.303}{D} \tag{2.22}$$

Based on this relationship, historical data for Decimal Reduction Times (D) can be converted to first-order rate constants (k).

Many examples of microbial survivor curves are not described by the traditional first-order model. These non-log-linear survivor relationships have been observed and reported in the literature for many years. Stumbo (1965) provided several examples and explanations. In general, the variations from the first-order survivor model have been incorporated into process design. In practice, appropriate adjustments in the process time have been incorporated to ensure product safety.

More recently, there have been numerous recommendations to account for the influence of non-log-linear microbial survivor curves on process design. Sapru, Teixeira, Semerage, and Lindsay (1992); Anderson, McClure, Baird-Parker, and Cole (1996); Kormandy and Kormandy (1997); and Peleg and Penchina (2000) have presented alternatives to the traditional log-linear model for microbial survivor curves. Heldman and Newsome (2003) evaluated all of these alternatives in a review and recommended the following model to account for the variations observed in many non-log-linear microbial survivor curves:

$$\log N_o - \log N = [t/D']^n \tag{2.23}$$

where D' = a time constant similar to the Decimal Reduction Time (D) and n = a coefficient to account for deviations from a log-linear relationship.

When the coefficient (n) in the preceding model becomes 1.0, the relationship between the logarithm of microbial survivors and time is linear. When the relationship is non-log-linear, and the survivor curve is concave downward, the values of $n > 1.0$. Many microbial survivor curves are concave upward, and the values of $n < 1.0$. An illustration of the various types of survivor curves is presented in Figure 2.10.

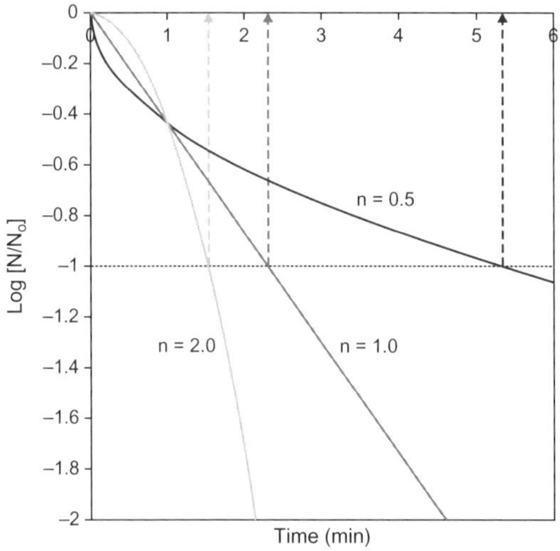

Figure 2.10 An illustration of non-log-linear survivor curves (adapted from Van Boekel, 1996).

Example 2.6

The survivor curve data for a microbial population are as follows:

Time (min)	Survivors (cfu)
0	9×10^8
0.6	2.25×10^8
1.0	9.0×10^7
1.5	9.0×10^6
1.7	4.5×10^6
2.0	5.67×10^5
2.2	9.0×10^4
2.6	1.43×10^3
3.0	28

Determine the parameters needed to describe the relationship between microbial survivors and time.

Given:
The populations of the microorganisms have been measured at time interval between 0 and 3 minutes.

Approach:
The first step is to determine if the relationship between survivors and time is log-linear. If the relationship is non-log-linear, Eq. (2.23) will be used to evaluate the constants (D′, n).

Solution:
1. The data can be transformed as follows:

Time	N/N_o	$\log(N/N_o)$	$\log[\log(N/N_o)]$
0	1.00	0	
0.6	0.25	−0.6	−0.22
1.0	0.10	−1.0	0.00
1.5	0.010	−2.0	0.30
1.7	0.005	−2.3	0.36
2.0	0.00063	−3.2	0.51
2.2	0.0001	−4.0	0.60
2.6	0.00000159	−5.8	0.76
3.0	0.00000003	−7.5	0.88

(Continued)

Example 2.6 (Continued)

2. Because it is evident that the relationship is not log-linear, Eq. (2.3) will be used. The expression can be rewritten in the following form:

$$\log [\log (N/N_o)] = n \log D' - n \log t$$

3. By plotting the data, the relationship presented in Figure 2.11 is obtained:

$$\log [\log (N/N_o)] = 1.5966 \log D' + 0.0554$$

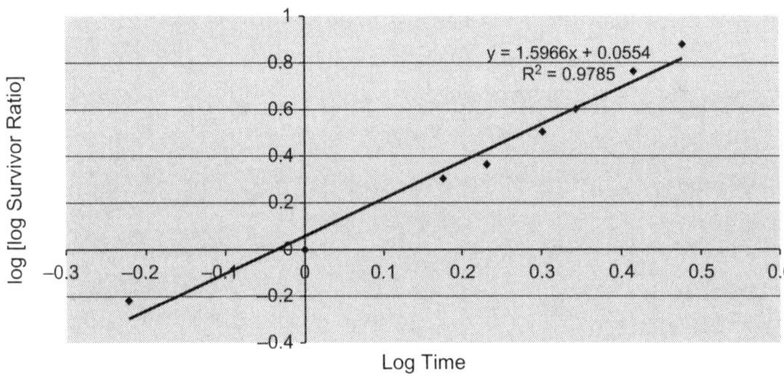

Figure 2.11 A plot of microbial survivors versus time, used to evaluate the parameter (n) for a non-log-linear model.

4. Based on the results from the analysis, the slope is 1.5944, and the intercept is 0.0554:

$$n = 1.6$$

and

$$D' = 10^{(0.0554)} = 1.083 \text{ min}$$

Recognize that the parameter "n" in Eq. (2.23) is not the same as "n" in Eq. (2.3). Both constants do account for the deviation of the survivor curve from log-linear. When n = 1 in Eq. (2.3), it becomes the first-order model. The same is true for Eq. (2.23). Equation (2.23) can be expressed as

$$\text{Ln } C - \text{Ln } C_o = (-k' t)^n \quad (2.24)$$

or

$$C/C_o = \exp [-k' t]^n \quad (2.25)$$

Clearly, Eq. (2.23) becomes the solution to the first-order model when $n = 1$ and is presented as Eq. (2.8). All of these relationships illustrate the applications of the parameter (n) to communicate the magnitude of the deviation of model from the first-order model. When the values of n are larger than 1.0, the survivor curve is concave downward, and larger values indicate greater deviations from the first-order model. When the values of n are less than 1.0, the survivor curve is concave upward, and smaller values represent greater deviations from the log-linear relationship.

A Thermal Resistance Coefficient (z) described the traditional relationship between the Decimal Reduction Time (D) and temperature. The magnitude of this temperature coefficient was determined by evaluating the temperature increase required to reduce the Decimal Reduction Time (D) by 90% (or one log cycle). An example of this relationship is presented in Figure 2.12. The following expression describes the relationship between the Decimal Reduction Time and temperature:

$$\mathrm{Log}\,[D_1/D_2] = (T_2 - T_1)/z \qquad (2.26)$$

where D_2 = Decimal Reduction Time at T_1, D_1 = Decimal Reduction Time at T_2, and z = Thermal Resistance Coefficient (Figure 2.12).

By comparing Eq. (2.23) with Eq. (2.15), the relationship between the Thermal Resistance Coefficient (z) and the Activation Energy Constant (E_A) can be expressed as

$$E_A = 2.303\, R\, \mathbf{T}^2/z \qquad (2.27)$$

where **T** = absolute temperature.

Although the range of temperature used in Eq. (2.26) is not specified, the range must be limited to the range of temperatures used to generate the Thermal Resistance Coefficient (z). The relationship between the two temperature coefficients (E_A and z) does not have theoretical significance. The Thermal Resistance Coefficient (z) describes the influence of temperature on the Decimal Reduction Time (D) over the range of temperatures used in experimental measurements. As suggested by Eq. (2.27), the relationship between the Activation Energy Constant (E_A) and the Thermal Resistance Coefficient (z) could be a function of temperature, but the influence should be small.

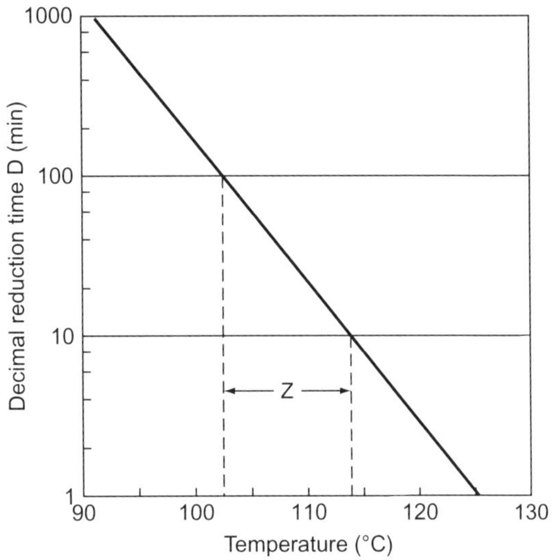

Figure 2.12 The relationship between Decimal Reduction Time and temperature.

Example 2.7

The results from measurement of survivor curves for a vegetative pathogen provided the following results:

Temperature	Decimal Reduction Time, D
80°C	105 min
90	10
95	4
100	1

Estimate the Activation Energy Constant needed to describe the influence of temperature on the inactivation rate of the microbial population.

Given:
The Decimal Reduction Times at five different temperatures have been provided.

Approach:
The relationship between the Decimal Reduction Time (D) and temperature can be determined by a plot of log D versus T to

evaluate the Thermal Resistance Constant (z). The Activation Energy Constant can be estimated by using Eq. (2.27).

Solution:

1. The plot of log D versus T is obtained as follows: The results from Figure 2.13 indicate that the slope is −0.1019.

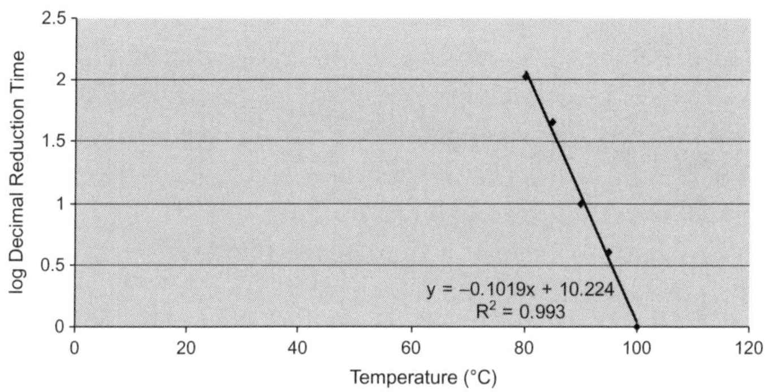

Figure 2.13 A plot of Decimal Reduction Times versus temperature, as used to evaluate the Thermal Resistance Coefficient.

2. The thermal resistance constant is

$$z = 1/0.1019 = 9.8\ \mathrm{C}$$

3. Using Eq. (2.27):

$$E_A = 2.303\,(8.31441)(90 + 273)/9.8 = 7.1 \times 10^5\ \mathrm{J/mole}$$
$$E_A = 710\ \mathrm{kJ/mole}$$

4. The same result could have been obtained by converting the Decimal Reduction Times to rate constants (k) using Eq. (2.22), followed by using the Arrhenius relationship to estimate the Activation Energy Constant.

2.6 Uniform parameters

As indicated by the origins and applications of kinetic parameters associated with preservation of foods, a variety of terms have evolved to describe the parameters. In most situations, the kinetic responses have defined the terms, but different terms have evolved for different components of the food product. For example, the rate constant for a first-order microbial survivor curve is the Decimal Reduction Time, while the rate constant for the first-order curve for retention of a vitamin is a first-order rate constant. As has been illustrated in this chapter, the kinetic parameters used to describe the changes in food product quality attributes during a preservation process are based on traditional reaction kinetics. Most of the parameters used to describe the changes in microbial populations during a preservation process are based on first-order reaction kinetics, but different parameters and terms have been used to express the constants. The relationships among the typical parameters have been described in this chapter.

As discussed in Chapter 1, one of the important goals of preservation process design is to optimize the process; that is, achieve the maximum retention of product quality attributes while ensuring microbiological safety. To pursue this goal, the best method is to use the same kinetic models, parameters, and terms for both microbial populations and food quality attributes. The following descriptors, symbols, and units are proposed for use in preservation process design and will be used whenever possible throughout this book.

k = rate constant to describe survivors of a microbial population or the retention of a food product quality attribute; 1/s.

For microbial survivor curves, this parameter can be converted from traditional Decimal Reduction Times (D) by using Eq. (2.22), as long as the survivor curves are first-order or log-linear. For non-log-linear survivor curves, the parameter (k′) may be used in place of the time constant (D′) from Eq. (2.23). In survivor curve equations, the symbol (N) designates the magnitude of the microbial population.

When describing changes in food quality attributes, the parameter (k) is used to express the rate constant for the retention food quality attributes during a preservation process. The majority of the constants are first-order constants; for situations other than first-order, the reaction order (zero, second) is noted by superscript (0, 2, 3, etc.). In some cases, an expression similar to Eq. (2.25) may be used to describe the change in quality attributes, and the reaction rate constant k′ along with the index (n) are used. In expressions used to describe changes in the magnitude of a food product quality attribute, the concentration is expressed as C.

n = index parameter used to account for deviations from non-log-linear survivor curves or non-first-order reaction models.

The magnitudes of all parameters associated with the model used to describe the change in microbial population or change in product quality attributes are functions of the preservation process. The Arrhenius equation relationship is used to describe the influence of the temperature on the rate constants.The primary parameter is defined in the heading of the following section.

E_A = Activation Energy Constant to describe the influence of temperature on the microbial inactivation rate or on the reaction rate constants for retention of a food quality attributes; kJ/mole.

These parameters may be generated directly from experimental data or converted from traditional thermal resistance constants (z) using Eq. (2.27). This approach allows for the influence of temperature on microbial inactivation rates to be described by Eq. (2.15). The same parameter can be used to describe the temperature influence on the non-log-linear rate constant (k′). Most data from the literature used to describe the influence of temperature on rate constants for retention of product quality attributes during preservation processes have been expressed as Activation Energy Constants (E_A). These same parameters can be used for situations when the rate constants are for the kinetic models other than first order.

When the preservation process uses high pressure, the influence of pressure on the rate constants is described by Eq. (2.20) and the following parameter.

ΔV = Activation Volume Constant for description of the influence of pressure on microbial inactivation rates or on reaction rate constants for food quality.

This parameter appears in Eq. (2.20), for reaction rate constants (k), as well as for parameters used to describe microbial inactivation rates. Some of current literature has used pressure coefficients (z_p) to describe the influence of pressure on inactivation rates for microbial populations. These constants can be converted to Activation Volume Constants using

$$\Delta V = -2.303\, R\, \mathbf{T}/z_p \qquad (2.28)$$

where

$$R = 8.31441\ m^3\ Pa/mole\ K$$

Although the amounts of data available to describe the influence of pressure on rate constants in food products is limited, most of the data have been analyzed to generate Activation Volume Constants.

These are the primary kinetic constants associated with preservation processes for food products. Other parameters associated with process design for food preservation will be presented and discussed in the appropriate chapters, along with discussion of process design.

List of symbols

A = intensity of reactant in kinetic reaction model
A_o = initial intensity of reactant
C = concentration of component in kinetic model
C_o = initial concentration of component
D = decimal reduction time; min
D' = decimal reduction time for non-log-linear microbial survivor curve; min
E_A = activation energy constant; kJ/mole
$\mathbf{E}$ = enzyme concentration
k = first-order reaction rate constant; 1/s
k_o = zero-order reaction rate constant; Conc./s
k_2 = second-order reaction rate constant; 1/Conc. s.
k^o = pre-exponential constant in Arrhenius relationship; 1/s
k_{ref} = reference rate constant; 1/s
k' = rate constant for non-log-linear survivor curve; 1/s
K_M = constant in Michaelis-Menten equation
n = parameter defining deviation from first-order kinetic relationship
n_c = constant defining deviation from first-order reaction due to initial concentration
n_t = constant defining deviation from first-order reaction due to time

N = microbial population
N_o = initial microbial population
P = pressure; MPa
P_{ref} = reference pressure; MPa
$\mathbf{P}$ = byproducts concentration in enzyme reaction
R = gas constant; kJ/mole K
$\mathbf{S}$ = substrate concentration in enzyme reaction
t = time; s or min
T = temperature; C
$\mathbf{T}$ = absolute temperature; K
ΔV = activation volume; cm^3/mole
V = enzyme reaction rate
V_{max} = reaction rate constant from Michaelis-Menten equation
z = thermal resistance constant; C
z_p = pressure change for one log change in rate constant; MPa

Bibliography

Anderson, W. A., McClure, P. J., Baird-Parker, A. C., & Cole, M. B. (1996). The application of a log-logistic model to describe the inactivation of *Clostrididium botulium 213B* at temperatures below 121.1°C. *Journal of Applied Bacteriology*, *80*, 283–290.

Barrow, G. M. (1996). *Physical chemistry* (6th ed.). New York: McGraw-Hill.

Ball, C. O., & Olson, F. C. W. (1957). *Sterilization in food technology*. New York: McGraw-Hill.

Fachin, D., van Loey Indrawati, A., Ludikhuyze, L., & Hendrickx, M. (2002). Thermal and high-pressure inactivation of tomato polygalacturonase: A kinetic study. *Journal of Food Science*, *67*(5), 1610–1615.

Heldman, D. R., & Newsome, R. L. (2003). Kinetic models for microbial survival during processing. *Food Technology*, *57*(8), 7–12.

IFT/FDA. (2001). Kinetics of microbial inactivation for alternative food processing technologies. *Journal Food Science* (Special Supplement), 1–108.

Kormandy, I., & Kormandy, L. (1997). Considerations for calculating heat inactivation processes when semilogarithmic thermal inactivation models are nonlinear. *Journal of Food Engineering*, *34*, 33.

Peleg, M., & Penchina, C. M. (2000). Modeling microbial survival during exposure to a lethal agent with varying intensity. *Critical Reviews in Food Science and Nutrition*, *40*(2), 159–172.

Sapru, V., Teixeira, A. A., Semerage, G. H., & Lindsay, J. A. (1992). Predicting thermophilic spore population dynamics for UHT sterilization processes. *Journal of Food Science*, *57*, 1248–1257.

Stumbo, C. R. (1965). *Thermobacteriology in food processing*. New York: Academic Press.

Teixeira, A. A. (2007). Thermal processing on canned foods. In D. R. Heldman & D. B Lund (Eds.), *Handbook of food engineering* (2nd ed., pp. 745–797). Boca Raton, FL: CRC Press, Taylor & Francis Group.

Villota, R., & Hawkes, J. G. (2007). Reaction kinetics in food systems. In D. R. Heldman & D. B. Lund (Eds.), *Handbook of food engineering* (2nd ed., pp. 125–286). Boca Raton, FL: CRC Press, Taylor & Francis Group.

Van Boekel, M. A. J. S. (1996). Statistical aspects of kinetic modeling for food science problems. *Journal of Food Science*, *61*(3), 477–489.

3

Kinetics of Inactivation of Microbial Populations

3.1 Characteristics of microbial survivor curves 50
3.2 Kinetic parameters for microbial populations. 54
3.3 Applications of kinetic parameters 59
3.4 Definition of microbial inactivation 76
3.5 Kinetic parameters for alternative preservation technologies 77
List of symbols 77
Bibliography. 78

The kinetics of microbial survival during a thermal process has a significant history. As documented by Ball and Olsen (1956) and Stumbo (1965), experimental measurements of the numbers of survivors as a function of time, at a constant temperature, have occurred continuously over a significant period of time. Although a significant focus of these measurements has been on pathogenic spores, information has also been assembled on the survivor curves for microbial populations causing spoilage in shelf-stable foods and for microbial pathogens in refrigerated foods. As indicated in Chapter 2, in most of the investigations, the survivor curves for microbial populations have been described by first-order models, and the rates have been expressed as Decimal Reduction Times (D). For a given microbial population, the Thermal Resistance Constant (z) has been used to describe the influence of temperature on D-values.

Food Preservation Process Design
ISBN: 978-0-12-372486-1

This chapter discusses the characteristics of microbial survivor curves. These discussions include distinctions between survivor curves for microbial spores as compared to vegetative cells, as well as the influence of product environment on the survivor curves. We will assemble typical kinetic parameters for both microbial pathogens and microbial populations causing product spoilage. These parameters include those collected during thermal treatments, as well as parameters describing survivor curves during treatments by alternative technologies, such as ultra-high pressure (UHP) and pulsed electric fields (PEF). We will also demonstrate the use of the parameters in predicting the fate of the microbial populations during typical preservation processes.

3.1 Characteristics of microbial survivor curves

The shape of the survivor curve for a microbial population has been debated from the early experiments associated with the development of a thermal process for commercial sterilization of foods. The characteristics of these curves were documented by Ball and Olsen (1956) and Stumbo (1965), and they have been highlighted in most recent reviews by IFT/FDA (2001). The origins of the traditional logarithmic (first-order) shape of the survivor curve can be traced to the early experiments of Esty and Meyer (1922). These experiments attempted to establish the "thermal death times," or the time required to inactivate a given number of *Cl. botulinim* spores. These experiments used 10^6 spores as an initial population, and "inactivation" was established as a 10^{-6} probability of survivors. The outcomes from the Esty and Meyer (1922) experiments were thermal death time curves as illustrated in Figure 3.1. These curves are thermal death times versus temperature, with the slope being expressed as the thermal resistance coefficient (z). These experiments established that "z" for *Cl. botulinum* was 10°C (18°F).

The early experiments for establishing the inactivation of spore populations did not include the measurement of survivor curves. The shape of the curve between the initial population and the probability of survivors was generally accepted to be log-linear and described by a first-order model. Later research indicated that many deviations from the log-linear relationship occur, but the logarithm

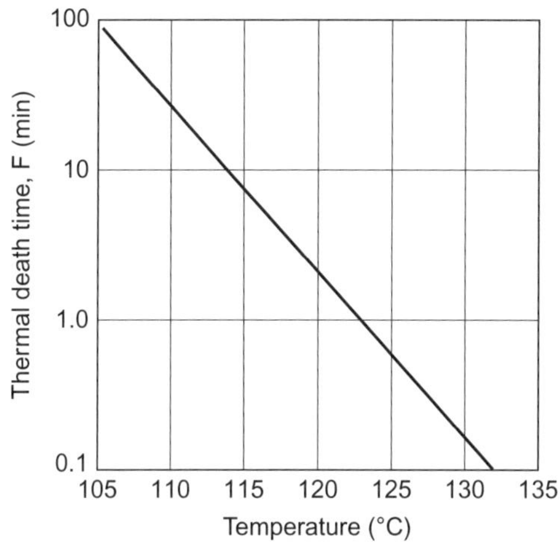

Figure 3.1 The thermal death time versus temperature relationship (from Heldman & Hartel, 1997).

of numbers versus time was accepted as the approach to describing the survivor curves for all microbial populations during exposure to the elevated temperatures. An additional outcome of the early experiments of Esty and Meyer (1922) was the concept of "12D"; the 12 log cycle reduction in population from the 10^6 to 10^{-6} was accepted as a safe process for commercially sterilized foods. The concept of the time for each log cycle of reduction in population became a Decimal Reduction Time (D).

Stumbo (1965) discussed the deviations from log-linear survivor curves in depth by presenting three typical variations from the first-order curve for spore populations. The survivor curve in Figure 3.2 illustrates an apparent increase in population within short periods of time after exposure to elevated temperature. This type of curve has been explained by spore germination during the early stages of heating and a corresponding increase in numbers of vegetative cells measured by assay methods. The increase occurs prior to the normal decrease in population caused by exposure to an elevated temperature. The concept of "shoulders" is not expected to cause deviations from log-linear survivor curves for a population of vegetative cells. Sapru, et al. (1992) have recommended more complex models to describe these unique characteristic of survivor curves for spore populations.

The second type of deviation from the log-linear survivor curve is illustrated in Figure 3.3. Stumbo (1965) suggests that a "lag"

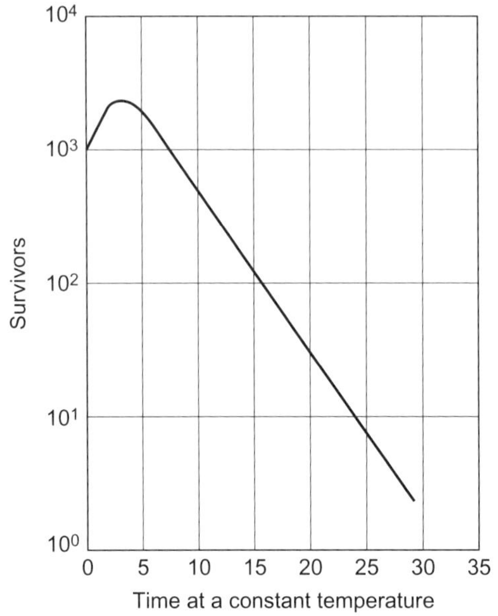

Figure 3.2 One type of non-log-linear survivor curve for microbial spores (from Stumbo, 1965).

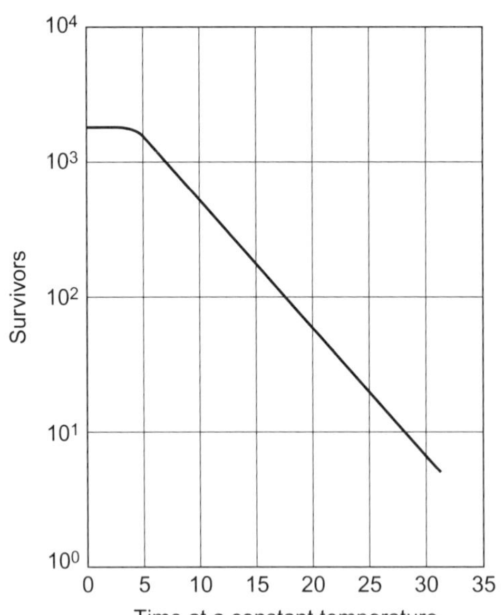

Figure 3.3 An example of a survivor curve with a lag during the initial exposure to high temperatures (from Stumbo, 1965).

at the beginning of the exposure to elevated temperatures occurs when the rate of spore germination is equal to the rate of inactivation for the microbial population. Following the deviation at the beginning of the experimental measurement of spore survival, the survivor curve is log-linear.

A third type of survivor curve is illustrated in Figure 3.4. Following a log-linear reduction in population during the first portion of exposure to an elevated temperature, a second portion of the survivor curve has the reduction in population occurring at a slower rate. Stumbo (1965) suggests this type of response occurs when the microbial population used in the experiment has a mixed culture. In the example (Figure 3.4), there are two different species of spores in the population; each with a different response to the elevated temperature. As indicated by Van Boekel (2002), these types of deviations have occurred in vegetative cell populations, and may be associated with different strains within the microbial population, with each having a different response to the elevated temperature.

The significance of the shape of the survivor curve needs to be explored in depth. As previously indicated, the establishment of the Decimal Reduction Time (D) has historical impact on the research literature associated with the response of microbial populations to

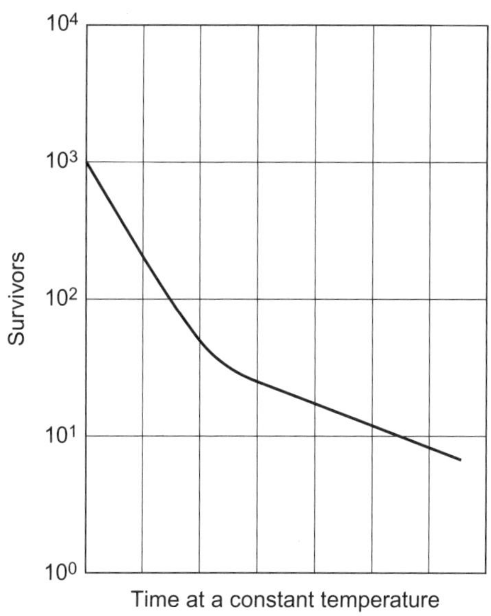

Figure 3.4 An example of a microbial survivor curve for a mixed culture population. (from Stumbo, 1965).

elevated temperatures. Although the impact was not intentional, the measurement and publishing of data in terms of Decimal Reduction Times (D) automatically assumes that the survivor curves are log-linear and can be described by first-order kinetic models. These assumptions, in turn, impact the process design and the capability to accurately predict the outcomes from the process. To provide better interpretation of historical data and to create a more refined set of parameters for the future, the parameters used to describe survivor curves must accommodate typical deviations from log-linear survivor curves.

3.2 Kinetic parameters for microbial populations

To assemble kinetic parameters for use in preservation processes, a consistent set of definitions has been adopted. The uniform definitions introduced in Section 2.6, Uniform Parameters, in Chapter 2, provide the basis for development. The kinetic parameters include those needed to describe survivor curves for an array of preservation technologies, along with the parameters required to describe the influence of the preservation agent intensity on the rate of microbial population inactivation.

Some of the typical kinetic parameters for microbial populations associated with preservation of food products are presented in Tables 3.1, 3.2, 3.3 and 3.4. A significant portion of the experimental data have been published in the past 15 to 20 years. These kinetic parameters for microbial survivor curves are based on measurements using alternatives technologies (ultra-high pressure [UHP] and pulsed electric fields [PEF]), as well as the kinetic parameters for similar microbial populations for traditional thermal processes. The parameters presented in these tables are the rate constant (k), a deviation coefficient (n), the intensity of the preservation agent for the rate constant, the agent intensity coefficient (E_A, z, ΔV, z_p), and the substrate carrier for the microbial population. The data presented focus on differences in the magnitudes of the kinetic parameters for vegetative pathogens as compared to pathogenic spores. In addition, we can evaluate the variability in kinetic constants due to the measurement's methods. Finally, we

Table 3.1 Typical kinetic parameters for *Salmonella spp.* survivor curves

Preservation process	Rate constant	Deviation parameter (n)	Agent intensity	Agent coefficients	Substrate
Thermal	3.569/min		60°C		Turkey
Thermal	2.075–4.113/min		65.5°C	392–499 kJ/mole	Various foods
Thermal	0.144/min	0.667	60°C		66% sugar
Thermal	0.002–0.006/min		70–71°C		Milk chocolate
UHP	0.384/min		345 MPa		Buffer
UHP	0.14/min		200 MPa	42.27 cm^3/mole	Tryptic soy broth
PEF	0.5758/microsec		83 kV/cm		NaCl

Table 3.2 Typical kinetic parameters for *E. coli spp.* survivor curves

Preservation process	Rate constant	Deviation parameter (n)	Agent intensity	Agent coefficients	Substrate
Thermal	0.45–1.77/min		57.2°C		Dairy products
Thermal	1.12–1.92/min		58°C	553.5–582.7 kJ/mole	Ground beef
Thermal	1.11–5.62/min		50°C	80.1–353.9 kJ/mole	Apple cider
UHP	0.25/min		250 MPa	45.3 cm^3/mole	Apple juice
UHP	0.768/min		400 MPa		Milk
PEF	0.0724/microsec	0.6	25 kV/cm		20% carrot juice
PEF	0.0236–0.245. microsec		30 kV/cm	10.33–24.91 kJ/mole	Liquid egg whites

Table 3.3 Typical kinetic parameters for *Listeria spp.* survivor curves

Preservation process	Rate constant	Deviation parameter (n)	Agent intensity	Agent coefficients	Substrate
Thermal	3.97–10.47/min		63.3°C	386 kJ/mole	Milk
Thermal	0.41/min		60°C	358.7 kJ/mole	Ground pork
Thermal	0.092/min		60°C	433.3 kJ/mole	Turkey breast
UHP	0.552–1.219/min		414 MPa		Ground pork
UHP	0.768/min		450 MPa		Eggs
PEF	0.0043/microsec		20 kV/cm		Buffer
PEF	0.086/microsec		50 kV/cm		Skim milk

Table 3.4 Typical kinetic parameters for *Clostridium spp.* survivor curves

Preservation process	Rate constant	Deviation parameter (n)	Agent intensity	Agent coefficients	Substrate
Thermal	0.929–3.775/min		110°C	242.1 kJ/mole	Vegetable products
Thermal	0.186–4.7.min		110°C	379.6 kJ/mole	Vegetable products
Thermal	0.731/min		104.1°C		Phosphate buffer
Thermal	0.08/min	0.99	105°C		Water
Pressure	0.218/min		689 MPa	4.39 cm^3/mole	Buffer
Pressure	0.723/min		700 MPa	7.42 cm^3/mole	
Pressure	4.69/min	0.26	700 MPa		Water

illustrate the influence of the substrate carrying the microbial population during the measurement of the kinetic parameters.

The kinetic parameters presented in Table 3.1 are for several different populations of *Salmonella*, including *senftenberg, typhimurium, enteritus, eastbourne,* and *dublin*. These parameters are representative of parameters available for microorganisms implicated in many food-borne infection outbreaks. The parameters presented in the table include rate constants (k) for several different substrates, along with the intensity of the preservation agent for each situation. The table includes activation energy constants, along with the pressure coefficient for an ultra-high pressure process. In addition, an example of the deviation coefficient (n) for a non-log-linear survivor curve is included in the table. A more detailed list of parameters and related information for these microbial populations are presented Table A.3.1 in the appendices. In addition, the table in the appendices includes kinetic data for *Enterobacter sakazakii*.

The kinetic parameters presented in Table 3.2 are for various species of *E. coli*, including *E. coli* ATCC and O157:H7. Many serious outbreaks of food-borne illness have been associated with these microbial populations. The kinetic parameters include rate constants and agent coefficients for thermal, pressure and PEF processes, and for several different substrate environments. A more detailed list of the kinetic parameters, and related information, for *E. coli* are presented in Table A.3.2 in the appendices.

Typical kinetics parameters to describe survivor curves for *Listeria species* are presented in Table 3.3. These parameters have been measured to assist in controlling food-borne infection outbreaks caused by products containing significant populations of *L. monocytogenese*. The table includes parameters for *L. innocus*, an organism identified as a potential surrogate for *L. monocytogenese*.

In addition to kinetic parameters to be used in the design of thermal processes, Table 3.3 provides parameters for UHP and PEF processes. Additional parameters for these microbial populations, along with additional background information, are presented in Table A.3.3 in the appendices. Finally, Table A.3.3 presents kinetic parameters for *Staphylococcus aureas*, which is another vegetative pathogen associated with food-safety concerns.

The kinetic parameters presented in Table 3.4 are for various *Clostridium* spores. The food safety importance of *Clostridium botulinum* has brought significant attention to the kinetic parameters of these pathogens and the factors influencing the magnitude of the parameters. In addition, several surrogate microorganisms have been identified—all with similar magnitudes of the kinetic parameters to describe survivor curves. More detailed information and background on the various parameters are provided in Table A.3.4 in Appendix A. In addition, the table includes kinetic parameters for the pathogenic spores of *Bacillus subtilis* and *Bacillus cereas*.

Tables A.3.1 to A.3.4 present additional and more specific kinetic parameters than are provided in Tables 3.1 to 3.4. Each table presents the type of preservation technology used during the experimental measurements: thermal, UHP or PEF. The specific microbial populations for each of the kinetic parameters have been identified. The rate constants are the Decimal Reduction Time (D) or a rate constant (k), and a deviation constant (n) to express the deviation from a first-order model. The tables identify the intensity of the agent (thermal, UHP, PEF) used to measure the rate constant. In addition, the tables present the appropriate coefficients (z, E_A, ΔV, etc.) to describe the influence of agent intensity on the rate constant. The substrate carrying the microbial population has been presented for each situation. Finally, each table includes additional information relevant to the measurements and the reference for the source of kinetic parameters included in the tables.

3.2.1 Microbial population variability

As is evident with kinetic parameters in Tables 3.1 to 3.4, the magnitudes of parameters for a given microbial population vary considerably. The factors influencing this variability are obvious in some cases, but the influence of experimental procedures seems to be a significant factor. One of the more significant procedural factors is the substrate carrying the microbial population during exposure to the preservation process conditions. The influence of the substrate is evident when evaluating the kinetic parameters for each of the different processes (thermal, UHP, PEF), and for different microbial population (vegetative cells, spores). The substrates range from water to complex food systems. In general, rate constants decrease as the substrate becomes more complex, indicating that the effectiveness of the process decreases as the composition of the food becomes more complex. The microbial populations seem to be protected by insoluble or soluble (or both) product components during the preservation process. The specific impact of the proteins, carbohydrates, fats, and ash on the response of the microbial population to a preservation process is not clear. Higher concentrations of these components seem to protect the microorganisms, and create lower rate constants. The coefficients used to express the influences of the intensity of the preservation agent indicate that the microbial population is more difficult to inactivate when the microorganism is carried in a food product.

3.2.2 Vegetative populations versus spores

The kinetic parameters for survivor curves of several vegetative pathogens have been presented in Tables 3.1, 3.2 and 3.3. In general, the magnitudes of these parameters clearly indicate that vegetative pathogens are less resistant to preservation processes than pathogenic spores, as presented in Table 3.4. There are no direct comparisons among the various parameters due to the intensity of the process agent used during measurements, but process requirements for pathogenic spores may be more than double the magnitude for vegetative pathogens. The coefficients used to describe the intensity of the process agent also indicate that spores are more resistant to preservation processes.

3.2.3 Process technology

The results in Tables 3.1 through 3.4 include kinetic parameters for thermal, UHP, and PEF processes. The most obvious difference in the effectiveness of these technologies is the significantly higher rate constants for the PEF process when applied to vegetative pathogens, as compared to thermal and UHP processes. These differences are at least 1000-fold, suggesting that similar magnitudes of reductions in microbial populations can be achieved in microseconds using PEF, as compared to several seconds or minutes when using thermal or

UHP processes. Note that many of the rate constants for the PEF measurements are based on only two points to describe the survivor curve: initial and final population. The results from one investigation (Rodrigo, Barbosa-Canovas, Martinez, & Rodrigo, 2003) indicate that the survivor curves during PEF processes may follow first-order model or the two-parameter model presented as Eq. (2.22 or 2.25) Kinetic parameters for applications of the PEF process to populations of microbial spores are limited to results for *B. cereus* and *B. subtilus* spores. These results indicate that rate constants for the PEF process are higher than for thermal or UHP processes.

The kinetic parameters for thermal and UHP processes have similar magnitudes. The results presented in Tables 3.1 through 3.4 have been assembled for typical ranges of temperatures and pressures. For vegetative pathogens, the results suggest that rate constants to describe microbial survivor curves measured at temperatures of approximately 60°C are similar to rate constants measured at pressures of 300 to 400 MPa. The kinetic parameters for pathogenic spores indicate that rate constants are lower than for vegetative populations. For these processes, the rate constants for thermal processes at 120°C are similar to UHP processes at 700 to 800 MPa. The magnitudes of the coefficients used to describe the influence of temperature and/or pressure on the rate constants suggest that vegetative pathogens and pathogenic spores respond to thermal and UHP processes in a similar manner.

3.3 Applications of kinetic parameters

Kinetic parameters presented in Tables 3.1 through 3.4 (and Tables A.3.1 to A.3.4) are typical of the information needed when designing processes for food product preservation. Because these parameters are for pathogenic microorganisms, the primary purpose of the process design is to ensure the microbiological safety of the products. When the populations of spoilage microorganisms happen to have lower rate constants, the processes need to incorporate the appropriate kinetic parameters to ensure the desired product shelf life. The following examples and discussions illustrate using kinetic parameters in the first step to designing the preservation process for a food product.

The various models presented in Chapter 2 are used to demonstrate the use of kinetic parameters for microbial pathogens in preservation process design. The model for the microbial survivor curve is the primary expression for preservation process design. This expression describes the reduction in microbial population as a function of time and is used to predict the time required to reach the target number of survivors.

Because the preservation process must be designed to ensure the safety of the product, the number of survivors must be expressed in terms of the probability of survivors. The probability of survivors becomes an expression of risk associated with a pathogen surviving the preservation process. An example probability is one survivor in one million units of product processed.

The survivor curve in Figure 3.5 illustrates the concept of survivor probability. In the illustration, the initial microbial population is 10^6 per product package unit. For this example, the survivor curve is assumed to follow a first-order model or log-linear model. The target number of survivors is 0.01 or a probability of 1 survivor in 100 product package units. Ultimately, this probability becomes the target population for the preservation process. In practice, this magnitude must be established based on a variety of issues associated with the product attributes, consumer expectations, and regulations.

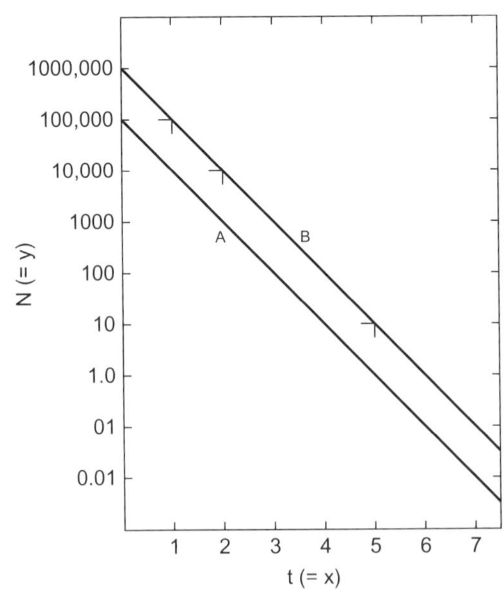

Figure 3.5 Microbial survivor curves illustrating probability of survival (from Ball & Olson, 1957).

Example 3.1

A ground beef product is receiving a mild thermal process to ensure product safety by reducing the population of *Salmonella typhimurium*. Measurements have indicated that populations of the pathogen may be as high as 10 in the product per package unit. Establish the process time needed to achieve an acceptable level of product safety.

Given:

1. Ground beef is the product.
2. Vegetative pathogen is *Salmonella typhimurium*.

Approach:

1. Select the appropriate kinetic parameters from Table 3.1 or Table A.3.1.
2. Establish the acceptable level of product safety.

Solution:

1. By reference to Table A.3.1, there are two potential rate constants for *Salmonella typhimurium* in ground beef: k = 8.8/min or 10.6/min at 60°C.
2. The acceptable risk associated with this product has been established as 1 survivor in 100,000 packages of the product.
3. Because the available kinetic parameters have been determined by using a first-order model, the following survivor curve equation, based on Eq. (2.11) of Chapter 2, is used:

$$N = N_o \exp(-k\,t)$$

4. To determine the process time, the preceding equation can be expressed as follows:

$$t = -\mathrm{Ln}\,(N/N_o)/k = \mathrm{Ln}\,(N_o/N)/k$$

5. Using the preceding expression and the input data

$$t = \mathrm{Ln}\,(10/0.00001)/8.8 = 1.57 \text{ min}$$

where N = 1/100,000 = 0.00001, the results indicate that a process time of 1.57 min at 60°C is needed to ensure the established level of product safety.

The preceding example can be used to illustrate several important aspects of preservation process design:

- The initial population of the pathogen in the food product must be established as accurately as possible. If measurements indicate a range of populations have been determined, the highest magnitude should be selected.
- Although two rate constants were discovered for the pathogen in ground beef, the lower of the two rates was selected for the computation.
- The accepted target probability of survivors was 1 in 100,000 product package units. Given the initial population, a 6 log cycle reduction in the population would be accomplished by the process.
- The process of 1.57 min at 60°C assumes that the microbial population, at any and all locations within the product package unit, is exposed to this time and temperature. In addition, that the product temperature is increased to 60°C instantaneously, held for 1.57 min, and then cooled instantaneously.

Example 3.1 is an illustration of predicting process time using the log-linear or first-order survivor model. Variations from the log-linear survivor curve model will impact the process time, and the influence must be considered. A frequently used model is Eq. (2.23 or 2.25) from Chapter 2. This model accommodates variations by incorporating an additional parameter (n), and several entries in Tables A.3.1 through A.3.4 in Appendix A have used this model to evaluate kinetic parameters for vegetative pathogens and pathogenic spores.

Example 3.2

A preservation process for a food product known to contain populations of *Salmonella typhimurium* is being developed. Measurements indicate that initial populations may be as high as 100 per product package. Recommend a process to ensure that the food safety risk from this product is negligible.

Given:

1. The specific composition of the product is not provided, but it is a high-carbohydrate product.
2. Kinetic parameters for *Salmonella typhimurium* are provided in Table A.3.1 in Appendix A.

Approach:

1. The kinetic parameters in Table A.3.1 for a high-carbohydrate product (high sugar content) are selected.
2. Because the kinetic parameters are for the non-log-linear model, Eq. (2.23) is used to predict the process time.
3. The negligible risk requirement is generally viewed as a 1 in 1 million risk. For this application, 1 survivor in 1 million product packages is used, or a probability of 0.000001.

Solution:

1. The kinetic parameters in Table A.3.1 include k = 0.144/min (D′ = 6.94 min) and n = 0.667 for 66% sugar at 60°C, and k = 0.1/min (D′ = 10 min), and n = 1.2 for 48% sugar at 60°C. The process time for both sets of kinetic parameters will be determined.
2. Using Eq. (2.23):

$$\log N_o - \log N = [t/D']^n$$

$$[\log (N_o/N)]^{1/n} = t/D'$$

$$t = D' \, [\log (N_o/N)]^{1/n}$$

3. By incorporating the kinetic parameters for 66% sugar at 60°C, and the magnitudes of microbial populations, the following result is obtained:

$$t = (6.94) \, [\log (100/0.000001)]^{1/0.667}$$
$$t = 156.8 \text{ min at } 60^\circ\text{C}$$

4. The process time for 48% sugar would be

$$t = (10) \, [\log (100/0.000001)]^{1/1.2}$$
$$t = 58.6 \text{ min at } 60^\circ\text{C}$$

5. The results for the same product substrate are significantly difference due to the difference in sugar concentration, as reflected in the differences in kinetic parameters.

The results for preservation process times in Example 3.2 provide interesting insights about the survivor curve model to be used for prediction of the process time. First of all, the significant difference in the process time in Examples 3.1 and 3.2 for the same vegetative pathogen is obvious. A portion of this difference is due to higher initial population in Example 3.2, as well as the lower probability of survivors associated with negligible risk. These influences account for a portion of the difference, but an additional portion of the difference must be attributed to the influence of the substrate on the response of the microorganism to the thermal process.

The results in Example 3.2 indicate that the process time for 66% sugar substrate is significantly greater than for 48% sugar. This difference is due to the differences in the kinetic parameters for the microorganism in the two substrates. The primary parameter is "n"; the factor describing the extent of deviation of the survivor curve model from the first-order model. For 66% sugar, the parameter is less than 1.0 (n = 0.667), suggesting that the survivor curve is concave upward, as compared to a log-linear survivor curve. The shapes of the curves were illustrated in Figure 2.10 of Chapter 2. The process time for this situation is expected to be longer than would be predicted by the first-order model. When the parameter is greater than 1.0 (n = 1.2), as indicated for 48% sugar, the survivor curve is concave downward, and the process time is lower than predicted by a log-linear survivor curve model. The example indicates that the process time for the concave upward survivor curve is more than double the process time when the survivor curve is concave downward.

A review of the kinetic parameters presented in Tables 3.1 through 3.4 indicates that rate constants are usually presented at a single temperature. If the process of interest is to be conducted at a temperature other than the temperature indicated in the table, the rate constant at a different temperature is needed. During measurements of kinetic parameters, rate constants are usually measured at more than one temperature, and the results are expressed as an agent intensity coefficient (E_A or z). These coefficients provide the information needed to predict process times at temperatures other than the temperature associated with the rate constant presented in the table. Caution should be used when computing rate constants outside the range of temperatures used to determine the agent intensity coefficient because the relationship between the rate constant and the agent intensity may change outside the range of measurements.

Example 3.3

A process for a shelf-stable vegetable product is being established. The pathogenic spore of concern is *Clostridium botulinum*, and measurements indicate that initial populations may be as high as 10 per product container. Recommend a process time at 121 °C sufficient to ensure negligible risk associated with distribution of the shelf-stable product.

Given:

1. The product of concern is a shelf-stable vegetable product, containing initial populations of *Cl. botulinum* of as high as 10 per product container.
2. A typical negligible risk expectation is 1 survivor per 1 million containers.
3. Typical kinetic parameters for the pathogenic spore are presented in Table 3.4.

Approach:

1. Because the available kinetic parameters are based on a first-order model, an expression based on Eq. (2.11) is used. Eq. (2.16) is used to compute the rate constant at the desired temperature.
2. The kinetic parameters in Table 3.4 for vegetable products are presented as ranges of values: k = 0.929 to 3.775/min or k = 0.186 to 4.7/min at a temperature of 110 °C. The temperature coefficient is the Activation Energy Constant, with values ranging from 242.1 to 375.4 kJ/mole, or 260.1 to 379.6 kJ/mole.

Solution:

1. The first step is to select the appropriate rate constant for *Cl. botulinum*. The lowest rate constant presented in Table 3.4 is 0.186 for *Cl. botulinum B*, which would provide the most conservative process time.
2. The Activation Energy Constants for *Cl. botulinum B* range from 260.1 to 379.6 kJ/mole, and the lower magnitude would provide the most conservative coefficient for computing the adjustment in the rate constant.
3. Using Eq. (2.16), the rate constant at 121 °C is determined as

$$\text{Ln } k = -[E_A/R\mathbf{T}] + \text{Ln } k_{ref}$$

(*Continued*)

Example 3.3 (Continued)

or

$$\text{Ln } k = -E_A/R\,[1/\mathbf{T} - 1/\mathbf{T}_{ref}] + \text{Ln } k_{ref}$$

$$\text{Ln } k = -(260{,}100/8.3144)[1/(121 + 273) - 1/(110 + 273)] + \text{Ln } (0.186)$$

$$\text{Ln } k = -(31{,}283)\,[2.538 \times 10 - 3 - 2.591 \times 10 - 3] + \text{Ln } (0.186)$$

$$\text{Ln } k = 0.7195 - 1.682 = -0.9625$$

$$k = 0.382/\text{min at } 121^\circ\text{C}$$

4. Using the first-order survivor curve expression:

$$t = \text{Ln } (10/0.000001)/0.382$$

$$t = 42.2 \text{ min at } 121^\circ\text{C}$$

The process time for the vegetable product is influenced by the process temperature; the time would be much higher at 110°C. The process time determined in Example 3.3 assumes that all portions of the product are exposed to the temperature of 121°C for the entire process time. The influence of time to reach the process temperature and the distribution of temperature within the container will be discussed in Chapter 6. The predicted process time is influenced by the temperature coefficient or Activation Energy Constant, as well. If magnitudes higher than 260.1 kJ/mole are used, the rate constants would be lower, and the process times would be longer.

The kinetic parameters presented in Tables 3.1 through 3.4 include data for populations of vegetative pathogens and pathogenic spores when exposed to UHPs. The use of UHP as a preservation process for a food product depends on the application of an elevated pressure to the product for a specified period of time. The steps in process design are similar to those for a thermal process.

Example 3.4

An ultra-high pressure preservation process is being developed to eliminate the food safety risk associated with *Listeria monocytogenes* in ground pork. The process, at 414 MPa, is expected to produce a product with negligible risk, while minimizing thermal impacts on product quality attributes. Populations of *Listeria monocytogenes* in the ground pork product have been measured and indicate that the population in the package prior to the process could be as high as 15. Determine the time for an UHP process to meet the process specifications described.

Given:

1. The target microbial pathogen is *Listeria monocytogenes* in ground pork.
2. An ultra-high pressure process at 414 MPa is used.
3. An initial population of 15 per product package and a negligible risk have been specified.

Approach:

1. The kinetic parameters needed, at 414 MPa, for the process design are presented in Table 3.3.
2. The negligible risk requirement requires a probability of 0.000001, or 1 survivor in 1 million packages.
3. The first-order survivor curve model is used.

Solution:

1. The kinetic parameters from Table 3.3 indicate that the rate constant is 0.552/min at 414 MPa and a temperature of 25°C. The rate constant selected is lowest in a range of measured parameters and would provide the conservative result. In addition, the lower temperature should minimize the influence of temperature on the product quality.
2. The first-order survivor curve model, based on Eq. (2.11), will be used as

$$N = N_o \exp[-k\,t]$$

or

$$t = -\mathrm{Ln}\,[N/N_o]/k = \mathrm{Ln}\,(N_o/N)/k$$

(Continued)

Example 3.4 (Continued)

3. These input parameters are used:

$$t = \text{Ln}\,[15/0.000001]/0.552$$

$$t = (16.524)/0.552$$

$$t = 29.9\ \text{min at } 414\ \text{MPa and } 25^{\circ}\text{C}$$

4. The results indicate that a high pressure process of 29.9 min at 414 MPa and 25°C would meet the expectations of negligible risk for the preservation process.

As previously discussed for thermal processes established by using the first-order or log-linear survivor models, there are several observations about the UHP process that deserve similar discussion:

- Meeting the expectation of the negligible risk is dependent on the initial population of the pathogen. The accurate measurement of this magnitude is a very important step in process design.
- The lowest rate constant from a range of measured values from the published literature should be used to ensure the most conservative result from the process design.
- The process time assumes that pressure (414 MPa) is achieved instantaneously at the beginning of the process, followed by a holding period of 29.9 min, followed by an instantaneous release of the pressure. The result also assumes that the pressure of 414 MPa is uniform throughout the product in the package during the preservation process.
- It has been established that the product temperature increases during a UHP process. For the process in Example 3.4, it is assume that product temperature does not exceed 25°C during the process. If the product temperature is increased to a magnitude where the temperature influences the rate constant for the pressure process, appropriate adjustments in the kinetic parameters would be required.

Example 3.5

Determine the influence of product temperature on the preservation process recommended for conditions presented in Example 3.4.

Given:

1. The same conditions as described in Example 3.4 are used.
2. A higher product temperature is evaluated.

Approach:

1. The only parameter to be changed from the solution given for Example 3.4 is temperature. The kinetic parameters presented in Table A.3.3 include a first-order rate constant for *Listeria monocytogeneses* at 50°C and at 414 MPa, as compared to the rate constant at 25°C used in the previous example.

Solution:

1. The first-order rate constant at 50°C for *Listeria monocytogeneses* is 3.656/min at 414 MPa. Note that the lowest magnitude rate constant is selected from the range of rate constants measured for these conditions.
2. Using the rate constant for 50°C, the following result is obtained:

$$t = \text{Ln}\,[15/0.000001]/3.656$$
$$t = 4.52 \text{ min at } 414 \text{ MPa and } 50^\circ\text{C}$$

3. The preservation process time at 414 MPa and 50°C is significantly lower than the process time at 25°C; 4.52 min as compared to 29.9 min. If the impact of the higher temperature on product quality attributes is not significant, the process at 50°C would be recommended.

The influence of pressure magnitude on the rate's contacts for reducing microbial populations is evident in the parameters presented in Tables 3.1 through 3.4 (as well as Tables A.3.1 to A.3.4). The magnitude of this influence is described by pressure coefficients: a pressure z-value or the Activation Volume Coefficient (ΔV). These coefficients are needed to evaluate the influence of pressure magnitude on process times.

Example 3.6

A UHP process is being developed for a pork product, with focus on ensuring that populations of *Listeria monocytogenes* in the product are reduced to negligible levels. The populations in the product may be as high 25 per product package. A system capable of providing a pressure of 375 MPa at 30°C has been selected. Determine the process time required.

Given:

1. The vegetative pathogen is *Listeria monocytogenes,* and kinetic parameters are provided in Table 3.3.
2. An initial population of 25 per product package has been established.
3. A negligible risk of survivors from the process requires 1 survivor per 1 million packages from the process.
4. The process is conducted at 375 MPa.

Approach:

1. The kinetic data from Table 3.3 include a rate constant at 400 MPa and the Activation Volume Constant for *Listeria monocytogenes*.
2. Equation (2.20) is used to compute the rate constant at 375 MPa.
3. The first-order rate constant at 375 MPa and 30°C are used in Eq. (2.11) to compute the process time.

Solution:

1. The parameters obtained from Table 3.3 are k = 0.652/min at 400 MPa and

$$V = -3.43 \times 10 - 5\ \text{m}^3/\text{mole}$$

2. Use Eq. (2.20):

$$\text{Ln } k = \text{Ln } k_{ref} - (\Delta V/R\mathbf{T})\,[P - P_{ref}]$$

$$\text{Ln } k = \text{Ln } (0.652) - [-3.43 \times 10^{-5}/(8.31441)(303)]\,[(375 - 400) \times 10^{6}]$$

$$\text{Ln } k = -0.4277 + [13.615 \times 10^{-9}]\,[-25 \times 10^{6}]$$

$$\text{Ln } k = -0.768$$

$$k = 0.464/\text{min at } 375\ \text{MPa}$$

3. Use Eq. (2.11):

$$t = \mathrm{Ln}\,[25/0.000001]/0.464$$

$$t = 36.7 \text{ min at } 375 \text{ MPa and } 30^{\circ}\text{C}$$

4. The process required at 375 MPa and 30°C is 36.7 min to ensure a negligible risk of survivors of *Listeria monocytogenes.*

These examples illustrate that the process times to be used for UHP preservation processes are established in the same manner as for thermal processes. The additional steps required to design preservation processes will be presented and discussed in Chapter 6.

Limited amounts of kinetic data are available for microbial survivor curves during exposure to PEF technologies. Although the kinetic parameters indicate that the rate constants are much higher than for thermal or UHP processes, the parameters are used in the same manner I process design. As previously indicated, the kinetic parameters to be used in examples are found in Tables 3.1 through 3.3 (and Tables A.3.1 to A.3.4).

Example 3.7

Liquid whole eggs are to be processed using PEF as a preservation technology. The process is expected to reduce the population of *Listeria enteritidis* to negligible levels, given that initial populations are as high as 100 for the amount of product placed in a final package. Determine the process time, when a field intensity of 30 kV/cm is to be used at 20°C.

Given:

1. A preservation process for liquid whole eggs is to be established using PEF with a field intensity of 30 kV/cm and 20°C.
2. The vegetative pathogen of concern is *Listeria enteritidis*, and the initial population is 100 per product package.

(*Continued*)

Example 3.7 (Continued)

Approach:

1. The kinetic parameters in Tables 3.3 and A.3.3 indicated that rate constants for the pathogen range from 0.0303 to 0.0546/microsec at 20 kV/cm and 20 °C.
2. Because the rate constants are for a first-order survivor curve model, a log-linear model is used for computing the process time.
3. For this product, a probability of 1 survivor in 10,000 product packages is acceptable.

Solution:

1. Equation (2.10) is the model to use for this application:

$$\mathrm{Ln}\,[N/N_o] = -k\,t$$

or

$$t = \mathrm{Ln}\,[N_o/N]\,/k$$

2. Use the input parameters:

$$t = \mathrm{Ln}\,[100/0.0001]/0.0303$$
$$t = 456 \text{ microsec at } 20\text{ kV/cm and } 20^{\circ}\mathrm{C}$$

3. The results indicate that the liquid whole egg needs to be exposed to the PEF of 20 kV/cm and 20 °C for 456 microseconds. This is a fraction of a second. The challenge is to achieve uniform exposure of the product as it flows through the PEF cell.

The commercial applications of PEF as a preservation process are still limited. The challenges include achieving uniform distribution of the field within a solid food product and evaluating the effectiveness in the inactivation of pathogenic spores. The appropriate kinetic parameters for use in process design are limited. In addition, there is some evidence that the survivor curves for microbial populations may not be first-order.

Example 3.8

A preservation process is being developed for a juice product. PEF is being considered as a technology to ensure negligible risk of *E. coli* survivors in the product. Populations of the pathogen in the product may be as high as 55 per product package. Determine the process time for PEF.

Given:

1. The pathogen of concern is *E. coli,* and the initial population is 55 per product package.
2. The product is juice, and PEF is the preservation process.

Approach:

1. The kinetic parameters are selected from Table 3.2 or A.3.2, where parameters for a carrot juice product have been recorded.
2. The appropriate survivor curve model is selected, and the process time is determined.
3. The negligible risk requirement indicates that a probability of survivors must be 0.000001.

Solution:

1. The kinetic parameters available from Tables 3.2 and A.3.2 include rate constants for first-order and non-log-linear survivor curves. The parameters are $k = 0.0138$/microseconds for the first-order model, and $D' = 31.8$ microseconds and $n = 0.6$ for the non-log-linear model.
2. For the first-order model, Eq. (2.10) is used:

$$t = \text{Ln}\,[55/0.000001]/0.0138$$
$$t = 1291.5 \text{ microsec} = 0.00129 \text{ sec}$$

3. When using the non-log-linear survivor curve model from Eq. (2.23):

$$t = (31.8)\,[\log{(55/0.000001)}]^{1/0.6}$$
$$t = 963.2 \text{ microsec} = 0.00096 \text{ sec}$$

4. The process times from the two models are not significantly different, and the longer of the two is selected. The recommended process time is 0.0013 sec with an electric field intensity of 25 kV/cm and at product temperature that does not exceed 40°C.

As indicated earlier, most applications of the PEF technology to food preservation involve liquid foods and the application of the electric field during flow through a treatment cell. In these types of applications, the process time is used to establish the product flow rate through the process treatment cell. As long as the application of the PEF is applied uniformly to the product during flow through the cell, the product flow rate (and velocity) is inversely proportional to the process time.

As illustrated in Example 3.3, pathogenic spores are a significant challenge in the development of the preservation process. When considering thermal processes, the process times and temperature are much higher than required for vegetative pathogens. In general, the kinetic parameters in Tables 3.1 through 3.4 (and Tables A.3.1 to A.3.4) indicate that the same challenges exist when developing processes for UHP preservation. The published literature does not contain kinetic parameters to describe the survivor curves for microbial spores using PEF treatments.

Example 3.9

A new shelf-stable food product is being developed, and safety from the spores of *Cl. botulinum* needs to be established. Populations of the pathogenic spores, equivalent to 10 per product package, must be considered to achieve negligible food safety risk for the product. A UHP processing system is being considered for the product.

Given:

1. The pathogen of concern is *Cl. botulinum,* with an initial population of 10 per product package.
2. A negligible risk criterion has been established for the product.
3. A high pressure processing system has been selected.

Approach:

1. The kinetic parameters for the pathogenic spore are selected from Tables 3.4 or A.3.4.

2. The appropriate survivor curve model is selected based on the kinetic parameters.
3. The negligible risk criterion requires a probability of 0.000001 survivors.

Solution:

1. Based on the kinetic parameters provided in Table A.3.4, the parameters for *Cl. sporogenes* have been selected. These parameters include k = 2.556/min at a pressure of 700 MPa and 108°C for a first-order survivor curve model. In addition, parameters for a non-log-linear model include D′ = 0.19 min and n = 0.25 for the same pathogenic spore at 700 MPa and 105°C.
2. Use the first-order survivor curve model:

$$t = \text{Ln}\,[N_o/N]/k$$

$$t = \text{Ln}\,[10/0.000001]/2.556$$

$$t = 6.3 \text{ min at } 700 \text{ MPa}$$

3. Use the non-log-linear model:

$$t = D' \log\,[N_o\,/N]^{1/n}$$

$$t = (0.19)\,[\log\,(10/0.000001)]^{1/0.25}$$

$$t = 456.2 \text{ min at } 700 \text{ MPa}$$

4. The process results suggest a significant difference between the two models. The process time from the non-log-linear survivor curve is much longer, most likely due to the concave upward shape of the curve. The low magnitude of the "n" parameter indicates the significant deviation from the first-order model and the longer process time.

Both of the process times obtained in Example 3.9 require a very high pressure (700 MPa) and represent a significant operational challenge. This challenge become most evident for pathogenic spores and is likely to limit immediate applications of UHP processes for shelf-stable foods.

3.4 Definition of microbial inactivation

The use of kinetic parameters to predict appropriate process times for food preservation processes depends on the survivor curve being an actual representation of pathogen survival during the preservation process. This is a primary concept of any preservation process and has been a guiding principle in the application of thermal processes for food preservation. In general, previous developments have demonstrated that the survivor curve data used to generate kinetic parameters are equivalent to the microbial survival occurring during application of the thermal processing of the food product in a commercial operation.

Research by Teo, Ravishankar, and Sizer (2001) indicates that survivor curve data for UHP processes may need more careful analysis. These results suggest that some portion of the survivors from UHP processes for vegetative pathogens in foods may be injured and may recover under appropriate conditions for growth. The focus of these observations was on the assay used to determine the numbers of microbial survivors during experiments for quantification of kinetic parameters. These experiments may require an extra step during recovery of survivors in the medium after the process treatment. The purpose of this step is to ensure that injured cells of the vegetative pathogen are recovered and included in survivor numbers used to quantify the kinetic parameters. Unless these precautions are followed, the kinetic parameters may result in predicted process times that are inadequate to achieve the desired targets in safety of the final product.

There is no evidence that the kinetic parameters for UHP processes of pathogenic spores may be influenced by the survival of injured cells. The results for vegetative cells suggest that the same precaution should be applied to the development of processes for pathogenic spores. This precaution should be considered during evaluation of survivor curve data for vegetative pathogens or pathogenic spores from PEF processes. Although there is no evidence of cell injury from PEF processes, the potential impact on survivor curves from PEF processes must be considered. These precautions need to be considered for other alternative technologies considered for preservation of food products.

3.5 Kinetic parameters for alternative preservation technologies

Most of the illustrations used to predict process times for UHP processes indicate that the kinetic parameters are very similar to those used for thermal processes. The rate constants (k, D, D′) for UHP processes are measured and used in the same manner as for thermal processes. As consideration is given to the uniformity of parameters, the focus should be on first-order rate constants (k) or multiple-order rate constants (k′,n).

The kinetic parameters used to describe the influence of pressure magnitude on rate constants should be uniform as well. Several investigations have used the traditional Thermal Inactivation Coefficient (z) as a parameter for describing the influence of pressure magnitude on the rate constants. Fortunately, these coefficients can be converted to the preferred parameter—the Activation Volume Coefficient (ΔV). Hopefully, future investigations will consider using the Activation Volume Coefficient on a consistent basis for describing the influence of pressure magnitude on rate constants.

The kinetic parameters for PEF processes present several different challenges. Many investigations have reported rate constants for either first-order or multiple-order survivor curves. The measurement of these constants presents unique challenges due to the very short time frame (microseconds) for reducing microbial populations. The application of these parameters can be used to predict process times at the same electric field intensity as used to measure the rate constant. The parameters for describing the impact of field intensity on rate constants are not uniform. Although a coefficient similar to thermal and pressure (z) has been used in some reports, an alternative model may be most appropriate.

List of symbols

D = Decimal Reduction Time

D′ = Decimal Reduction Time for non-log-linear microbial survivor curve

E_A = Activation Energy Constant

k = first-order reaction rate constant

k_o = zero-order reaction rate constant
k_2 = second-order reaction rate constant
k^o = pre-exponential constant in Arrhenius relationship
k_{ref} = reference rate constant
N = microbial population
N_o = initial microbial population
P = pressure
P_{ref} = reference pressure
R = gas constant
t = time
T = temperature
T = absolute temperature
ΔV = activation volume
z = thermal resistance constant
z_p = pressure change for one log change in rate constant

Bibliography

Ahmed, M. N., Connor, D. E., & Huffman, D. L. (1995). Heat-resistance of *Escherichia coli* O157:H7 in meat and poultry as affected by product composition. *Journal of Food Science*, *60*, 606–610.

Ahn, J., Balasubramaniam, V. M., & Yousef, A. E. (2007). Inactivation kinetics of selected aerobic and anaerobic bacterial surrogate spores by pressure-assisted thermal processing. *International Journal of Food Microbiology*, *113*, 321–329.

Alvarez, I., Pagan, R., Condon, S., & Raso, J. (2003). The influence of process parameters for the inactivation of *L. monocytogenes* by pulsed electric fields. *International Journal of Food Microbiology*, *87*, 87–95.

Alvarez, I., Raso, J., Palop, A., & Sala, F. J. (2000). Influence of factors on the inactivation of *Salmonella senftenberg* by pulsed electric field. *International Journal of Food Microbiology*, *85*, 143–146.

Amiali, M., Ngadi, M. O., Smith, J. P., & Raghaven, V. G. S. (2006). Inactivation of *Escherichia coli* O157:H7 and *Salmonella enteritidis* in liquid egg white using pulsed electric field. *Journal of Food Science*, *71*(3), M88–M94.

Ananth, V., Dickson, J. S., Olson, D. G., & Murano, E. A. (1998). Shelf-life extension, safety and quality of fresh pork loin treated with high hydrostatic pressure. *Journal of Food Protection*, *61*(12), 1649–1656.

Ball, C. O., & Olson, F. C. W. (1957). *Sterilization in food technology*. New York: McGraw-Hill.

Bolton, D. J., McMahon, C. M., Doherty, A. M., Sheridan, J. J., McDowell, D. A., & Blair, J. S. (2000). Thermal inactivation of *Listeria monocytogenes* and *Yersinia enterocolitica* in minced beef under laboratory conditions and in sous-vide prepared minced and solid beef cooked in a commercial retort. *Journal of Applied Microbiology*, *79*, 626–632.

Calderon-Miranda, M. L., Barbosa-Canovas, G. V., & Swanson, B. G. (1999a). Inactivation of *Listeria innocua* in liquid whole egg by pulsed electric fields and nisin. *International Journal of Food Microbiology*, *51*, 7–17.

Calderon-Miranda, M. L., Barbosa-Canovas, G. V., & Swanson, B. G. (1999b). Inactivation of *Listeria innocua* in skimmed milk by pulsed electric fields and nisin. *International Journal of Food Microbiology*, *51*, 19–30.

Carlson, T. R., Marks, B. P., Booren, A. M., Ryser, E. T., & Orta-Ramirez, A. (2005). Effect of water activity on thermal inactivation of *Salmonella* in ground turkey. *Journal of Food Science*, *70*(7), M363–M366.

Chen, H., & Hoover, D. G. (2003). Modeling the combined effect of high hydrostatic pressure and mild heat on the inactivation of *Listeria monocytogenes* Scott A in whole milk. *Innovative Food Science & Emerging Technology*, *4*, 25–34.

Cho, H.-Y., Sastry, S. K., & Yousef, A. E. (1999). Kinetics of inactivation of *Bacillus subtilis* spores by continuous or intermittent ohmic and conventional heating. *Biotechnology and Bioengineering*, *62*(3), 368–372.

Cserhalmi, Z., Vadac, I., Beczner, J., & Czukor, B. (2002). Inactivation of *Saccharomyces cerevisiae* and *Bacillus cereus* by pulsed electric field technology. *Innovative Food Science & Emerging Technology*, *3*(1), 41–45.

Dunn, J. E., & Pearlman, J. S. (1987). Methods and apparatus for extending the shelf-life of fluid food products. Maxwell Laboratories Inc. U.S. Patent 4,695,472.

Edelson-Mammel, S. G., & Buchanan, R. L. (2004). Thermal inactivation of *Enterobacter sakazakii* in rehydrated infant formula. *Journal of Food Protection, 67*(1), 60–63.

Enache, E., Chen, Y., Awuah, G., Economides, A., & Scott, V. (2006). Thermal resistance parameters for pathogens in white grape juice concentrate. *Journal of Food Protection*, *69*(3), 564–569.

Esty, J. R., & Meyer, K. R. (1922). The heat resistance of the spores of *B. botulinus* and allied anaerobes. *The Journal of Infectious Diseases*, *31*, 650–663.

Evrendilek, G. A., & Zhang, Q. H. (2003). Effects of ph, temperature and pre-pulsed electric field treatment on pulsed electric field and heat inactivation of *Escherichia coli* 0157:H7. *Journal of Food Protection*, *66*, 755–759.

Farber, J. M. (1989). Thermal resistance of *Listeria* monocytogenes in foods. *International Journal of Food Microbiology*, *8*, 285–291.

Fernandez-Molina, J. J., Barkstrom, E., Torstenssen, P., Barbosa-Canovas, G. V., & Swanson, B. G. (1999). Shelf-life extension of raw skim milk by combining heat and pulsed electric fields. In: The 6th Conference of Food Engineering (CoFE99). G.V. Barbosa-Canovas and S. Lombardo (Eds.), AIChE, Dallas. pp. 349–355.

Gaze, J. E., Brown, G. D., Gaskell, D. E., & Banks, J. G. (1989). Heat resistance of *Listeria* monocytogenes in homogenates of chicken, beef steak and carrot. *Food Microbiology*, *6*, 251–259.

Gervilla, R., Capellas, M., Ferragut, V., & Guamis, B. (1997). Effect of high hydrostatic pressure on *Listeria innocua* 910 CECT inoculated in ewes' milk. *Journal of Food Protection*, *60*(1), 33–37.

Gervilla, R., Felipe, X., Ferragut, V., & Guamis, G. (1997). Effect of high hydrostatic pressure on *Escherichia coli* and *Pseudomonas fluorescens* strains in bovine milk. *Journal of Dairy Science*, *80*, 2297–2303.

Grahl, T., Sitzman, W., & Marki, H. (1992). Killing of microorganisms in fluid media by high voltage pulses. *DECHEMA Biotechnology Conferences*, 675–679.

Gupta, R. P., & Murray, W. (1988). Pulsed high electric field sterilization. *IEEE Pulsed Power Conference Record National Research Council*, 58–64.

Heinz, V., Phillips, S. T., Zenker, M., & Knorr, D. (1999). Inactivation of *Bacillus subtilis* by high intensity pulsed electric fields. *Food Biotechnology*, *13*, 155–168.

Heldman, D. R., & Hartel, R. W. (1997). *Principles of food processing*. New York: Chapman & Hall, Inc.

Heldman, D. R., & Newsome, R. L. (2003). Kinetic models for microbial survival during processing. *Food Technology*, *57*(8), 7–12.

Hulsheger, H., & Nieman, E. G. (1980). Lethal effect of high voltage pulsed on *E. coli* K12. *Radiation and Environmental Biophysics*, *18*(4), 281–288.

Huang, L., & Juneja, V. K. (2003). Thermal inactivation of *Escherichia coli* O157:H7 in ground beef supplemented with sodium lactate. *Journal of Food Protection*, *66*(4), 664–667.

ICMSF. (1996). *Microorganims in foods. Microbiological specification of food pathogens*. London: Blackie Academic & Professional Publishers (Book 5).

IFT/FDA. (2001). Kinetics of microbial inactivation for alternative food processing technologies. *Journal of Food Science* (Special Supplement), 1–108.

Juneja, V. K. (2003). Predictive model for the combined effect of temperature, sodium lactate, and sodium diacetate on the heat resistance of *Listeria monocytogenes* in beef. *Journal of Food Protection*, *66*(5), 804–811.

Juneja, V. K., Synder, O. P., & Manner, B. S. (1997). Thermal destruction of *Escherichia coli* O157:H7 in beef and chicken: Determination of D- and z-values. *International Journal of Food Microbiology*, *35*, 231–237.

Kotrola, J. S., & Connor, D. F. (1997). Heat inactivation *Escherichia coli* O157:H7 in turkey meat as affected by sodium chloride, sodium lactate, polyphosphate, and fat content. *Journal of Food Protection*, *60*, 898–902.

Line, J. E., Fain, A. R., Moran, A. B., Martin, L. M., Lechowich, R. V., & Carosella, J. M. (1991). Lethality of heat to *Escherichia coli* O157:H7: D-value and z-value determinations in ground beef. *Journal of Food Protection*, *54*, 762–786.

Mackey, B. M., Forestier, K., & Issacs, N. (1995). Factors affecting the resistance of *Listeria monocytogenes* to high hydrostatic pressure. *Food Biotechnology*, *9*, 1–11.

Marquez, V. O., Mittal, G. S., & Griffiths, M. W. (1997). Destruction and inhibition of bacterial spores by high voltage pulsed electric field. *Journal of Food Science*, *62*(2), 399–401, 409.

Martin-Belloso, O., Vega-Mercado, H., Qin, B. L., Chang, E. J., Barbosa-Canovas, G. V., & Swanson, B. G. (1997a). Inactivation of *Escherichia coli* suspended in liquid egg using pulsed electric fields. *Journal of Food Processing and Preservation*, *21*(3), 193–208.

Martin-Belloso, O., Qin, B. L., Chang, E. J., Barbosa-Canovas, G. V., & Swanson, B. G. (1997b). Inactivation of *Escherichia coli* in skim milk by high intensity pulsed electric fields. *Journal of Food Process Engineering*, *20*, 317–336.

Mattick, K. L., Legan, J. D., Humphrey, T. J., & Peleg, M. (2001). Calculating Salmonella inactivation in nonisothermal heat treatments from isothermal nonlinear survival curves. *Journal of Food Protection*, *64*(5), 606–613.

Metrick, C., Hoover, D. G., & Farkas, D. F. (1989). Effects of high hydrostatic pressure on heat-resistant and heat-sensitive strains of Salmonella. *Journal of Food Science*, *54*, 1547–1564.

Murphy, R. Y., Beard, R. L., Martin, E. M, Duncan, L. K., & Marcy, J. A. (2004). Comparative study of thermal inactivation of *Escherichia coli* O157:H7, salmonella, and *Listeria monocytogenes* in ground pork. *Journal of Food Science*, *69*(4), FSM97–FSM101.

Murphy, R. Y., Duncan, L. K., Johnson, E. R., Davis, M. D., & Smith, J. N. (2002). Thermal inactivation D- and z-values of Salmonella serotypes and *Listeria innocua* in chicken patties, chicken tenders, franks, beef patties, and blended beef and turkey patties. *Journal of Food Protection*, *65*(1), 53–60.

Murphy, R. Y., Duncan, L. K., Beard, B. L., & Driscoll, K. H. (2003). D and z values of salmonella, *Listeria innocua* and *Listeria monocytogenes* in fully cooked poultry products. *Journal of Food Science*, *68*(4), 1443–1447.

Noma, S., Shimoda, M., & Hayakawa, I. (2002). Inactivation of vegetative bacteria by rapid decompression treatment. *Journal of Food Science*, *67*(90), 3408–3411.

O'Bryan, C. A., Crandall, P. G., Martin, E. M., Griffis, C. L., & Johnson, M. G. (2006). Heat resistance of salmonella spp., *Listeria monocytogenes, Escherichia coli* 0157:H7, and *Listeria innoccua* M1, A potential surrogate for *Listeria monocytogenes*, in meat poultry: A review. *Journal of Food Science*, *71*(3), R23–R30.

Osaili, T. M., Griffis, C. L., Martin, E. M., Beard, B. L., Keener, A., & Marcy, J. A. (2006). Thermal inactivation studies of *Escherichia coli* O157:H7, Salmonella, and *Listeria monocytogenes* in ready-to-eat chicken-fried beef patties. *Journal of Food Protection*, *69*(5), 1080–1086.

Osaili, T. M., Griffis, C. L., Martin, E. M., Beard, B. L., Keener, A. E., & Marcy, J. A. (2007). Thermal inactivation of *Escherichia coli* O157:H7, Salmonella, and *Listeria monocytogenes* in breaded pork patties. *Journal of Food Science*, *72*(2), M56–M61.

Pagan, R., Esplugas, S., Gongora-Nieto, M. M., Barbosa-Canovas, G. V., & Swanson, B. G. (1998). Inactivation of *Bacillus subtilus* spores using high intensity pulsed electric fields in combination with other food conservation technologies. *Food Science & Technology International*, *4*(1), 33–44.

Patterson, M. F., Quinn, M., Simpson, R., & Gilmore, A. (1995). Sensitivity of vegetative pathogens to high hydrostatic pressure treatment in phosphate buffered saline and foods. *Journal of Food Protection*, *58*(5), 524–529.

Patterson, M. F., & Kirpatrick, D. J.. (1998). The combined effect of high hydrostatic pressure and mild heat on inactivation of pathogens in milk and poultry. *Journal of Food Protection*, *61*(4), 432–436.

Pina Perez, M. C., Aliaga, D. R., Reyes, D. S., & Martinez Lopez, A. (2007). Pressure inactivation of *Enterobacter sakazakii* in infant formula milk. *Journal of Food Protection*, *70*(10), 2281–2289.

Ponce, E., Pia, R., Mor-Mur, M., Gervilla, R., & Guamis, B. (1998). Inactivation of *Listeria innocua* inoculated into liquid whole egg by high hydrostatic pressure. *Journal of Food Protection*, *61*(1), 119–122.

Qin, B. L., Chang, E. J., Barbosa-Canovas, G. V., & Swanson, B. G. (1995a). Nonthermal inactivation of *S. cerevisiae* in apple juice using pulsed electric fields. *Lebensmittel Wissenschaft Technology*, *28*(60), 564–568.

Qin, B. L., Pothakamury, U. R., Vega, H., Barbosa-Canovas, G. V., & Swanson, B. G. (1995b). Food pasteurization using high intensity pulsed electric fields. *Journal of Food Technology*, *49*(12), 55–60.

Qin, B. L., Zhang, Q. H., Barbosa-Canovas, G. V., Swanson, B. G., & Pedrow, P. D. (1995c). Pulsed electric field treatment chamber design for liquid food pasteurization using a finite element method. *Transactions ASAE*, *38*(2), 557–565.

Ramaswamy, H. S., Riahi, E., & Idziak, E. (2003). High-pressure destruction kinetics of *E. coli* (29055) in apple juice. *Journal of Food Science*, *68*(5), 1750–1756.

Reddy, N. R., Solomon, H. M., Fingerhut, G. A., Rhodehamel, E. J., Balasubramaniam, V. M., & Palaniappan, S. (1999). Inactivation of *Clostridium botulium* Type E spores by high pressure processing. *Journal of Food Safety*, *19*, 277–288.

Reina, L. D., Jin, Z. T., Yousef, A. E., & Zhang, Q. H. (1998). Inactivation of *Listeria monocytogenes* in milk by pulsed electric field. *Journal of Food Protection*, *61*(9), 1203–1206.

Rodrigo, D., Barbosa-Canovas, G. V., Martinez, A., & Rodrigo, M. (2003). Weibull distribution function based on an empirical mathematical model for inactivation of *Escherichia coli* by pulsed electric fields. *Journal of Food Protection*, *66*(60), 1007–1012.

Rovier, P., Carpi, G., Dall'Aglio, G., Gola, S., Maggi, A., Miglioli, I., & Scaramuzza, N. (1996). High pressure heat treatments: Evaluation of the sterilizing effect and of thermal damage. *Industrial Conservation*, *71*, 473–483.

Sapru, V., Teixeira, A. A., Semerage, G. H., & Lindsay, J. A. (1992). Predicting thermophylic spore population dynamics for UHT sterilization processes. *Journal of Food Science*, *57*, 1248–1257.

Sela, S., Pinto, R., Merin, U., & Rosen, B. (2003). Thermal inactivation of *Escherichia coli* in camel milk. *Journal of Food Protection*, *66*(9), 1708–1711.

Sensoy, L., Zhang, Q. H., & Sastry, S. K. (1997). Inactivation kinetics of *Salmonella dublin* by pulsed electric fields. *Journal of Food Process Engineering*, *20*, 367–381.

Smith, S. E., Maurer, J. L., Orta-Ramirez, A., Ryser, E. T., & Smith, S. M. (2001). Thermal inactivation of *Salmonella spp. Salmonella typhimurium* DT104 and *Escherichia coli* O157:H7 in ground beef. *Journal of Food Science*, *66*, 1164–1168.

Steenstrup, L. L., & Floros, J. D. (2002). Statistical modeling of D- and z-value of *E. coli* O157:H7 and pH in apple cider containing preservatives. *Journal of Food Science*, *67*(2), 793–796.

Stopforth, J. D., Surslim, R., Kottapalli, B., Hill, W. E., & Samadpour, M. (2008). Thermal inactivation D- and z- values of multidrug-resistant and non-multidrug-resistant Salmonella serotypes and survival in ground beef exposed to consumer-style cooking. *Journal of Food Protection*, *71*(3), 509–515.

Stumbo, C. R. (1965). *Thermobacteriology in food processing*. New York: Academic Press.

Teo, A. Y., Ravishankar, S., & Sizer, C. E. (2001). Effect of low-temperature, high-pressure treatment on the survival of *Escherichia coli* O157:H7 and *Salmonella* in unpasteurized fruit juices. *Journal of Food Protection*, *62*(1), 40–45.

Vega-Mercado, H., Martin-Belloso, O., Chang, E.-J., Barbosa-Canovas, G. V., & Swanson, B. G. (1996a). Inactivation of *Escherichia coli* and *Bacillus subtilis* suspended in pea soup using pulsed electric fields. *Journal of Food Processing and Preservation*, *20*(6), 501–510.

Vega-Mercado, H., Pothakamury, U. R., Chang, E.-J., Barbosa-Canovas, G. V., & Swanson, B. G. (1996b). Inactivation of *Escherichia coli* by combining pH, ionic strength and pulsed electric field hurdles. *Food Research International*, *29*(2), 117–121.

Van Boekel, M. A. J. S. (1996). Statistical aspects of kinetic modeling for food science problems. *Journal of Food Science*, *61*(3), 477–489.

Zhang, Q., Qin, B. L., Barbosa-Canovas, G. V., & Swanson, B. G. (1995). Inactivation of *Esterichia coli* for food pasteurization by high-strength pulsed electric fields. *Journal of Food Processing and Preservation*, *19*, 103–118.

Zhang, H. Q., Chang, E. J., & Barbosa-Canovas, G. V. (1994). Inactivation of microorganisms in a semisolid model food using high voltage electric fields. *Lebensmittel Wissenschaft Technology*, *27*(6), 538–543.

Zhang, Q. H., Monsalve-Gonzales, A., Barbosa-Canovas, G. V., & Swanson, B. G. (1994). Inactivation of *E. coli* and *S. cerevisiae* by pulsed electric fields under controlled temperature conditions. *Transactions ASAE*, *37*(2), 581–587.

4 Kinetics of Food Quality Attribute Retention

4.1 Characteristics of quality retention kinetics. 88
4.2 Kinetic parameters for product quality retention 91
4.3 Applications of kinetic parameters for quality attributes. 95
4.4 Impacts of preservation processes on quality attributes 105
List of symbols 106
Bibliography. 107

Most preservation processes are expected to have some impact on quality attributes of the food product. For most thermal processes, the influence is significant, and the quality attributes of the food after the process may be different from those of the product prior to the process. Much of the interest in alternative preservation technologies is to reduce negative impacts of the preservation process while still achieving the desired reduction in microbial populations.

Investigations into the changes in product quality attributes during a preservation process reveal that the changes may be described by the models for reaction kinetics presented in Chapter 2. The retention of many heat-sensitive quality attributes has been described by first-order reactions and first-order rate constants. The kinetic parameters available to describe the retention of food quality attributes are presented and discussed in this chapter. In addition, the kinetic parameters are used to demonstrate the quality retention that occurs during a preservation process in a quantitative manner. These illustrations provide quantitative insights into differences

Food Preservation Process Design
ISBN: 978-0-12-372486-1

in preservation processes and provide a framework for evaluating processes when attempting to determine the process with minimum impact on the quality attributes of the product.

Quantifying quality retention during preservation processes for food products has developed over a considerable period of time. The extent to which this information has been used in process design and optimization of processes is limited. The examples presented in this chapter demonstrate the opportunities for predicting the magnitudes of quality retention during preservation processes and potentially minimizing the changes in quality attributes.

4.1 Characteristics of quality retention kinetics

The interest in retention of food quality attributes during preservation processes is evident in the early research of Feliciotti and Esselen (1957). After measuring thiamine concentration as a function of time at several elevated temperatures, these researchers developed time–temperature relationships to describe thiamine retention, as illustrated in Figure 4.1. Incorporating the thermal death time curve for a commercial sterilization process illustrated the advantages associated with using higher temperature processes for shorter time. These relationships were among the first attempts to quantify preservation processes and improve the retention of quality attributes in a food product. These results and applications provided the basis for measurement of the kinetic parameters published over the past 50 years.

Villota and Hawkes (2007) provide an excellent review of the reaction kinetics associated with changes in quality attributes in foods. The changes in vitamin concentration during preservation processes, storage, and distribution have been described in terms of first-order kinetics, and the kinetic parameters include first-order rate constants and activation energy constants. Although the published literature includes a few references to the use of non-first-order models, there are limited numbers of kinetic parameters for models other than first order.

Villota and Hawkes (2007) assembled kinetic parameters for retaining color pigment intensity during preservation processes, storage, and distribution. These parameters include first-order rate

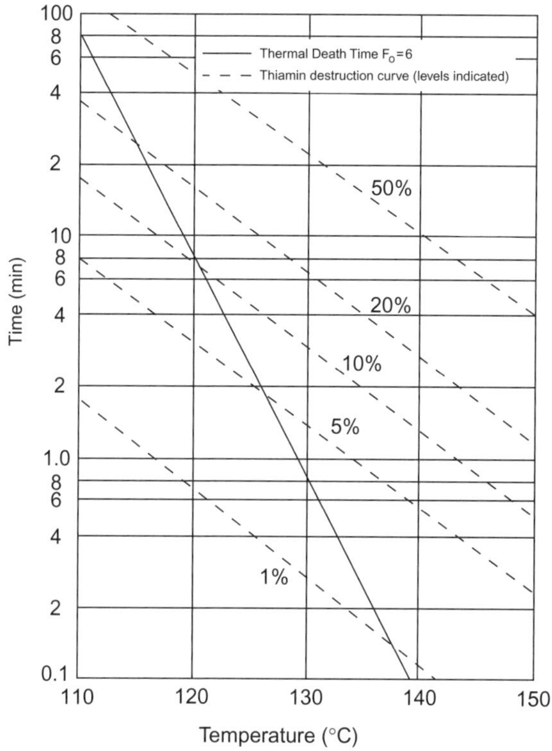

Figure 4.1 The improvements in thiamine retention achieved with high temperature, short time thermal processes (from Lewis & Heppel, 2000).

constants for Maillard browning and nonenzymatic browning. There is limited evidence of kinetic parameters for non-first-order models.

Villota and Hawkes (2007) present a detailed review of the chemical reactions occurring during changes in quality attributes. The review clearly illustrates that the kinetics of the reactions are far more complex than indicated by first-order rate constants. In most situations, several intermediate reactions occur during the change in the initial concentration or in the intensity of the quality attributes. These reactions create intermediate reaction products and may or may not define an end product from a series of reactions. Careful interpretation of these reactions is required to ensure that the kinetic parameters describe the quality change of primary interest. There may be situations where intermediate products represent an enhancement in the quality attribute of interest. The best example of this type of situation is the series of reactions involving lycopene in tomatoes or tomato products.

The kinetic parameters for retention of quality attributes in foods, as assembled by Villota and Hawkes (2007), are representative of the retention of the more prominent product quality attributes. The use of first-order rate constants seems to be appropriate for describing the change in the concentration or intensity of that attribute as a function of time. In addition, the use of activation energy constants to describe the influence of temperature on the magnitude of the rate constants has been recognized in most references. The activation energy constants presented in Table 4.1 present the magnitudes for typical reactions occurring in food products.

The magnitude of the constants is an excellent indicator of the sensitivity of the reaction to temperature. The activation energy constants for most product quality attributes have relatively low magnitudes (<200 kJ/mole) as compared to the constants for inactivation of spores and vegetative cells (250 to 625 kJ/mole). The only product quality attribute with activation energy constants in the same range as the pathogenic spores or vegetative pathogens is protein denaturation. The differences in the magnitudes provide a basis and an opportunity for improvements in quality retention through the use of higher temperature processes for shorter periods of time.

Table 4.1 Activation energy constants for quality attributes in food products

Reaction	Activation energy (kJ/mole)
Enzyme reactions	0–34
Chlorophyll degradation	20–115
Ascorbic acid retention	20–165
Anthocyanins	30–125
α-Tocopherol	40–55
trans-Retinol	40–120
Betalains	40–120
Nonenzymatic browning	40–165
Hydrolysis of disaccharides	40–65
Lipid oxidation	40–105
Spore destruction	250–335
Vegetative cell inactivation	210–625
Protein denaturation	335–500

Source: Adapted from Villota and Hawkes (2007).

4.2 Kinetic parameters for product quality retention

Tables 4.2 and 4.3 present the typical kinetic parameters for retention of several different product quality attributes. These tables include the first-order rate constants and activation energy constants for the different quality attributes. Most of parameters describe retention of product quality attributes during a thermal process, while a few parameters are for retention during ultra-high pressure (UHP) processes. Additional information, including the specific product applications, unique aspects of the measurements, and the reference source of the kinetic parameters, is presented in Tables A.4.2 and A.4.3.

4.2.1 Retention of heat-sensitive vitamins

The kinetic parameters presented in Table 4.2 are for heat-sensitive vitamins in food products. The first-order rate constants and activation energy constants provide quantitative information on the reductions in vitamin concentrations in various products during preservation processes. In general, the magnitudes of the rate constants depend on the temperature of the process, and the magnitudes of activation energy constants are within the anticipated ranges previously shown in Table 4.1. In a few situations, the influence of product composition (as indicated by sugar concentration [Brix] or pH) is more evident when comparing parameters, as illustrated in Table A.4.2.

Table 4.2 Typical kinetic parameters for retention of heat-sensitive vitamins in foods

Preservation process	Vitamin	Rate constant	Agent intensity	Agent coefficients	Substrate
Thermal	Ascorbic Acid	0.00900/min	132.2°C	164.4 kJ/mole	Canned Peas
Thermal	Ascorbic Acid	0.0967/min	150°C	117.6 kJ/mole	Orange Juice
Thermal	Thiamine	0.002511/min	98°C	113.4 kJ/mole	Meat Loaf
Thermal	Thiamine	0.0435/min	138°C	97.1 kJ/mole	Peas
Thermal	5-Methyl	0.249/min	70°C	33.1 kJ/mole	Apple Juice
Pressure	Ascorbic Acid	0.005744/min	850 MPa	74.6 kJ/mole[a]	Tomato Juice
Pressure	Ascorbic Acid	0.010289/min	850 MPa	84.1 kJ/mole[a]	Orange Juice

[a]Agent coefficient represents impact of temperature on rate constant.

The magnitude of kinetic parameters for retention of ascorbic acid (Vitamin C) is influenced by the product substrate, as well as product composition. The first-order rate constants seem to fall in the same range, although magnitudes for grapefruit juice and peas are lower than for orange juice and tomato juice. Results in Table A.4.2 indicate that activation energy constants for grapefruit juice are considerably lower than the other products, and it appears that sugar concentration (Brix) has a significant influence on both activation energy constants and rate constants.

The folates include several nutrients of importance found in different food products. The two heat-sensitive nutrients of concern are 5-methyltetrahydrofolate and folic acid. The magnitudes of the first-order rate constants for 5-methyltetrahydrofolate are consistent from product to product and very close to magnitudes for phosphate buffer (at pH = 7.0). The influence of the product (apple juice, tomato juice) as compared to phosphate buffer is dependent on pH, and there is a corresponding influence on the magnitudes of the activation energy constant. The influence of pH on rate constants for changes in folic acid is relatively small, and the magnitudes of the constants increase slightly as pH decreases. There appears to be a slight influence of pH on the magnitudes of the activation energy constant.

First-order rate constants and activation energy constants for thiamin (Vitamin B_1) retention have been measured for a variety of food products, including vegetables and meats. The magnitudes of these published values are very consistent with limited variation in constants from one product to another. Although there is limited published data for retention of riboflavin (Vitamin B_2), the kinetic parameters indicate the impact of preservation processes. The rate constants are similar for two different products, but the magnitudes of the activation energy constants are significantly different for the same two products.

The published literature presents kinetic constants for reactions associated with Vitamin B_6 retention in foods. In general, the kinetic parameters (both first-order rate constants and activation energy constants) for the various compounds appear to be similar. The two rate constants for pyridoxine are significantly different but have been measured in two dramatically different products: breakfast cereal versus a liquid casein model system. The published literature contains only one reference with kinetic parameters for retention of Vitamin B_{12} (Cyanocobalamin) concentration during a thermal process. The rate constant indicates that retention is relatively high during thermal processing of milk, but the influence of temperature on rate constant was not determined.

Several key kinetic parameters are presented in Table 4.2 and Table A.4.2 to quantify retention of Vitamin A concentrations during a thermal process. The magnitudes of both first-order rate constants and activation energy constants are influenced by the food product substrate. Rate constants for retention of Vitamin A in beef liver are considerably higher than the magnitudes of rate constants for three different vegetable products. In addition, the magnitudes of the activation energy constants for vegetables vary considerably from one investigation to another.

There are limited numbers of references to measurements of retention of vitamin concentrations during UHP preservation processes. The results indicate that retention of ascorbic acid concentrations (expressed as first-order rate constants) is significantly better during a UHP process (850 MPa) than during a comparable thermal process (150°C). The activation energy constants for UHP processes describe the influence of temperature on vitamin retention at a specific pressure magnitude. Measurements of folate (5-methyltetrahydrofolate) retention during UHP processes indicate that kinetic constants are similar to constants for thermal processes. The activation energy constants for UHP process are considerably lower than the magnitude at normal pressure.

4.2.2 Retention of product color

The kinetic parameters to describe the impact of a preservation process on retention of color in various food products are published in the references presented at the end of this chapter. As illustrated in Table 4.3 and Table A.4.3, the influence of various thermal processes on chlorophyll, anthocyanins, betalains, carotenoids, and browning have been assembled. There are limited amounts of kinetic data to describe the influence of UHP processes on chlorophyll retention.

In general, the kinetic parameters for retention of chlorophyll seem consistent from one product to another, even though different methods of measurement have been used. The first-order rate constants range from 0.01575/min at 80°C to 0.2666/min at 126°C, with the latter value describing the retention of chlorophyll A. A more significant range of activation energy constants, from 38.5 kJ/mole (for green beans) to 114.2 kJ/mole (for

Table 4.3 Typical kinetic parameters for retention of heat-sensitive colors in foods

Preservation process	Attributes	Rate constant	Agent intensity	Agent coefficients	Substrate
Thermal	Chlorophyll	0.154/min	120°C	84.9 kJ/mole	Peas
Thermal	Chlorophyll	0.0943/min	120°C	69.0 kJ/mole	Broccoli Juice
Thermal	Anthocyanins	0.02666/min	108°C	105.0 kJ/mole	Citrus Buffer
Thermal	Betalains	0.113/min	100°C	76.2 kJ/mole	Beet Juice
Thermal	Carotenoids	0.024105/min	130°C	88.3 kJ/mole	Tomato Juice
Thermal	Browning	0.017/min	130°C	100.4 kJ/mole	Apple Juice
Thermal	Browning	0.1643/min	96°C	127.2 kJ/mole	Grapefruit Juice
Pressure	Chlorophyll	0.10239/min	850 MPa		Broccoli Juice

chlorophyll A in peas), has been reported. Potential reasons for these differences include variability in products and measurement methods. Obvious variability exists in the magnitude of kinetic parameters for peas, due to constants based on total chlorophyll measurements, as compared to chlorophyll A or B.

The kinetic parameters for retention of the anthocyanins have been measured in fruit juices and similar citrus products. The first-order rate constants range from 0.0005661/min at 80°C to 0.02666/min at 108°C. The variability in kinetic parameters suggests that the magnitudes are influenced by sugar concentration (Brix) and pH. The range of activation energy constants are from 54.8 kJ/mole for a pH = 3.4 grape buffer to 104.6 kJ/mole for a citrus buffer. These parameters do not appear to be influenced by sugar concentration (Brix).

The first-order rate constants for retention of the carotenoids include higher magnitudes in blue crab and much lower values for lycopene in tomato products. The activation energy constants are similar for blue crab and lycopene in tomato juice, while this parameter is much lower for lycopene in tomato puree. There is no apparent reason for the variability in these parameters. The chemical reactions associated with changes in lycopene during a thermal process are complex, and the reactions influence the magnitude of the kinetic parameters.

The published literature provides a variety of different types of kinetic parameters for browning in foods during thermal processes. The parameters measured during a thermal process for apple juice are similar, for either Maillard browning or nonenzymatic browning. The magnitudes of kinetic parameters for apple juice are similar to those for peaches and pears, even though slightly different measurement criteria have been used. Considerable variability exists among the kinetic parameters for grapefruit juice, due to the measurement methods. The first-order rate constants for browning of flour dough during baking are influenced by the moisture content of the dough. The magnitudes of activation energy constants based on browning during the baking of flour dough are similar.

Limited amounts of kinetic data for color change during UHP processes have been reported. The first-order rate constants for chlorophyll retention in broccoli indicate that the magnitudes at 800 MPa (at 80°C) are similar to magnitudes based on retention measured during a thermal process at 120°C. These results suggest that UHP causes a degradation of chlorophyll, independent of the influence of temperature.

4.3 Applications of kinetic parameters for quality attributes

The kinetic parameters for changes in quality attributes of foods can be used to predict the retention of the attribute during a preservation process. The appropriate model from Chapter 2 must be selected for this purpose. The steps involved in the prediction begin with the process time (at a given temperature) required to accomplish the target preservation process. The design of these processes was presented and illustrated in Chapter 3. Using process times, the appropriate kinetic model is then used to predict the change in magnitude of the quality attribute.

Example 4.1

The ground beef product described in Example 3.2 of Chapter 3 is receiving a mild thermal process of 1.57 min at 60°C to ensure product safety associated with *Salmonella typhimurium*. Estimate the loss of thiamine due to this thermal process.

Given:

1. The product is ground beef.
2. The quality attribute of concern is thiamine.
3. The process is established as 1.57 min at 60°C.

(*Continued*)

Example 4.1 (Continued)

Approach:

1. The kinetic parameters for thiamine that change in ground beef during a thermal process are given in Table 4.2.
2. Because the kinetic parameters are based on first-order rate constants, the appropriate prediction model can be selected.
3. Composition tables indicate that the thiamine concentration of lean ground beef (10% fat) is 0.042 mg/100 g product.

Solution:

1. The appropriate prediction model for Thiamine in ground beef is Eq. (2.11) from Chapter 2, as follows:

$$C = C_o \exp[-k\,t]$$

   C is the concentration of thiamine, and k is the first-order rate constant.

2. By reference to Table 4.2, the kinetic parameters for thiamine in ground beef include k = 0.002511/min at 98°C, and $E_A = 113.4$ kJ/mole.
3. To use the prediction model, the rate constant (k) must be expressed at the same temperature as the process time (60°C), using Eq. (2.16) from Chapter 2 as follows:

$$\text{Ln } k = \text{Ln } k_{ref} + [E_A/R\ (1/\mathbf{T}_{ref} - 1/\mathbf{T})]$$

Or:

$$\text{Ln } k = \text{Ln } (0.002511) + [(113{,}400/8.314441)\ (1/371 - 1/333)]$$
$$\text{Ln } k = -5.987 - 4.1952 = -10.1822$$
$$k = 0.0000378/\text{min at } 60^\circ\text{C}$$

4. Using the prediction model:

$$C/C_o = \exp[-(0.0000378)\ (1.57)]$$
$$= 0.99994$$

5. The model predicts 99.994% retention of the thiamine after the mild thermal process. Using this retention level, the thiamine concentration in the hamburger following the process is 0.0419975 mg per 100 g of product.

The previous example illustrates the steps needed to quantify the retention of a nutrient from a food product during a thermal process, when the process required to eliminate a microbial safety risk has been specified. Thiamine is sensitive to thermal processes, but the modest process in the example does not create a significant change in the concentration of thiamine.

Example 4.2

If the vegetable in Example 3.3 of Chapter 3 is peas, estimate the retention of chlorophyll during the process. A thermal process of 42.2 min at 121°C has been determined to ensure elimination *Cl. botulium* as a food safety hazard.

Given:

1. The product is peas, and the quality attribute is chlorophyll.
2. The thermal process is 42.2 min at 121°C.

Approach:

1. The kinetic parameters for chlorophyll retention are obtained from Table 4.2: k = 0.154/min at 120°C, and E_A = 84.9 kJ/mole. Note that parameters represent retention of total chlorophyll, and the rate constants were determined over a range of temperatures similar to the conditions described in the example.
2. The model used for prediction of changes in the quality attribute is a first-order kinetic model, which is the same as the model used to measure the rate constants for chlorophyll in Table 4.2.

Solution:

1. The first-order rate expression is obtained from Eq. (2.11) of Chapter 2:

$$C/C_o = \exp(-k\,t)$$

   The ratio (C/C_o) represents the retention fraction of chlorophyll after the process.
2. Because the rate constant presented in Table 4.2 was measured at 120°C, and the process is conducted at 121°C, a modest change is required:

$$\mathrm{Ln}\,k = \mathrm{Ln}\,k_{ref} + [E_A/R\,(1/\mathbf{T}_{ref} - 1/\mathbf{T})]$$
$$\mathrm{Ln}\,k = \mathrm{Ln}\,(0.154) + [(84{,}900/8.31441)\,(1/393 - 1/394)]$$
$$\mathrm{Ln}\,k = -1.8708 + [(10{,}211.2)\,(0.000006458)]$$
$$k = 0.1645/\text{min at } 121^{\circ}\text{C}$$

(*Continued*)

Example 4.2 (Continued)

3. Using the first-order rate constant and the process time:

$$C/C_o = \exp\left[-(0.1645)\,(42.2)\right]$$
$$= 0.000966$$

4. This prediction indicates that chlorophyll retention is less than 0.1%, or the loss of chlorophyll is 99.9%.

The sensitivity of chlorophyll to thermal processes and the highly detrimental influence on the color of green vegetables is illustrated in the previous example. The objective of process design is to modify the preservation process to improve retention of chlorophyll and similar quality attributes.

Example 4.3

The thermal process for commercial sterilization of peas has a significant influence on quality attributes. Using information presented in Examples 3.3 and 4.2, evaluate the impact of the preservation process on the ascorbic acid (Vitamin C) and thiamine concentration of shelf-stable peas.

Given:

1. The preservation process required for shelf-stable peas is 42.2 min at 121°C.
2. The quality attributes are ascorbic acid and thiamin.

Approach:

1. The kinetic parameters for ascorbic acid are k = 0.009/min at 132.2°C and E_A = 164.4 kJ/mole.
2. The kinetic parameters for thiamine are k = 0.0435/min at 138°C and E_A = 97.1 kJ/min.
3. Composition data for raw peas indicate that the ascorbic acid content is 40 mg/100 g, and thiamine content is 0.266 mg/100 g product.

Solution:

1. Compute the rate constants at 121°C:

$$\text{Ln } k = \text{Ln } k_{ref} + [E_A/R\,(1/\mathbf{T}_{ref} - 1/\mathbf{T})]$$

For ascorbic acid:

$$\text{Ln k} = \text{Ln}(0.009) + [(117{,}600/8.31441)(1/405.2 - 1/394)]$$
$$\text{Ln k} = -4.711 - 0.9923 = -5.703$$
$$k = 0.00335/\text{min at } 121^{\circ}\text{C}$$

For thiamine:

$$\text{Ln k} = \text{Ln}(0.0435) + [(97{,}100/8.31441)(1/411 - 1/394)]$$
$$\text{Ln k} = -3.135 - 1.226 = -4.361$$
$$k = 0.01276/\text{min at } 121^{\circ}\text{C}$$

2. Use the first-order kinetic model to predict retention of the vitamins:

$$C/C_o = \exp[-k\,t]$$
$$C/C_o = \exp[-(0.00335)(42.2)]$$
$$= 0.868 \text{ for ascorbic acid}$$
$$C/C_o = \exp[-(0.01276)(42.2)]$$
$$= 0.584 \text{ for thiamin}$$

3. Using the predicted retention fraction for ascorbic acid, the amount in the peas following the process is (0.868) (40) = 34.72 mg/100 g product. Composition measurements for canned peas indicate a content of 9.6 mg/100 g. The potential reasons for the differences between the predicted retention and the measured amount are discussed in a later chapter.
4. Using the predicted retention fraction for thiamine, the amount in the peas following the process is (0.584) (0.266) = 0.156 mg/100 g product. Composition measurements for canned peas indicate a content of 0.121 mg/100 g, an amount relatively close to the prediction.

The previous example provides insight into the potential use of prediction models for product quality attributes. In addition, Examples 4.2 and 4.3 illustrate that the impact of a given preservation process on multiple quality attributes can be evaluated using appropriate kinetic models.

Example 4.4

A process for pasteurization is being developed for apple juice. The process must ensure negligible risk from *E. coli* O157:H7 in the product delivered to the consumer. In addition, the process to be developed should provide the optimum retention of 5-methyltetrahydrofolate in the product. Initial populations of the pathogen on the raw apple juice may be as high as 1000 per consumer package size.

Given:

1. The product is apple juice.
2. The pathogen of concern is *E. coli* O157:H7.
3. The quality attribute of concern is 5-methyltetrahydofolate,

Approach:

1. The best available kinetic parameters for inactivation of *E. coli* O157:H7 in apple juice are selected.
2. The kinetic parameters for 5-methyltetrahydrofolate in apple juice have been measured.
3. The appropriate kinetic models for prediction of the pathogen survivors during a thermal process are determined.
4. The appropriate kinetic models for prediction of 5-methyltetrahydrofolate retention during a thermal process are selected.

Solution:

1. The best available kinetic parameters for *E. coli* O157:H7 in a product similar to apple juice are for apple cider in Table 3.2 of Chapter 3. Because ranges are provided, the most conservative parameters are selected. The first-order rate constant (k) is 1.11/min at 50°C, and the activation energy constant (E_A) is 350 kJ/mole.
2. The kinetic data for 5-methylterahydrofolate in apple juice are presented in Table 4.2: k = 0.249/min at 70°C, and $E_A = 33.05$ kJ/mole.
3. Two different processes are evaluated: a low temperature (55°C) process for a longer time and a high temperature (65°C) process for a short time period.
4. The first-order rate constants for *E. coli* O157:H7 at 55°C is

$$\text{Ln } k = \text{Ln } (1.11) + [(350{,}000/8.31441)\ (1/323 - 1/328)]$$
$$\text{Ln } k = 0.10436 + 1.9867 = 2.091$$
$$k = 8.0934/\text{min at } 55^{\circ}\text{C}.$$

and

$$\text{Ln } k = \text{Ln } (1.11) + [(350{,}000/8.31441)\ (1/323 - 1/338)]$$
$$\text{Ln } k = 0.10436 + 5.7837 = 5.888$$
$$k = 360.7/\text{min at } 65\ ^\circ\text{C}$$

5. The process is based on negligible risk, or a probability of one surviving pathogen in 1 million containers, or a probability of 10^{-6}. Using the first-order survivor curve model,

$$N = N_o \exp\ (-k\ t)$$

or

$$0.000001 = (1000)\ \exp\ [-(8.0934)\ t]$$

$$1 \times 10^{-9} = \exp\ [-8.0934\ t]$$
$$t = 2.56 \text{ min is process time at } 55^\circ\text{C}$$

and

$$1 \times 10^{-9} = \exp\ [-(360.7)\ t]$$
$$t = 0.0575 \text{ min is process time at } 65^\circ\text{C}$$

6. To evaluate the impact of the process on retention of 5-methyltetrahydrofolate, the first-order rate constants must be computed at 55°C and 65°C:

$$\text{Ln } k = \text{Ln } (0.249) + [(33{,}050/8.31441)\ (1/343 - 1/328)]$$
$$\text{Ln } k = -1.3903 - 0.7176 = -2.1079$$
$$k = 0.1466/\text{min at } 55^\circ\text{C}$$

and

$$\text{Ln } k = \text{Ln } (0.249) + [(33{,}050/8.31441)(1/343 - 1/358)]$$
$$\text{Ln } k = -1.3903 - 1.71435 = -3.105$$
$$k = 0.2098/\text{min at } 65^\circ\text{C}$$

7. Given the process times and the rate constants, the following steps can be completed:

For a process time of 2.56 min at 55°C, the retention of 5-methyltetrahydrofolate is

$$C/C_o = \exp\ [-0.1466\ (2.56)] = 0.687 \text{ or } 68.7\% \text{ retention}$$

For a process time of 0.0575 min at 65°C, the retention of 5-methyltetrahydrofolate is

$$C/C_o = \exp\ [-0.2098\ (0.0575)] = 0.988 \text{ or } 98.8\% \text{ retention}$$

(*Continued*)

Example 4.4 (Continued)

8. The computations indicate that the nutrient loss is 31.3% for the low temperature and the longer time process, and 1.2% at the high temperature and shorter time process.

The previous example illustrates the impacts of a thermal process on a microbial population and a sensitive food product component. The sensitivity of key product components to a thermal process is evident in the magnitudes of the rate constants at different temperatures. In most situations, the use of high temperature, short time processes improves the retention of nutrients or similar product quality attributes. These improvements can be evaluated by using appropriate kinetic parameters.

Example 4.5

A UHP process is being considered for orange juice. The impact of the process on ascorbic acid content of the product is being evaluated. The pathogen of concern in the product is *E. coli*; initial populations, as high as 5 per product package, will be used to establish the process. The ascorbic acid content of the product following the process will be estimated.

Given:

1. A UHP preservation process for orange juice is being developed.
2. The impact of the process on ascorbic acid is being evaluated.
3. The pathogen of concern is *E. coli*, with initial populations as high as 10 per package.

Approach:

1. The best available kinetic parameters for ascorbic acid in orange juice during a pressure process include k = 0.010289/min at 850 MPa and an activation energy constant of 84.1 kJ/mole at 80°C are obtained from Table 4.3 and Table A.4.3. The first-order rate constant at atmospheric pressure is 0.0967/min at 150°C, and E_A = 117.57 kJ/mole.

2. The best available kinetic parameters for *E. coli* during a pressure process were reported for apple juice in Table 3.2. The parameters include k = 0.25/min at 250 MPa, with a pressure coefficient of 126 MPa, at 25°C.
3. Although the process could be conducted at one of several pressures between atmospheric and 850 MPa, the evaluation is conducted at an intermediate range pressure of 400 MPa. The available kinetic information is used as inputs to the appropriate models for process design.
4. Another adjustment for UHP processes is the elevation in product temperature as pressure is increased. This adjustment is normally 10°C for each 100 MPa, or an increase of 40°C for the process being developed.

Solution:

1. To establish the process, the kinetic parameters for *E. coli* are used as inputs for the first-order survivor curve model. The rate constant at 400 MPa is determined from

$$[\log k - \log (0.25)]/(400 - 250) = 1/126$$

Then

$$k = 3.876/\text{min at } 400 \text{ MPa}$$

2. Use the first-order survivor curve model

$$N = N_o \exp(-k\, t)$$

and a negligible risk to establish the final population of 0.000001:

$$0.000001 = (10) \exp[-(3.876)\, t]$$
$$t = 4.158 \text{ min at } 400 \text{ MPa}$$

3. To evaluate the impact of the preceding process on ascorbic acid retention, the appropriate kinetic parameters must be determined. The first-order rate constant for ascorbic acid retention at 400 MPa is estimated by using the relationship between rate constants and pressure. The rate constant at 850 MPa is at 80°C, so the rate constant at atmospheric pressure must be adjusted from 150°C to the same temperature:

$$k = (0.0967) \exp[-(117{,}570/8.31441)(1/353 - 1/423)]$$
$$k = 0.0001278/\text{min at } 80°\text{C}$$

(Continued)

Example 4.5 (Continued)

Then

$$[\mathrm{Ln}\,(0.010289) - \mathrm{Ln}\,(0.0001278)]/(850 - 0)$$
$$= [\mathrm{Ln}\,(0.010289) - \mathrm{Ln}\,k]/(850 - 400)$$
$$k = 0.001008/\mathrm{min} \text{ at } 400 \text{ MPa and } 80°\mathrm{C}$$

4. If the product enters the process at 40°C, the temperature during the process is 80°C, and the impact on ascorbic acid retention is

$$C/C_o = \exp\,[-(0.001008)\,(4.158)] = 0.9958$$

or

Ascorbic Acid Retention = 99.58% or 0.42% loss

5. The ascorbic acid retention for a thermal process designed to accomplish the same reduction in pathogen population can be estimated using the kinetic parameters for apple cider at 80°C; k = 1.11/min or D = 2.075 min and z = 26.5°C:

$$[\log\,(2.075) - \log D]/(50 - 80) = 1/26.5$$
$$D = 28.152 \text{ min}$$
$$k = 0.0819/\mathrm{min}$$

Then

$$0.000001 = (10)\exp\,[-(0.0819)\,t]$$
$$t = 196.8 \text{ min at } 80°\mathrm{C} \text{ for thermal process}$$

Using the kinetic parameters for ascorbic acid,

$$C/C_o = \exp\,[-(0.0001278)\,(196.8)]$$
$$= 0.975 \text{ or } 97.5\% \text{ retention}$$

The ascorbic acid loss is 2.5% for the equivalent thermal process.

The previous example illustrates the potential value of the UHP process on retention of quality attributes.

4.4 Impacts of preservation processes on quality attributes

The impact of a preservation process on the quality attributes of a food is a critical concern in designing the process and selecting the preservation technology. The retention of quality during a thermal process has been a concern, but the inability to quantify the extent of quality change has prevented evaluation and adjustments. In general, the availability of kinetic parameters to quantify the impact of the process on specific quality attributes provides opportunities for improvements in preservation process design for current and future technologies.

The ideal kinetic parameters for quality attributes quantify the retention of quality attributes for a product using the most appropriate kinetic model. Although most changes can be described by first-order kinetic models, future investigations should ensure that alternative models (second-order, etc.) are available and considered. As has been illustrated in this chapter, the retention of specific quality attributes can be predicted for each preservation technology. This important observation demonstrates the opportunity to evaluate processes and to accomplish quantitative comparisons of different processes. The examples and illustrations in this chapter emphasize the need for more complete kinetic data, including parameters for an array of quality attributes as well as alternative preservation technologies. Consideration should be given to defining a more specific protocol for collection and assembly of kinetic parameters to ensure that the most appropriate information is assembled for use in the future.

The importance of relative magnitudes of kinetic parameters must be emphasized. The classic relationship between the activation energy constants for microbial inactivation, as compared to the same parameter for product quality attributes, provides a tool for rapid evaluation of preservation processes. In general, the higher magnitudes of the activation energy constants for quality attributes, as compared to lower magnitudes for microbial populations, provide guidance in the evaluation of processes, as well as the quality attributes to be considered in the evaluation of the process.

More detailed evaluation of preservation processes consider the following:

- The establishment of the most appropriate process to be used to ensure maximum retention of a given quality attribute, while still meeting the food safety expectations. Although models for evaluation of multiple quality attributes are available, the complexity of such models and the lack of kinetic parameters for specific quality attributes may limit the applications of these models.
- The selection of the most appropriate preservation process technology for a given product, as well as the product quality attributes to be evaluated.
- An evaluation of several product quality attributes when a given preservation process is considered. The differences in the kinetic parameters for different quality attributes can be used to demonstrate the impacts of the process on the retention of different quality attributes.

In summary, the quantity of kinetic parameters for food quality attributes needs to be expanded. Quality attribute parameters are influenced by the product structure, so unique characteristics of the product impact the kinetic parameters for product quality attributes. Often, the method of measurement of the concentration or intensity of the quality attribute impacts the magnitude of the kinetic parameters and the predicted retention of the quality attributes. In addition, evaluating alternative preservation technologies requires consideration of additional and unique variables. For example, the influence of UHP on quality attribute retention cannot be measured without considering the influence of temperature. Future measurements of kinetic parameters for quality attributes need to expand on the kinetic parameters available for these types of applications.

List of symbols

C = concentration of component in kinetic model
C_o = initial concentration of component
D = decimal reduction time
E_A = activation energy constant
k = first-order reaction rate constant
k^o = pre-exponential constant in Arrhenius relationship

k_{ref} = reference rate constant
n = parameter defining deviation from first-order kinetic relationship
N = microbial population
N_o = initial microbial population
P = pressure
P_{ref} = reference pressure
R = gas constant
t = time
T = temperature
T_{ref} = reference temperature
T = absolute temperature
ΔV = activation volume
z = thermal resistance constant
z_p = pressure change for one log change in rate constant

Bibliography

Avila, I. M. L. B., & Silva, C. L. M. (1981). Photochemical degradation of betanine and selected anthocyanins. *Journal of Food Science*, *46*(6), 1934–1937.

Bendix, G. H., Herberlein, D. G., Ptak, L. R., & Clifcorn, L. E. (1951). Factors influencing the stability of thiamine during heat sterilization. *Food Research*, *46*(6), 494–503.

Cemeroglu, B., Velioglu, S., & Isik, S. (1994). Degradation kinetics of anthocyanins in sour cherry juice and concentrate. *Journal of Food Science*, *59*(6), 1216–1218.

Dietrich, W. C., & Neumann, H. J. (1965). Blanching brussel sprouts. *Food Technology*, *19*(7), 1174–1177.

Dietrich, W. C., Olsen, R. L., Nutting, M. D., Nuemann, H. J., & Boggs, M. M. (1959). Time-temperature tolerances of frozen foods. XVIII. Effect of blanching condition on color stability of frozen beans. *Food Technology*, *13*(5), 258–261.

Evan, S. R., Gregory, J. F., & Kirk, J. R. (1981). Thermal degradation kinetics of pyridoxine hydrochloride in dehydrated model food systems. *Journal of Food Science*, *46*(2), 555–558, 563.

Feliciotti, E., & Esselen, W. B., Jr. (1957). Thermal destruction rates of thiamine in pureed meats and vegetables. *Food Technology*, *11*(2), 77–84.

Gregory, J. F., & Hiner, M. E. (1983). Thermal stability of vitamin B_6 compounds in liquid model food systems. *Journal of Food Science*, *48*(4), 1323–1327, 1339.

Herrmann, J. (1970). Calculation of the chemical and sensory alterations in food during heating and storage processes. *Ernaehrungsforschung*, *15*(4), 279–299.

Herrmann, J., & Nour, S. (1977). Modelluntersuchungen zur Bildung von Carbonylverbindungen und Braunung durch die Mailard-Reaktion beim Backprozeb. *Die Nahrung*, *21*(4), 319–330.

Himelbloom, B. H., Rutledge, J. E., & Biede, S. L. (1983). Color changes in blue crabs (Callinectes sapidus) during cooking. *Journal of Food Science*, *48*(2), 652–653.

Ibarz, A., Pagan, J., & Garza, S. (1999). Kinetic models for color changes in pear puree during heating at relatively high temperatures. *Journal of Food Engineering*, *39*(4), 415–422.

Lathrop, P. J., & Leung, H. K. (1980). Rates of ascorbic acid degradation during thermal processing of canned peas. *Journal of Food Science*, *45*(1), 152–153.

Lau, M. H., Tang, J., & Swanson, B. G. (2000). Kinetics of texture and color changes in green asparagus during thermal treatments. *Journal of Food Engineering*, *45*(4), 23–236.

Lenz, M. K., & Lund, D. B. (1977). The lethality-Fourier number method: Experimental verification of a model for calculating average quality factor retention in conduction heating canned foods. *Journal of Food Science*, *42*(4), 997–1001.

Lewis, M., & Heppel, N. (2000). *Continuous thermal processing of foods*. Gaithersburg, MD: Aspen Publishers, Inc.

Miki, N., & Akatsu, K. (1970). Effect of heating sterilization on color of tomato juice. *Nippon Shokuhn Kogyo Gakkai-Shi [Japan]*, *17*(5), 175–181.

Mishkin, M., & Saguy, I. (1982). Thermal stability of pomegranate juice. *Zeitschrift fur Lebensmittel-Untersuchung und – Forschung*, *175*(6), 410–412.

Mnkeni, A. P., & Beveridge, T. (1982). Thermal destruction of pteroy glutamic acid in buffer and model food system. *Journal of Food Science*, *47*(6), 2038–2063.

Mulley, E. A., Stumbo, C. R., & Hunting, W. M. (1975). Kinetics of thiamine destruction by heat. A new method for studying reaction rates in model systems and food products at high temperatures. *Journal of Food Science*, *40*(5), 985–988.

Mulley, E. A., Stumbo, C. R., & Hunting, W. M. (1975). Kinetics of thiamine destruction by heat. Effect of pH and form of vitamin on its rate of rate of destruction. *Journal of Food Science*, *40*(5), 989–992.

Navankasattusas, S., & Lund, D. B. (1982). Thermal destruction of vitamin B_6 viamers in buffer solution and cauliflower puree. *Journal of Food Science*, *47*(5), 1512–1518.

Nguyen, M. T., Indrawati, & Hendrickx, M. (2003). Model studies on the stability of folic acid and 5-methyltetrahydrofolic acid degradation during thermal treatment in combination with high hydrostatic pressure. *Journal of Agricultural and Food Chemistry*, *51*(11), 3352–3357.

Saguy, I. (1979). Thermostability of red beet pigments (betanine and vulgaxanthin-1): Influence of pH and temperature. *Journal of Food Science*, *44*(5), 1554–1555.

Saguy, I., Kopelman, I. J., & Mizarhi, S. (1978a). Extent of nonenzymatic browning in grapefruit juice during thermal and concentration processes: Kinetics and prediction. *Journal of Food Processing and Preservation*, *2*(3), 175–184.

Saguy, I., Kopelman, I. J., & Mizrahi, S. (1978b). Simulation of ascorbic acid stability during heat processing and concentration of grapefruit juice. *Journal of Food Process Engineering*, *2*(3), 213–225.

Sanchez, A. H., Rejano, L., & Montano, A. (1991). Kinetics of destruction by heat of color and texture pickled green olives. *Journal of the Science of Food and Agriculture*, *54*(3), 379–385.

Sastry, L. V. L., & Tischer, R. G. (1952). Behavior of anthocyanin pigments in Concord grapes during heat processing and storage. *Food Technology*, *6*(3), 82–86.

Schwartz, S. J., & von Elbe, J. H. (1983). Kinetics of chlorophyll degradation to pyropheophytin in vegetables. *Journal of Food Science*, *48*(4), 1303–1306.

Shi, J., LeMaguer, M., Bryan, M., & Kakuda, Y. (2003). Kinetics of lycopene degradation in tomato puree by heat and light irradiation. *Journal of Food Process Engineering*, *25*(6), 485–498.

Skjoldebrand, C., Anas, A., Oste, R., & Sjodin, P. (1983). Prediction of thiamine content in convective heated meat products. *Journal of Food Technology*, *18*(1), 61–83.

Steet, J. A., & Tong, C. H. (1996a). Degradation kinetic of green color and chlorophylls in peas by colorimetry and HPLC. *Journal of Food Science*, *61*(5), 924–927, 931.

Steet, J. A., & Tong, C. H. (1996b). Quantification of color change resulting from pheophytinzation and non-enzymatic browning reactions in thermally processed green peas. *Journal of Agriculture and Food Chemistry*, *44*(6), 1531–1537.

Stefanovich, A. F., & Karel, M. (1982). Kinetics of beta-carotene degradation at temperatures typical of air drying of foods. *Journal of Food Processing and Preservation*, *6*(4), 227–242.

Tanchev, S. (1972). Kinetics of the thermal degradation of anthocyanins of the raspberry. *Zeitschrift fur Lebensmittel-Untersuchung und –Forschung*, *150*(1), 28–30.

Tanchev, S. (1983). Kinetics of thermal degradation of anthocyanins. Research in food science and nutrition. Vol. 2. Basic Studies in Food Science. *Proceedings of 6th International Congress of Food Science and Technology*, 96.

van den Broeck, I., Ludikuyze, L., Weemaes, C., van Loey, A., & Hendricks, M. (1998). Kinetics for isobaric–isothermal degradation of L-ascorbic acid. *Journal of Agriculture and Food Chemistry*, *46*(5), 2001–2006.

Viberg, U., Jagerstad, M., Oste, R., & Sjoholn, I. (1997). Thermal processing of 5-methyltetrahydrofolic acid in the UHT region in the presence of oxygen. *Food Chemistry*, *59*(3), 381–386.

Villota, R., & Hawkes, J. G. (2007). Reaction kinetics in food systems. In D. R. Heldman & D. B. Lund (Eds.), *Handbook of food engineering* (2nd ed., pp. 125–286). Boca Raton, FL: CRC Press, Taylor & Francis Group.

von Elbe, J. H., Maing, I. Y., & Amundson, C. H. (1974). Color stability of betanine. *Journal of Food Science*, *39*(2), 334–337.

Watanabe, F., Abe, K., Fujita, T., Goto, M., Hiemori, M., & Nakano, Y. (1998). Effect of microwave heating on the loss of vitamin B_{12} in foods. *Journal of Agriculture and Food Chemistry*, *46*(1), 206–210.

Weemans, C. A., Ooms, V., Indrawati, I. L., van den Broeck, A., van Loey, A. M., & Hendrickx, M. E. (1999). Pressure-temperature degradation of green color in broccoli juice. *Journal of Food Science*, *64*(3), 504–508.

Weemans, C. A., Ooms, V., van Loey, A. M., & Hendrickx, M. E. (1999). Kinetics of chlorophyll degradation and color loss in broccoli juice. *Journal of Agriculture and Food Chemistry*, *47*(6), 2404–2409.

Wilkinson, S. A., Earle, M. D., & Cleland, A. C. (1982). Effects of food composition, pH, and copper on the degradation of vitamin A in beef liver puree during heat processing. *Journal of Food Science*, *47*(3), 844–848.

5 Physical Transport Models

5.1 Physical properties 112
5.2 Heating and cooling in containers 125
5.3 Ohmic heating 138
5.4 Microwave heating 139
5.5 Ultra-high pressure applications 141
List of symbols 143
Bibliography 144

The design of a preservation process depends on several parameters, and many involve physical phenomenon. The application of a thermal process depends on the transport of thermal energy within the product structure to achieve the temperature increase required, as well as the cooling of the product to complete the process. An ultra-high pressure (UHP) process requires the application and uniform distribution of the pressure to the product structure. In addition, the thermal energy associated with UHP results in a rise in temperature, and the resulting temperature distribution within the product should be analyzed. Similar physical phenomena are associated with other preservation technologies.

This chapter focuses on the significant base of information on thermal energy transport models, as well as the more limited number of models for transport phenomenon occurring during the use of alternative technologies for food preservation. The transport models for processes involving thermal energy provide an excellent basis for developing and applying transport models for alternative processes.

Food Preservation Process Design
ISBN: 978-0-12-372486-1

The information to be developed and presented includes physical properties of foods, as influenced by temperature, pressure, and other variables associated with the preservation process. The appropriate and available transport models are presented and illustrated in this chapter. The models referenced are for thermal energy transport, but information on pressure distribution and electric field distribution are also explored. As in previous chapters, examples are used to illustrate the applications of the physical transport models.

5.1 Physical properties

Applying models to predict temperature distribution histories within a food product structure depend on access to reliable physical properties data. Although significant amounts of data have been published in recent years, the challenge lies in identifying the most appropriate data for a specific food product. Knowing that these properties are dependent on product composition has provided the opportunity to use the relationship of the properties to product composition in achieving very good estimates of property magnitudes. The relationships provided by Choi and Okos (1986) are significant components of this approach.

5.1.1 Density

The density of a food is defined as mass per unit volume (kg/m^3). The general relationship becomes

$$\rho = \frac{1}{\Sigma(m_i/\rho_i)} \tag{5.1}$$

where m_i = the mass fraction of each product component; ρ_i = the density of each product component.

This basic physical property is influenced by structural characteristics of the product and by the water content of the product. In addition, the air space within a given mass of product has significant influence on the density. Most often, there is a distinction between bulk density and particle density. Bulk density accounts for the void space (air) between the individual food particles, while particle density is based on the mass and volume of the individual food particles. Most often, preservation processes are applied to

Table 5.1 Relationships between density and temperature for individual food components

Property	Component	Temperature function	Standard error	Standard % error
ρ (kg/m^3)	Protein	$\rho = 1.3299 \times 10^3 - 5.1840 \times 10^{-1}T$	39.9501	3.07
	Fat	$\rho = 9.2559 \times 10^2 - 4.1757 \times 10^{-1}T$	4.2554	0.47
	Carbohydrate	$\rho = 1.5991 \times 10^3 - 3.1046 \times 10^{-1}T$	93.1249	5.98
	Fiber	$\rho = 1.3115 \times 10^3 - 3.6589 \times 10^{-1}T$	8.2687	0.64
	Ash	$\rho = 2.4238 \times 10^3 - 2.8063 \times 10^{-1}T$	2.2315	0.09
	Water	$\rho = 9.9718 \times 10^2 + 3.1439 \times 10^{-3}T - 3.7574 \times 10^{-3}T^2$	2.1044	0.22
	Ice	$\rho = 9.1689 \times 10^2 - 1.3071 \times 10^{-1}T$	0.5382	0.06

Source: Choi and Okos (1986).

foods with relatively high water contents and without air or void spaces. Based on this observation, Eq. (5.1) can be applied without considering an air or gas fraction within the product structure. The relationships between individual food product component density and temperature are presented in Table 5.1.

As illustrated, the magnitude of density of each food component is influenced by temperature. The composition of many foods is available from the USDA Food Composition database (http://www.nal.usda.gov/fnic/foodcomp/search/).

As might be expected, pressure has a direct influence on the density of a food product. The results of an investigation (Min, Sastry, &

Example 5.1

Estimate the density of tomato puree at 30°C, based on composition of the product.

Given:

1. Product is tomato puree at 30°C.

Approach:

1. Use the USDA Food Composition database to determine product composition.
2. Use the relationships in Table 5.1 as a source of density for each product component.

(Continued)

Example 5.1 (Continued)

Solution:

1. Based on the USDA Food Composition database, the following values are used:

Water	87.88 g/100 g
Protein	1.65
Total Lipids	0.21
Carbohydrate	7.08
Fiber	1.9
Ash	1.28

2. Based on the relationships in Table 5.1, the density of individual components become the following:

Water	997.6 kg/m^3
Protein	1314.5
Total Lipids	913.1
Carbohydrate	1589.8
Fiber	1300.5
Ash	2415.4

3. Using Eq. (5.1), the density of the tomato puree is predicted:

$$\rho = 1/((0.8788/997.6) + (0.0165/1314.5) + (0.0021/913.1) + (0.0708/1589.8) + (0.019/1330.5) + (0.0128/2425.4))$$

$$\rho = 1055.6 \text{ kg/m}^3 \text{ for tomato puree at } 30°\text{C}$$

Balasubramaniam, 2010) of the influence of UHP (up to 600 MPa) on the density of several food products suggest that the magnitude of the density increases by about 25 kg/m^3 for each 100 MPa of pressure. There are specific product composition factors that could influence this general guideline.

5.1.2 Viscosity

Another physical property of a liquid food is viscosity. Viscosity is a measure of the resistant forces within a layer of the fluid in

Table 5.2 Viscosities of typical Newtonian foods

Product	Composition	Temperature (°C)	Viscosity (Pa s)
Cream	10% fat	40	0.00148
	10% fat	60	0.00107
	10% fat	80	0.00083
Cream	20% fat	60	0.00171
	30% fat	60	0.00289
	40% fat	60	0.00510
Homogenized milk	-	20	0.0020
	-	40	0.0015
	-	60	0.000775
	-	80	0.0006
Raw milk	-	0	0.00344
	-	10	0.00264
	-	20	0.00199
	-	30	0.00149
	-	40	0.00123
Corn oil	-	25	0.0565
	-	38	0.0317
Cottonseed oil	-	20	0.0704
	-	38	0.0306
Peanut oil	-	25	0.0656
	-	38	0.0251
Safflower oil	-	25	0.0522
	-	38	0.0286
Soybean oil	-	30	0.04
Honey, buckwheat	18.6% T.S.	24.8	3.86
Sage	18.6% T.S.	25.9	8.88
White clover	18.2% T.S.	25.0	4.80
Apple juice	20° Brix	27	0.0021
	60° Brix	27	0.03
Grape juice	20° Brix	27	0.0025
	60° Brix	27	0.11
Corn syrup	48.4% T.S.	27	0.053

Source: Steffe (1983).

relation to the forces causing movement of the fluid. Many liquids exhibit a direct proportionality between shear rate and shear stress and are called Newtonian fluids. Water is a Newtonian liquid, along with foods such as honey, fluid milk, and fruit juices. The properties presented in Table 5.2 are examples of Newtonian viscosity coefficients. The viscosity of a food depends on temperature and the concentration of product components, as illustrated for several of the products in Table 5.2. As temperature increases, the viscosity of a food decreases.

Many food products are not Newtonian and require more complex models to describe the relationship between shear stress and rate of shear. These models and appropriate parameters are described in Singh and Heldman (2009). Pressure influences the viscosity of a food product.

5.1.3 Specific heat

Specific heat is the quantity of heat that is gained or lost by a unit mass of a food to accomplish a unit change in temperature, with no change of state within the product:

$$c_p = Q/m\,(T_1 - T_2) \tag{5.2}$$

where Q = quantity of thermal energy, kJ; m = product mass, kg; $T_1 - T_2$ = temperature change, C.

The specific heat (c_p) is expressed in appropriate units (kJ/kg K) and is one of several physical properties needed for a complete analysis of the heating and/or cooling of a food product. The magnitude of specific heat values depends on the composition of the food. The relationships presented by Choi and Okos (1986) provide the basis for predicting specific heat magnitudes for food products. The following model can be used for these predictions:

$$c_p = \Sigma[c_{pi}\,(m_i)] \tag{5.3}$$

where c_{pi} = specific heat values for individual product components; m_i = mass fractions for individual product components.

The specific heat values for the individual food components depend on temperature as predicted by the relationships in Table 5.3.

As indicated, these relationships can predict the dependence of each food component. Equation (5.3) is used to find the specific heat of a food product.

Table 5.3 Relationships between specific heat and temperature for individual food components

Property	Component	Temperature function	Standard error	Standard % error
c_p (kJ/kg°C)	Protein	$c_p = 2.0082 + 1.2089 \times 10^{-3} T - 1.3129 \times 10^{-6} T^2$	0.1147	5.57
	Fat	$c_p = 1.9842 + 1.4733 \times 10^{-3} T - 4.8008 \times 10^{-6} T^2$	0.0236	1.16
	Carbohydrate	$c_p = 1.5488 + 1.9625 \times 10^{-3} T - 5.9399 \times 10^{-6} T^2$	0.0986	5.96
	Fiber	$c_p = 1.8459 + 1.8306 \times 10^{-3} T - 4.6509 \times 10^{-6} T^2$	0.0293	1.66
	Ash	$c_p = 1.0926 + 1.8896 \times 10^{-3} T - 3.6817 \times 10^{-6} T^2$	0.0296	2.47
	Water[a]	$c_p = 4.1289 - 5.3062 \times 10^{-3} T + 9.9516 \times 10^{-4} T^2$	0.0988	2.15
	Water[b]	$c_p = 4.1289 - 9.0864 \times 10^{-5} T + 5.4731 \times 10^{-6} T^2$	0.0159	0.38
	Ice	$c_p = 2.0623 + 6.0769 \times 10^{-3} T$	0.0014	0.07

Source: Choi and Okos (1986).
[a]For the temperature range of −40°C to 0°C
[b]For the temperature range of 0°C to 150°C

Example 5.2

Predict the specific heat of 90% lean raw ground beef at 5°C.

Given:

1. The product is raw ground beef with 10% fat.
2. The product temperature is 5°C.

Approach:

1. The composition of the ground beef is obtained from the USDA Food Composition database.
2. The relationships from Table 5.3 are used to predict the specific heat of individual product components at 5°C.
3. Equation (5.3) is used to predict the specific heat of the product.

(*Continued*)

Example 5.2 (Continued)

Solution:

1. The composition of 90% lean ground beef is obtained from the USDA Food Composition database:

Water	69.5 g/100 g
Protein	20.0
Total Lipids	9.52
Ash	0.98

2. Using the relationships from Table 5.3, the specific heat of each component is as follows:

Water	4.1309 kJ/kg K
Protein	2.0142
Fat	1.9915
Ash	1.102

3. Using Eq. (5.3),

$$c_p = (0.695)(4.1309) + (0.2)(2.0142) + (0.0952)(1.9915) + (0.0098)(1.102)$$

$$c_p = 3.4742 \text{ kJ/kg K}$$

According to Nguyen, Balasubramaniam, & Sastry (in press), pressure has a modest influence on specific heat magnitude for food products. When evaluating a 600 MPa range of UHPs, the specific heats decreased about 15%, or approximately 2.5% for each 100 MPa of pressure above atmospheric pressure.

5.1.4 Thermal conductivity

The thermal conductivity of a food is the fourth property that needs to be considered when analyzing the transfer of thermal energy. This property describes the magnitude of thermal energy transfer per unit time through a unit thickness of a material for a unit of temperature gradient across that thickness.

The thermal conductivity of a high-moisture food should be close to the thermal conductivity of water at the same temperature. Alternatively, the thermal conductivity of a dried, porous food is influenced by the amount of air within the product structure and will have a much lower magnitude. Temperature influences the thermal conductivity of a given product, and the property value increases with increasing temperature.

Thermal conductivity of a food can be predicted based on composition of the product, using the following relationship:

$$k = \Sigma [k_i \ (Y_i)] \tag{5.4}$$

where k_i = thermal conductivity of an individual product component; Y_i = volume fraction of the individual product component.

The volume fraction of a given product can be predicted based on the following relationship:

$$Y_i = \frac{m_i/\rho_i}{\Sigma(m_i/\rho_i)} \tag{5.5}$$

where m_i = mass fraction of an individual product component; ρ_i = density of the individual product component.

Equations (5.4) and (5.5) can be used to predict the thermal conductivities of homogeneous food products. For foods with anisotropic structures, the thermal conductivity values depend on the direction of thermal energy transfer. For example, if the product is beef, and the structure has fiber components that influence the rate of heat transfer, the thermal conductivity value will be different when measured parallel to the component, as compared to perpendicular to the component. For beef, this difference in thermal conductivity magnitudes can be as much as 10%. Kopelman (1966) developed mathematical models to predict the thermal conductivity of foods with anisotropic structures, and applications are discussed in Heldman and Singh (1981) and Heldman (2007).

The thermal conductivities of individual food components can be predicted as a function of temperature have been provided by Choi and Okos (1986), and are presented in Table 5.4.

These relationships, along with Eq. (5.4) and (5.5), can be used to predict thermal conductivities of food products.

Table 5.4 Relationships between thermal conductivity and temperature for individual food components

Property	Component	Temperature function	Standard error	Standard % error
k (W/mX)	Protein	$k = 1.7881 \times 10^{-1} + 1.1958 \times 10^{-3}T - 2.7178 \times 10^{-6}T^2$	0.012	5.91
	Fat	$k = 1.8071 \times 10^{-1} - 2.7604 \times 10^{-4}T - 1.7749 \times 10^{-7}T^2$	0.0032	1.95
	Carbohydrate	$k = 2.0141 \times 10^{-1} + 1.3874 \times 10^{-3}T - 4.3312 \times 10^{-6}T^2$	0.0134	5.42
	Fiber	$k = 1.8331 \times 10^{-1} + 1.2497 \times 10^{-3}T - 3.1683 \times 10^{-6}T^2$	0.0127	5.55
	Ash	$k = 3.2962 \times 10^{-1} + 1.4011 \times 10^{-3}T - 2.9069 \times 10^{-6}T^2$	0.0083	2.15
	Water	$k = 5.7109 \times 10^{-1} + 1.7625 \times 10^{-3}T - 6.7036 \times 10^{-6}T^2$	0.0028	0.45
	Ice	$k = 2.2196 - 6.2489 \times 10^{-3}T + 1.0154 \times 10^{-4}T^2$	0.0079	0.79

Source: Choi and Okos (1986).

Example 5.3

Use the composition of applesauce to predict the thermal conductivity of the product at 80°C.

Given:

1. The product is applesauce.
2. The product temperature is 80°C.

Approach:

1. The composition of the product is obtained from the USDA Food Composition database.
2. The relationships in Table 5.4 will be used to predict the thermal conductivity values for individual product components.
3. The thermal conductivity of applesauce is computed using Eq. (5.4) and Eq. (5.5).

Solution:

1. The composition of unsweetened applesauce is obtained from the USDA Food Composition database:

Water	88.22 g/100 g
Protein	0.17

Total Lipids	0.10
Carbohydrates	10.17
Fiber	1.10
Ash	0.24

2. The thermal conductivity values for individual food components are computed using the relationships from Table 5.4 at 80°C:

Water	0.6693 W/m K
Protein	0.2571
Fat	0.1537
Carbohydrates	0.2847
Fiber	0.2630
Ash	0.4231

3. To predict the thermal conductivity of the product, the volume fraction for each component must be computed:

Water	0.920101
Protein	0.001376
Fat	0.001169
Carbohydrates	0.067376
Fiber	0.008946
Ash	0.001042

4. Using Eq. (5.5), the thermal conductivity of the applesauce is

$$\begin{aligned} k = & (0.6639)(0.920101) + (0.2571)(0.001376) \\ & + (0.1537)(0.001169) + (0.2847)(0.067376) \\ & + (0.2630)(0.008946) + (0.4231)(0.001042) \\ k = & 0.6334\ \mathrm{W/mK} \end{aligned}$$

The thermal conductivity values for foods are influenced by pressure, as well. Based on results from Nguyen et al. (in press), the thermal conductivity of the product increases approximately 40% between atmospheric pressure and 600 MPa. The increase may not be linear with pressure and seems to vary with product composition.

5.1.5 Electrical conductivity of foods

Ohmic heating is an alternative technology for heating a food product and has been incorporated into preservation processes for foods.

This technology uses the movement of ions within an electrolyte and the corresponding heating associated with this action. The movement of ions within a food structure causes a temperature rise when the product is placed between two electrodes. This temperature increase occurs when an alternating electric current or similar waveform is passed through the product.

The primary product property associated with ohmic heating is electrical conductivity. The property indicates the ability of the product to conduct an electric current. Electrical conductivity can be defined by the following expression:

$$\sigma_E = K_E\ L/A \tag{5.6}$$

where K_E = electrical conductance, S; L = length, m; A = area, m.

By recognizing that electrical conductance is the inverse of electrical resistance,

$$K_E = 1/R_e \tag{5.7}$$

where R_e = electrical resistance, ohms.

Electrical conductivity (σ_E) is expressed in Siemens/m (S/m). The electrical conductivity of a food increases linearly with temperature:

$$\sigma_E = \sigma_o\ [1 + c\ T] \tag{5.8}$$

where σ_o = electrical conductivity at 0°C; c = coefficient.

The magnitude of electrical conductivity for a food depends on the composition of the product, the mass fraction of individual components, and the types of components (Table 5.5).

Table 5.5 Electrical conductivities and coefficients for selected food products

Product	σ_o (S/m)	c (1/C)
Potato	0.04	0.28
Carrot	−0.218	−0.064
Yam	−0.149	−0.07
Chicken	0.194	0.036
Beef	0.264	0.027

Source: Adapted from Palaniappan and Sastry (1991).

Example 5.4

Predict the electrical conductivity of potato at 30°C.

Given:
1. The product is potato.
2. The temperature is 30°C.

Approach:
1. Equation (5.8) will be used for the calculation.

Solution:
1. Equation (5.8) is applied as follows:

$$\sigma_E = (0.04)[1 + (0.28)(30)] = 0.376\ \text{S/m}$$

Foods containing electrolytes such as salts, acids, certain gums, and thickeners contain charged groups with significant influence on the magnitude of electrical conductivity.

5.1.6 Dielectric properties

During microwave heating of a food product, electrical properties of the product dictate the characteristics of heating. The relative dielectric constant (ε') and the relative dielectric loss (ε'') are the primary properties to be considered. Foods are not good electrical insulators, and they generally absorb a large fraction of the energy when placed in a microwave field. These situations result in instantaneous heating (Mudgett, 1995). The di-electric loss factor for a food (ε'') expresses the extent of conversion of an externally applied electrical field to heat and can be described by the following expression:

$$\varepsilon'' = \varepsilon' \tan \delta \qquad (5.9)$$

where $\tan \delta$ = loss tangent.

The loss tangent, $\tan \delta$, provides an indication of the extent of electrical field penetration by the food product and the extent of electrical energy dissipation as thermal energy. Typical dielectric properties of foods are presented in Table 5.6.

As illustrated in Table 5.6, the dielectric properties of foods are influenced by water content, temperature, and frequency. In addition, an attenuation factor or penetration depth (Z) influences the distribution of the microwave energy within the product. Singh and Heldman (2009) discuss the significance of these factors.

Table 5.6 Dielectric properties of several food products

			915 MHz			2450 MHz		
Product	**Water**	**Temp**	ε'	ε''	**Z**	ε'	ε''	**Z**
Beef	72.8	25	62	27	3.1	61	17	1.8
		50	55	39	2.1	55	18	1.6
Pork	67.2	25	59	26	3.2	58	16	1.9
		50	52	38	2.1	52	17	1.7
Ham	77.5	25	61	96	1.0	60	42	0.8
		50	54	140	0.8	53	55	0.6
Cod	80.0	25	66	34	2.6	65	20	1.6
		50	59	50	1.7	59	22	1.4
Potato	76.4	25	65	19	4.5	64	14	2.2
		50	58	27	3.0	58	13	2.3
Carrot	90.9	25	73	20	4.5	72	15	2.2
		50	65	28	3.1	65	14	2.3
Apple	87.0	25	71	11	8.0	70	11	3.0
		50	64	13	6.5	63	9	3.4

Source: Adapted from Mudgett (1995).

Example 5.5

Determine the relative dielectric loss and the loss tangent for potato at 20°C and at a frequency of 2450 MHz.

Given:

1. The product is potato.
2. The temperature is 60°C.
3. The frequency is 2450 MHz.

Approach:

1. The properties from Table 5.6 and Eq. (5.9) will be used to determine the relative dielectric constant (ε') and the loss tangent (δ).

Solution:

1. From Table 5.6, the dielectric loss factor (ε'') is 14, and the relative dielectric constant (ε') is 64, at a frequency of 2450 MHz and 25°C.
2. By using Eq. (5.9)

$$\tan\delta = 14/64 = 0.219$$

The dielectric properties of the food product become direct inputs to the expressions used to predict the rate of heating when using microwave energy.

5.2 Heating and cooling in containers

Heating and cooling are important components of many food-preservation processes. When the food product is placed in a container or package prior to the process, the heating and cooling characteristics are described by unsteady-state or transient heat transfer relationships. During preservation processes, the unsteady-state heat transfer relationships provide temperature distribution histories within the product.

During unsteady-state heat transfer, temperature is a function of time and location, and the following partial differential equation describes one-dimensional thermal energy transfer in a radial direction:

$$\frac{\partial T}{\partial t} = \frac{k}{\rho c_p r^n}\frac{\partial}{\partial r}\left(r^n \frac{\partial T}{\partial r}\right) \tag{5.10}$$

where T = temperature, C; t = time, s; r = distance from the center of the cylindrical geometry, m.

The exponent (n) in Eq. (5.10) allows the equation to be used for different geometries:

1. n = 1 for a slab.
2. n = 2 for a cylinder.
3. n = 3 for a sphere.

In most cases, thermal preservation processes for foods involve heated water or steam as a heating medium. For these situations, the thermal energy transfer at the surface of the product container is forced convection, and the following boundary condition applies:

$$k\frac{\partial T}{\partial r}\Big|_{r=R} = h(T_a - T_s) \tag{5.11}$$

Table 5.7 Typical convective heat transfer coefficients for applications in food preservation processes

Fluid	Convective heat-transfer coefficient (W/[m² K])
Air	
Free convection	5–25
Forced convection	10–200
Water	
Free convection	20–100
Forced convection	50–10,000
Boiling water	3000–100,000
Condensing water vapor	5000–100,000

Source: Singh and Heldman (2008).

where h = convective heat transfer coefficient, W/m^2 K; T_a = temperature of heating or cooling medium, C; T_s = temperature at the surface of the container or package, C.

The magnitude of convective heat transfer coefficients depends on many factors. Some typical values are presented in Table 5.7.

The analytical solutions of Eq. (5.10) are provided for standard geometric shapes: sphere, infinite cylinder, or infinite slab. The applications of these solutions require an evaluation of the relative contributions of heat transfer at the container surface, compared to heat transfer within the product inside the container. This ratio has been expressed by a dimensionless number called the Biot Number:

$$N_{Bi} = \frac{hd_c}{k} \tag{5.12}$$

where d_c = a characteristic dimension, m; k = thermal conductivity of the product inside the container, W/m K.

When convective heat transfer at the surface of container is much higher than the conduction heat transfer within the product, the Biot number will be high. When Biot numbers are above 40, a condition of negligible surface resistance to heat transfer exists. At Biot numbers less than 0.1, negligible resistance to heat transfer exists within the container. When Biot numbers are between 0.1 and 40, a condition of finite internal and external resistance to heat transfer is defined. The characteristic dimension (d_c) applies to all geometries and is the shortest distance from the geometric center to the surface.

The conditions for negligible internal resistance to heat transfer in a food container represents a special case. Because the thermal conductivity of a conduction-heating food is very low, it is unlikely that these conditions will occur. If the product in the container is a liquid and is mixed, the temperature distribution within the liquid can be relatively uniform during heating and cooling. For these situations, a thermal energy balance during unsteady-state heat transfer provides the following expression:

$$q = \rho c_p V \frac{dT}{dT} = hA(T_a - T) \tag{5.13}$$

where ρ = density of food product; c_p = specific heat of food product; V = volume of product in container; A = surface area of container.

Equation (5.13) can be solved to obtain the following expression:

$$\frac{T - T_a}{T_i - T_a} = \exp - \frac{h\,A\,t}{\rho\, c_p V} \tag{5.14}$$

where T = product temperature; T_i = initial product temperature; t = time.

The temperature ratio on the left side of Eq. (5.14) is a temperature ratio. At the beginning of the heating/cooling process, the temperature ratio is 1 and decreases with time. The right side of the expression is an exponential function that describes the decrease in product temperature with time.

Example 5.6

Calculate the temperature of tomato juice being thermal processed in a 307 × 604 can during heating in hot water at 100°C after 5 minutes of the process. The convective heat-transfer coefficient in the steam environment is 500 W/m^2 K. The initial temperature of tomato juice is 20°C.

Given:

1. Water temperature = 100°C.
2. Initial temperature = 20°C.
3. Time of heating = 10 min.

Approach:

1. Because the product containers are rotated continuously during the thermal process, it can be assumed that temperature gradients within the product do not exist. Equation (5.14) can be used to compute the product temperature

(Continued)

Example 5.6 (Continued)

after 5 minutes of the process. The density and specific heat of the product will be predicted based on composition.

Solution:

1. Based on the dimensions of the can, the surface area and volume of the can are computed.
   The diameter of the can is 307 (or $3\frac{7}{16}$ in.) or 0.0873 m.
   The can height is 604 ($6\frac{4}{16}$ in.) or 0.15875 m.
   $A = \pi dh + 2\pi r^2 = \pi(0.0873)(0.15875) + 2\pi(0.0873/2)^2 = 0.0555\,m^2$
   $V = \pi r^2 h = \pi(0.0873/2)^2(0.15875) = 0.00095\,m^3$
2. The density of the tomato juice is estimated using Eq. (5.1). The composition of tomato juice is obtained from the USDA Food Composition database, and the following densities of each product component at 60°C are obtained using the relationships in Table 5.1:

Water	93.90 g/100 g	998.7 kg/m3
Protein	0.76	1298.8
Total Lipids	0.05	900.5
Carbohydrate	3.84	1580.5
Fiber	0.40	1289.6
Ash	1.05	2407.0

Using Eq. (5.1), the density is computed:

$$\rho = 1/(0.939/998.7) + (0.0076/1298.8) + (0.0005/900.5) + (0.0384/1580.5) + (0.004/1289.6) + (0.0105/2407)$$

$$\rho = 1039.2\ kg/m^3$$

3. The specific heat of the tomato juice is computed using Eq. (5.3) and the composition of tomato juice. The following specific heats of each product component is determined from relationships in Table 5.2:

Water	93.90 g/100 g	3.6507 kJ/kg K
Protein	0.76	2.0760
Total Lipids	0.05	2.0560
Carbohydrates	3.84	1.6452
Fiber	0.40	1.9725
Ash	1.05	1.1927

4. Using Eq. (5.3), the specific heat of the product can be computed:

$$c_p = (0.939)(3.6507) + (0.0076)(2.076) + (0.005)(2.056) + (0.0384)(1.6452) + (0.004)(1.9725) + (0.0105)(1.1927)$$

$$c_p = 3.53765 \text{ kJ/kg K}$$

5. The temperature of the tomato juice is computed by using Eq. (5.14):

$$(T - T_a)/(T_i - T_a) = \exp[-(500)(0.0555)(5)(60)/(1039.2)(3537.65)(0.00095)]$$

$$(T - T_a)/(T_i - T_a) = 0.0922$$

$$T = 0.0922\,(20 - 100) + 100 = 92.6°C$$

6. The temperature of the tomato juice is expected to be 92.6°C after 5 minutes of heating.

When the conditions during heating or cooling are described by finite internal and surface resistance to heat transfer, and the Biot number is in the range of 0.1 to 40, the solution to Eq. (5.12) is more complex. These solutions are presented in temperature–time charts. The chart for a sphere is presented in Figure 5.1.

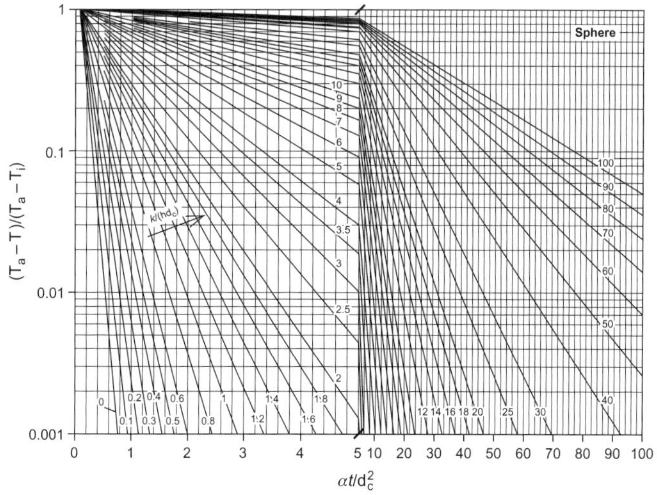

Figure 5.1 Unsteady-state heat transfer chart for a sphere (from Singh & Heldman, 2008).

The temperature–time chart in Figure 5.1 presents the relationship between the temperature ratio and the Fourier Number, at various magnitudes of the inverse Biot Number. The Fourier Number is defined as

$$N_{Fo} = \alpha t/d_c^2 = k\,t/\rho\, c_p d_c^2 \qquad (5.15)$$

where α = thermal diffusivity; $m^2/s = k/\rho\, c_p$.

As indicated earlier, the characteristic dimension (d_c) is the shortest distance from the surface to the center of the object, or the radius of the sphere. The temperature–time chart for an infinite slab is presented in Figure 5.2.

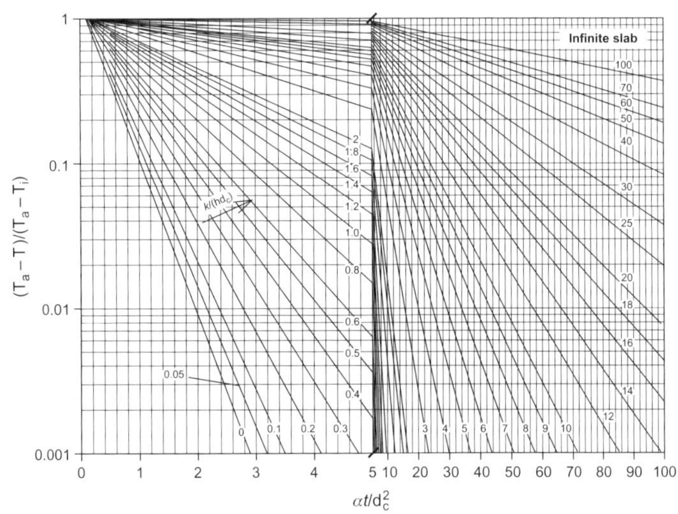

Figure 5.2 Unsteady-state heat transfer chart for an infinite slab (from Singh & Heldman, 2008).

The characteristic dimension (d_c) for an infinite slab is the shortest distance from geometric center to the surface of an object with infinite length in two dimensions. For this geometry, the characteristic dimension is the half-thickness of the slab.

The temperature–time chart for an infinite cylinder is presented in Figure 5.3.

An infinite cylinder is a rod-shaped object with finite length, and the characteristic dimension (d_c) is the radius of the cylinder.

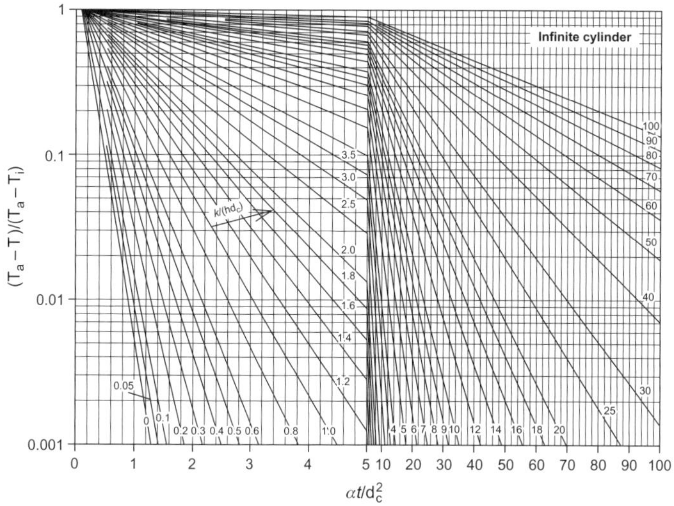

Figure 5.3 Unsteady-state heat transfer chart for an infinite cylinder (from Singh & Heldman, 2008).

All other parameters in Figures 5.2 and 5.3 are the same as in Figure 5.1.

When the Biot numbers are greater than 40, there is negligible resistance to heat transfer at the surface. For these situations, the temperatures can be predicted as a function of time using Figures 5.1, 5.2 and 5.3. When using these charts, the inverse of the Biot number will be very small, and the curves for k/h d_c = 0 should be used.

When evaluating unsteady-state heat transfer in a finite geometry, several relationships have been developed. Singh and Heldman (2009) describe and illustrate the applications of these expressions of typical food containers (cans, pouches, brick-shapes, and boxes).

5.2.1 Predicting temperature during transient heat transfer

In many food preservation situations, the temperature to be predicted occurs after the temperature ratio is less than 0.7. For these situations, Ball (1923) recognized a simplified approach, which has been incorporated into the prediction of temperatures during thermal process design. For this approach, two parameters are defined: a time factor (f_h) and a lag factor (j_c). When these parameters are introduced into the general solution to Eq. (5.10), the following expression is obtained:

$$\mathrm{Ln}\frac{(T_a - T)}{j_c(T_a - T_i)} = -\frac{2.303}{f_h}t \tag{5.16}$$

After conversion of the logarithms, the expression becomes

$$\log(T_a - T) = -\frac{t}{f_h} + \log[j_c(T_a - T_i)] \tag{5.17}$$

This expression illustrates that the magnitudes of the two parameters (f_h, j_c) can be evaluated experimentally by measurement of temperature as a function of time. The heating rate constant (f_h) is the time required for a one log cycle change in the temperature difference on the linear portion of the temperature–time relationship. The heating lag constant (j_c) describes the portion of the temperature–time curve at the beginning of heating (or cooling) and prior to the linear portion of the log-linear relationship between temperature and time.

Pflug, Blaisdell, and Kopelman (1965) have analyzed the relationships between the parameters in Eq. (5.17) and the key factors influencing heating and cooling rates for conduction-heating objects. The results of the analysis were presented in a series of charts to allow for prediction of heating or cooling times. One of the charts presented a relationship between a dimensionless number, incorporating the heating rate constant (f_h), and the Biot number, as presented in Figure 5.4.

The relationships presented in Figure 5.4 are for the three standard geometries. The range of Biot numbers considered is to conditions of finite internal and surface resistance to heat transfer, and illustrates that there is negligible change to the magnitude of the time factor (f_h) at Biot numbers above 40.

The heating/cooling lag constant varies with location within the object, and (j_c) represents the influence of the Biot number on the magnitude at the geometric center of the object. This relationship is presented in Figure 5.5.

The relationships in Figure 5.5 are for the three standard geometries, and indicate that the magnitude of the lag constants are

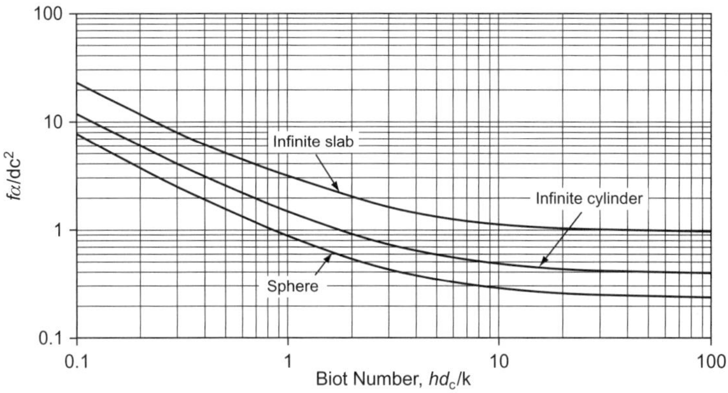

Figure 5.4 The relationship between the heating rate constant (f_h) and Biot number (from Pflug, Blaisdell, & Kopelman, 1965).

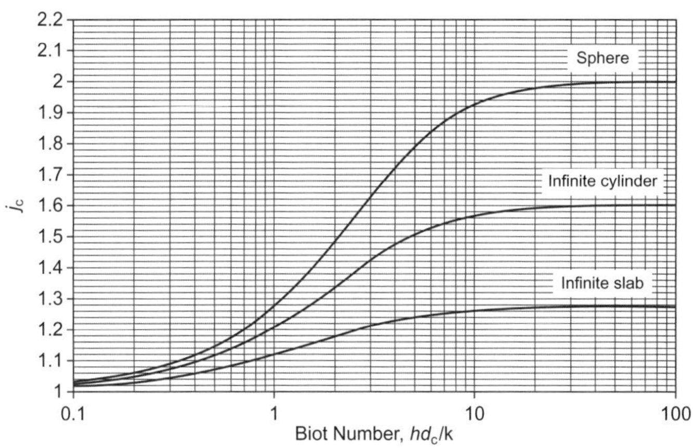

Figure 5.5 The relationship between the lag constant (j_c) at the geometric center and the Biot number (from Pflug, Blaisdell, & Kopelman, 1965).

near 1.0 at low Biot numbers, before increasing to higher constant values at Biot numbers above 40.

Equation (5.16) predicts the mass average temperature within the object by using the appropriate value of the lag constant. The relationships between the mass average lag constant (j_m) and the Biot number are presented in Figure 5.6.

As indicated by Figure 5.6, the mass average lag constant has magnitudes of 1.0 at low Biot numbers and decreases as the Biot number increases.

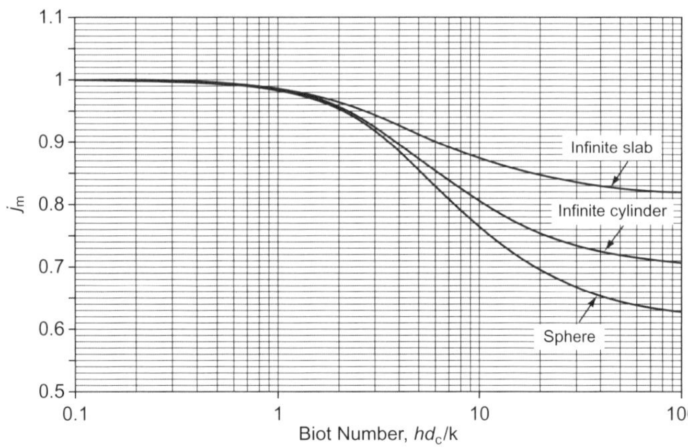

Figure 5.6 The relationship between the mass average lag constant and Biot number (from Pflug, Blaisdell, & Kopelman, 1965).

Equation (5.17) and the parameters from Figures 5.4, 5.5 and 5.6 can be used to predict temperatures within finite geometries that are more similar to food product containers and packages. When the shape of the object is a finite cylinder, the values of f_h and j_c are estimated from

$$\frac{1}{f_{\text{finite cylinder}}} = \frac{1}{f_{\text{infinite cylinder}}} + \frac{1}{f_{\text{infinite slab}}} \tag{5.18}$$

and

$$j_c(\text{finite cylinder}) = j_c(\text{infinite cylinder}) \times j_c(\text{infinite slab}) \tag{5.19}$$

A brick-shaped object gets the relationships

$$\frac{1}{f_{\text{brick}}} = \frac{1}{f_{\text{infinite slab}_1}} + \frac{1}{f_{\text{infinite slab}_2}} + \frac{1}{f_{\text{infinite slab}_3}} \tag{5.20}$$

and

$$j_{c,\text{ brick}} = j_{c,\text{ infinite slab}_1} \times j_{c,\text{ infinite slab}_2} \times j_{c,\text{ infinite slab}_3} \tag{5.21}$$

Although there are limitations to the use of these relationships, estimations of heating/cooling times or temperatures after short periods of heating or cooling are acceptable for most situations encountered during thermal processes for food.

Example 5.8

A canned mashed sweet potato product is being heated from 30°C in a steam environment at 120°C. The convective heat transfer coefficient is 3000 W/m^2 K. The product is in a 307 × 409 can, and the temperature at the center of the can is being estimated after 30 minutes of heating.

Given:

1. Initial temperature = 30°C.
2. Heating medium temperature = 120°C.
3. Convective heat transfer coefficient = 3000 W/m^2 K.
4. Can height = 0.1159 m.
5. Can diameter = 0.0873 m.
6. Heating time = 30 min.

Approach:

1. Use the charts in Figures 5.4, 5.5, and 5.6.
2. Compute the values for density, specific heat, and thermal conductivity of the product.
3. Calculate the Biot numbers for the radius and half-height of the can.
4. Find the values for the f_h and j_c factors, using Figures 5.4, 5.5, and 5.6.
5. Use Eq. (5.16) to compute the product temperature at the center of the can after 10 minutes.

Solution:

1. The density of the sweet potato is estimated at the mid-point temperature (75°C), based on the composition in the USDA Food Composition database. The composition and density of each component are as follows:

Water	0.7388	999.53 kg/m^3
Protein	0.0198	1291.02
Lipids	0.0020	894.27
Carbohydrates	0.2149	1575.82
Fiber	0.0170	1284.06
Ash	0.0075	2402.75

(*Continued*)

Example 5.8 (Continued)

Based on the preceding information and Eq. (5.1),

$$\rho = 1/[(0.7388/999.53) + (0.0198/1291.02) + (0.002/894.27) + (0.2149/1575.82) + (0.017/1284.06) + (0.0075/2402.75)]$$

$$\rho = 1099.74 \text{ kg/m}^3$$

2. The specific heat of the sweet potato is estimated based on the following composition and specific heat information:

Water	0.7388	3.5255 kJ/kg K
Protein	0.0198	2.0195
Lipids	0.0020	2.0688
Carbohydrates	0.2149	1.6626
Fiber	0.0170	2.0094
Ash	0.0075	1.2136

Based on the preceding information and Eq. (5.3),

$$c_p = (0.7388)(3.5255) + (0.0198)(2.0195) + (0.002)(2.0688) + (0.2149)(1.6626) + (0.017)(2.0094) + (0.0025)(1.2136)$$

$$c_p = 3.0492 \text{ kJ/kg K}$$

3. The thermal conductivity of the sweet potato is estimated using the following composition , volume fraction and thermal conductivity information:

Water	0.7388	0.81274	0.6656 W/m K
Protein	0.0198	0.01686	0.2532
Lipids	0.0020	0.00246	0.2024
Carbohydrates	0.2149	0.14995	0.2811
Fiber	0.0170	0.01456	0.2592
Ash	0.0075	0.00343	0.4188

Using the preceding information and Eq. (5.4),

$$k = (0.81274)(0.6656) + (0.01686)(0.2532) + (0.00246)(0.2024) + (0.14995)(0.2811) + (0.01456)(0.2592) + (0.00343)(0.4188)$$

$$k = 0.593 \text{ W/m K}$$

4. The Biot number for an infinite cylinder is computed using the radius of the can:

$$\begin{aligned} N_{Bi} &= (3000)(0.0873/2)/0.5935 \\ &= 220.8 \end{aligned}$$

5. The Biot number for the infinite slab is computed using the half-height of the can:

$$\begin{aligned} N_{Bi} &= (3000)(0.1159/2)/0.593 \\ &= 293.2 \end{aligned}$$

6. Using Figure 5.4, the equations for the infinite cylinder are

$$f_h\, k/\rho\, c_p d_c^{\,2} = 0.395$$

and

$$\begin{aligned} f_h &= 0.395\,(1099.74)(3049.2)(0.0873/2)^2/(0.593) = 4318.7 \text{ s} \\ f_h &= 71.98 \text{ min} \end{aligned}$$

7. Using Figure 5.4, the equations for an infinite slab are

$$f_h\, k/\rho\, c_p d_c^{\,2} = 0.97$$

and

$$\begin{aligned} f_h &= 0.97\,(1103.35)(3049.3)(0.1159/2)/(0.5935) = 18465.9 \text{ s} \\ f_h &= 307.8 \text{ min} \end{aligned}$$

8. Using Eq. (5.18),

$$\begin{aligned} 1/f_h &= (1/71.98) + (1/307.8) \\ f_h &= 58.3 \text{ min} \end{aligned}$$

9. The values for lag factor (j_c) are obtained from Figure 5.5:

$$\begin{aligned} j_c &= 1.602 \text{ for the infinite cylinder} \\ j_c &= 1.276 \text{ for the infinite slab} \end{aligned}$$

(*Continued*)

Example 5.8 (Continued)

10. Using Eq. (5.19),

$$j_c = (1.602)(1.276) = 2.044$$

11. With the given information and factors and Eq. (5.17),

$$\log (120 - T) = -(30/58.3) + \log [(2.044)(120 - 30)]$$
$$T = 63.75°C$$

As is evident, the time for heating a conduction-heating product in a container is relatively long.

5.3 Ohmic heating

The application of an alternating electric current to a food product creates thermal energy generation within the food. The increase in temperature within the food structure is more uniform and rapid than during heat transfer within a conduction-heating product. Ohmic heating depends on the movement of electric current through the product structure. As indicated in Section 5.1.3, the rate of heating depends on the electrical conductivity (σ_E) of the product.

The temperature within the food product during ohmic heating can be described by an expression based on a thermal energy balance. Singh and Heldman (2009) described the solution to the thermal energy balance expression for products flowing through an ohmic heating system as

$$(aT + b)/(aT_i + b) = \exp [(a\,\pi\, d_c L)/(M\, c_p)] \qquad (5.22)$$

where $a = \{[(\Delta V)^2\, d_c\, \sigma_o\, c]/4\} - U$; $b = \{[d_c\, (\Delta V)^2\, \sigma_o]/4\} - UT_a$; ΔV = voltage gradient, V/m; U = overall heat transfer coefficient, W/m^2 K; M = flow rate of product through heater, kg/s.

A typical application of ohmic heating provides for rapid and uniform heating of solid food particulates being conveyed in a liquid food carrier. The expressions presented can be used to determine the conditions needed for the product to reach the uniform temperatures required for a preservation process.

5.4 Microwave heating

Electromagnetic radiation is the basis for microwave heating of foods. Although microwaves are found in the range of frequencies from 300 MHz to 300 GHz, applications in the United States are confined to 915 and 2450 MHz.

Microwaves generate thermal energy within the food product structure, and the thermal energy causes an increase in temperature. The increase in temperature of a food product depends on composition of the food and the dielectric properties of the components as described in Section 5.1.5.

The conversion of microwave energy into thermal energy can be estimated by the relationship (Copson, 1975; Decareau & Peterson, 1986):

$$P = (55.61 \times 10^{-14})\, E^2 f\, \varepsilon' \tan\delta \qquad (5.23)$$

where P = power dissipation, W/cm^3; E = electric field strength, V/cm; f = frequency, Hz; ε' = relative dielectric constant, tan δ = loss tangent.

The electric field strength and the frequency are characteristics of the equipment being used, while the relative dielectric constant and the loss constant are properties of the product. The microwave energy source establishes the electrical field strength (E) and frequency (f) for a given system. Equation (5.23) illustrates that the electrical field strength has significant impact on the power dissipation.

The thermal energy required for a given increase in temperature was presented in Eq. (5.13) as

$$q = \rho\, c_p V\, dT/dt \qquad (5.24)$$

where q = thermal energy, W; ρ = density of product, kg/m^3; c_p = specific heat of the product, kJ/kg K; T = temperature, C; t = time, s.

By assuming that all energy dissipation from the microwave is converted into thermal energy and an increase in product temperature, Eq. (5.23) can be set equal to Eq. (5.24) to obtain

$$\frac{dT}{dt} = (55.61 \times 10^{-14})\, E^2\, f\, \varepsilon \frac{\tan \delta}{\rho\, c_p V} \tag{5.25}$$

This is the rate of temperature increase due to dissipation of microwave energy for each cubic centimeter of product volume. This expression provides a reasonable estimate of the rate of temperature increase and assumes that the product absorbs all microwave energy dissipation, and no thermal energy is lost. The lack of uniform temperature distribution is due to composition variability within the product structure. The impact of this factor is random and leads to the slowest heating point being at a random location within the product structure. Although this factor cannot be ignored in process design, the value of microwave heating in preservation can be demonstrated and compared to more traditional thermal processes.

An additional aspect of microwave heating that is influenced by the electrical properties of the food is penetration depth of the microwaves. The penetration depth is estimated from the attenuation factor, which is a function of the loss tangent, the relative dielectric constant, and the frequency of the microwave field. Because the frequency and wavelength are inversely related, it's evident that microwave energy at a frequency of 915 MHz penetrates more deeply than at a frequency of 2450 MHz. The background on factors influencing the application of microwave energy for heating of foods is presented in Singh and Heldman (2009).

Example 5.9

The heating of 10 cm × 10 cm × 5 cm pieces of potato in a microwave system is being estimated. The frequency of the system is 2450 MHz at 25°C. The electric field strength applied to the product is 12 V/cm. For this situation, the thermal properties of potato include a density of 900 kg/m^3, and the specific heat is 3.88 kJ/kg K. Estimate the temperature of the potato after 1 min.

Given:

1. The product being heated is potato.
2. Dimensions of the potato piece are 10 cm × 10 cm × 5 cm.
3. The frequency is 2450 MHz.
4. The temperature is 25°C
5. Potato density is 900 kg/m^3.
6. Specific heat of potato is 3.88 kJ/kg K.
7. The electric field is 12 V/m.

Approach:

1. The dielectric properties of the potato will be determined using Table 5.6.
2. Equation (5.25) will be used to estimate the rate of temperature increase.
3. The heating rate and time of heating will be used to calculate temperature after 1 min.

Solution:

1. The dielectric properties of potato have been determined in Example 5.4. The properties include $\varepsilon' = 64$, and the $\tan\delta = 0.219$.
2. Using Eq. (5.25),

$$dT/dt = (55.61 \times 10^{-14})(12)^2(2450 \times 10^6)(64)(0.219)/(900)(3880)(5 \times 10^{-4})$$

$$dT/dt = 1.575 \times 10^{-3}\,\text{C/s for a volume of 1 cm}^3$$

3. Given the volume of the potato piece (500 cm^3), the rate of temperature increase can be estimated:

$$dT/dt = 0.7875\ \text{C/s}$$

4. After 1 minute, the potato temperature is

$$T = 20 + (0.7875)(60) = 67.25°\text{C}$$

5.5 Ultra-high pressure applications

The applications of ultra-high pressures (UHPs) for preservation processes provide some unique opportunities. The influence of high pressure on the physical properties of food has been referenced in earlier parts of this chapter. In most cases, pressure influences thermal energy transfer in a significant manner.

When considering the specific application of UHP for food preservation, all evidence indicates that the application of pressure to a food results in uniform pressures throughout the product structure. Minor variations in product composition or structure may create minor differences for short periods of time, but these variations should not influence process design in a significant manner.

One of the impacts of pressure application to a food product is an increase in temperature within the product structure. The temperature increase is uniform throughout the products structure and is proportional to the magnitude of pressure applied. In addition, the increase in temperature depends on product composition, but an increase of 10°C per 100 MPa is a general expectation. During a preservation process based on UHP, the profiles of pressure and temperature illustrated in Figure 5.7 would be expected.

During the process, the pressure is increased to the desired amount and is held at that pressure for the required duration of the process. Finally, the pressure is released, and the product returns to atmospheric pressure. As the pressure is applied to the product, the temperature of the product is increased to some level associated

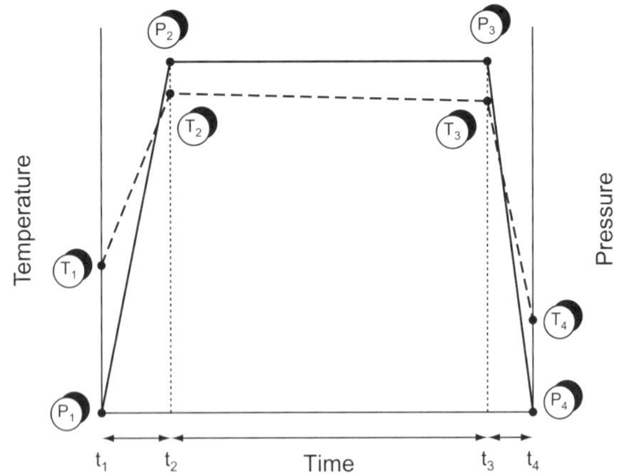

Figure 5.7 Pressure and temperature profiles during the UHP preservation process (Somerville & Balasubramaniam, 2009).

with the process pressure. During the holding period at high pressure, the temperature is expected to decrease slightly. When the pressure is released, the product temperature returns to a level near the same as before the process.

The influence of the temperature increase on the preservation process is a function of the initial temperature of the product. If the initial temperature is low, the increase in temperature will not be sufficient for the temperature to contribute to the reduction in microbial population. Alternatively, higher initial product temperatures result in a temperature rise to levels during holding with significant contribution to the preservation process. The specific applications of these processes are analyzed in the next chapter.

List of symbols

A = area, m^2
c = temperature coefficient for Eq. (5.10)
c_p = specific heat; kJ/kg K
d_c = characteristic dimension, m
E = electric field strength, V/m
E_V = voltage intensity, V
ε' = relative dielectric constant
ε'' = relative dielectric loss constant
f = frequency of microwave energy, Hz
f_h = heating rate constant, s
h = convective heat transfer coefficient, W/m^{2k}
I = electric current, amp
j_c = heating lag constant at center
j_m = mass average heating lag constant
k = thermal conductivity, W/m K
k_i = thermal conductivity of individual components
L = length, m
M = mass flow rate, kg/s
m = mass fraction
m_i = mass fraction of individual components
N_{Bi} = Biot number
n = coefficient in Eq. (5.)
P = power dissipation, W/cm^3
Q = thermal energy, kJ

q = thermal energy transfer, W
R_e = electrical resistance, ohms
r = radius, m
ρ = density, kg/m^3
ρ_i = density of individual components
σ = electrical conductivity
σ_o = electrical conductivity at 0°C
T = temperature,°C
T_a = surrounding temperature,°C
T_i = initial temperature,°C
T_s = surface temperature,°C
t = time, s
$\tan \delta$ = loss tangent
U = overall heat transfer coefficient, W/m^2 K
V = volume, m^3
ΔV = voltage gradient, V/m
Y = volume fraction
Y_i = volume fraction of individual components
Z = penetration depth, m

Bibliography

Ball, C. O. (1923). Thermal process time for canned foods Part 1. *Bulletin of the National Research Council*, *7*(37), 76 pages.

Choi, Y., & Okos, M. R. (1986). Effects of temperature and composition on thermal properties of food. In M. Le Maguer & P. Jelen (Eds.), *Food engineering and process applications, Vol. 1, Transport Phenomenon* (pp. 93–101). London: Elsevier Applied Science Publishers.

Copson, D. A. (1975). *Microwave heating*. Westport, CT: AVI Pub. Co, Inc.

Decareau, R. V., & Peterson, R. A. (1986). *Microwave processing and engineering*. Deerfield, FL: VCH Publ. Co.

Heldman, D. R. (2007). Food freezing. In D. R. Heldman & D. B. Lund (Eds.), *Handbook of food engineering* (2nd ed., pp. 427–470). Boca Raton, FL: CRC Press, Inc.

Heldman, D. R., & Singh, R. P. (1981). *Food process engineering* (2nd ed.). Westport, CT: AVI Publ. Co, Inc.

Kopelman, I. J. (1966). *Transient heat transfer and thermal properties of foods*. PhD Dissertation. Michigan State University, East Lansing.

Min, S., Sastry, S. K., & Balasubramaniam, V. M. (2010). Compressibility and density of select liquid and solid foods under pressures up to 700 MPa. *Journal of Food Engineering, 96*(4), 568–574.

Mudgett, R. E. (1995). Electrical properties of foods. In M. A. Rao & S. S. R. Rizvi (Eds.), *Engineering properties of foods* (2nd ed., pp. 389–455). New York: Marcel Dekker. Inc.

Nguyen, L. T., Balasubramaniam, V. M., & Sastry, S. K. (in press). Determination of in-situ thermal conductivity, thermal diffusivity, volumetric specific heat and isobaric specific heat of selected foods under pressure. *International Journal of Food Properties*.

Palaniappan, S., & Sastry, S. K. (1991). Electrical conductivity of selected solid during ohmic heating. *Journal of Food Process Engineering, 14*, 221–236.

Pflug, I. J., Blaisdell, J. L., & Kopelman, I. J. (1965). Developing temperature-time curves for objects that can be approximated by a sphere, infinite plate or infinite cylinder. *ASHRAE Transactions, 71*(1), 238–248.

Singh, R. P., & Heldman, D. R. (2009). *Introduction to food engineering* (4th ed.). San Diego, CA: Elsevier Academic Press.

Somerville, J., & Balasubramaniam, V. M. (2009). Pressure-assisted thermal sterilization of low-acid shelf-stable foods. *Resource: Engineering & Technology for a Sustainable World, 16*(7), 14–17.

Steffe, J. F. (1983). *Rheological properties of liquid foods*. ASAE Paper No. 83-6512. St. Joseph, MI: ASAE.

Steffe, J. F. (1996). *Rheological methods in food process engineering* (2nd ed.). East Lansing, MI: Freeman Press.

Process Design Models

6

6.1 The process design parameter 148
6.2 General approaches to preservation process design 150
6.3 Process design targets 166
6.4 Integrated impacts of preservation processes 171
6.5 Design of a microwave process 182
6.6 Design of an ohmic heating process 193
6.7 Design of ultra-high pressure processes 197
6.8 Design of pulsed-electric-field processes 202
6.9 Design of combined processes 209
List of symbols 212
Bibliography 214

The process for preserving a food product depends on the target microbial population and the physical phenomenon causing the reduction in the microbial population. When describing the preservation process, the models for the changes in microbial population must be integrated with the models for the distribution of the physical phenomenon within the product structure. The importance of this integration step is evident when considering the following concepts:

- The reduction in the microbial population is a function of time.
- In most situations, the distribution of the physical phenomenon causing reduction in microbial populations is also time dependent.
- The parameter used to define the process is time.

In this chapter, the appropriate models from previous chapters are integrated to develop process design models to accomplish the

Food Preservation Process Design
ISBN: 978-0-12-372486-1

desired results of the preservation process. The impact of the process on quality attributes of the food product is evaluated using the same integration steps.

The integration steps associated with a preservation process can be described in general terms. There are three basic expressions involved in the integration:

1. An expression to describe the microbial survivor curve, including parameters for the rate of reduction in microbial population: $N = f(\mathbf{k}, t)$
2. An expression to describe the influence of the physical phenomenon on the rate of reduction in the microbial population: $\mathbf{k} = f(PP)$
3. An expression to describe the distribution history of the physical phenomenon within the product mass during the process: $PP = f(x, y, z, t)$

The final integration expression combines all three of the previous expressions in the following manner:

$$N = f\{f[f(x, y, z, t)], t\} \tag{6.1}$$

As is evident from this very general expression, the microbial population is a function of time, both directly and indirectly. The research literature for food preservation processes provides numerous approaches to accommodate the mathematics of integration. The following sections illustrate these approaches and discuss the positive aspects of each approach. As might be expected, many of the approaches are based on using thermal energy as the physical phenomenon for preservation. Alternative preservation technologies are illustrated whenever possible. In addition, this chapter illustrates the influence of preservation processes on other product components and quality attributes.

6.1 The process design parameter

The design parameter for a preservation process is time. In general, this parameter is the time required to ensure the microbial population is reduced to a target level. Although this suggests a straightforward application of the survivor curve equation, the definition of the appropriate process time must account for the location within the product mass, and the variability in the intensity of the physical phenomenon used to cause the reduction in the microbial population.

Most of the quantitative guidance for development of process time for preservation processes has been derived from the design of thermal processes (commercial sterilization) for shelf-stable foods. Thc process time is important for the operator of the process and the specific process conditions associated with ensuring that the thermal process reduces the target microbial population to the target level. These conditions are illustrated in Figure 6.1.

The operator process time is defined in terms of temperature in the immediate environment of the product. In the illustration, the temperature (T_M) of the system increases to the desired level in some finite period of time (t_{cut}), and remains at that level until product cooling is initiated. Meanwhile, the product temperature (T) increases gradually during the process and decreases gradually after cooling begins. The differences between environment and product temperatures are most evident in food products with significant mass and when product heating is by conduction. For these types of preservation processes, the important location within the product is the slowest heating location. As indicated in Figure 6.1, the operator process time is

$$t_p = t_c - t_r \tag{6.2}$$

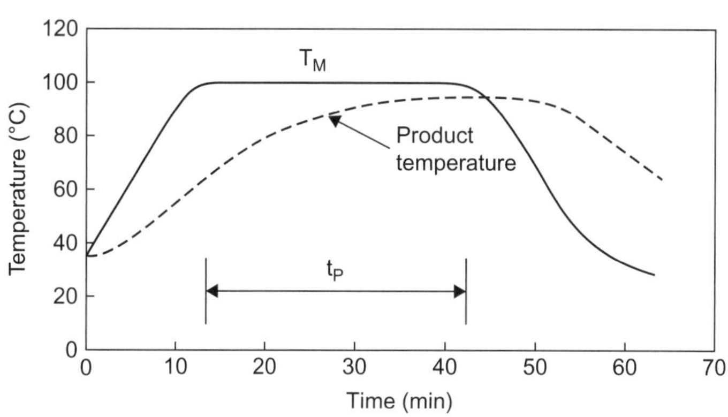

Figure 6.1 Temperature–time curves during a thermal process (from Singh & Heldman, 2009).

where t_c = time at the beginning of cooling; t_r = time when the retort reaches temperature.

The process time provides the operator of the process with a time that can be easily applied: the time after the system reaches temperature until the cooling portion of the process begins. Given this definition, the process design challenge is to ensure that the desired reduction in microbial population occurs at the slowest heating location when the operator process time is provided. Minor adjustments can be incorporated for the time required for the system to reach maximum temperature. The process design parameter can be applied for any type of preservation process, including continuous flow systems and alternative technology systems, as well as the system illustrated in Figure 6.1.

6.2 General approaches to preservation process design

The published literature for thermal processes provides the basis for integrating kinetic models with models for transport phenomena. Although the general approach is applied to alternative preservation technologies, the concepts are illustrated using the traditional thermal process.

6.2.1 Time-step calculation

The impact of a preservation process on any individual component of a food can be evaluated by time-step computation. There are three key expressions to be used:

$$N = N_o \exp(-\mathbf{k}t) \tag{6.3}$$

where N = microbial population; **k** = first-order rate constant; t = time.

This expression can be applied to any type of preservation process, as long as the survivor curve can be described by a first-order relationship. When the survivor curve is not described by a first-order model, an appropriate alternative model should be used. In addition, the same expression can be used to describe the changes in any other component (quality attribute) of the food product during the process.

The second key expression is

$$\mathbf{k} = \mathbf{k^o} \exp\left[-E_A/R_g\mathbf{T}\right] \tag{6.4}$$

where E_A = the Activation Energy Constant; $\mathbf{T}$ = absolute temperature; $\mathbf{k^o}$ = reference rate constant; R_g = the gas constant.

This expression describes the influence of temperature on the rate constant, or the rate the microbial population is decreased. For nonthermal preservation processes, appropriate expressions describing the influence of agent intensity on the rate constant are used.

The third key expression describes the temperature distribution history within the food product. A general expression is

$$T = f(x, \alpha, h, t) \tag{6.5}$$

where x = location within the product structure; α = thermal diffusivity of the food product; h = convective heat-transfer coefficient at the product surface; t = time.

A more specific expression for predicting the temperature distribution history within the product during a thermal process depends on the geometry of the product (and/or package/container). For liquid foods, the expressions are unique for the heat exchanger used for the process. For nonthermal preservation processes, appropriate alternative expressions are selected to describe the distribution of agent intensity within the product structure as a function of time.

The first step in the time-step calculation is selecting an appropriate time interval for the computation. Because many thermal processes are accomplished in time periods of minutes, one-minute intervals are appropriate. For short processes, one-second intervals are considered. Assuming the time interval is one minute, the first computation step uses Eq. (6.5) to calculate the mean temperature for the first minute of the process. This temperature is designated as ($T_{0.5}$), or the temperature at the midpoint in the first minute,

$$T_{0.5} = f(x, \alpha, h, t_{0.5}) \tag{6.6}$$

where $t_{0.5}$ is the first time interval. The thermophysical properties (α) of the food product and the convective heat-transfer coefficient (h) are dependent on temperature, and the most appropriate magnitude for these parameters is at the midpoint temperature for the first time interval. Because this is the temperature being computed, the temperature at the beginning of the time step can be used to establish the properties.

The second step in the time-step calculation is computing the rate constant at the temperature computed during the first step of the process using Eq. (6.4):

$$\mathbf{k_{0.5}} = \mathbf{k^{o}} \exp\left[-E_A/R_g\mathbf{T_{0.5}}\right] \tag{6.7}$$

The rate constant ($\mathbf{k_{0.5}}$) applies for the first time step in the process, and the temperature ($\mathbf{T_{0.5}}$) is the absolute temperature, based on the temperature obtained from Eq. (6.6).

The third step in the time-step calculation is the reduction in the microbial population during the first time step. Using Eq. (6.3), the change is obtained from

$$N_1 = N_o \exp\left(-\mathbf{k_{0.5}}\Delta t\right) \tag{6.8}$$

where N_1 becomes the microbial population after the first time interval of the process, when the rate constant is $\mathbf{k_{0.5}}$, and the time step (Δt) is one minute.

If the previous three steps, as described by Eq. (6.6, 6.7, 6.8) are repeated for the second time step, the temperature becomes

$$T_{1.5} = f(x, \alpha, h, t_{1.5}) \tag{6.9}$$

where the temperature at 1.5 time increments (1.5 minutes) into the process is calculated, and the thermophysical properties (α, h) are selected for the temperature at the beginning of the second time step. The rate constant for the second time step is computed as follows:

$$\mathbf{k_{1.5}} = \mathbf{k_o} \exp\left[-E_A/R_g\mathbf{T_{1.5}}\right] \tag{6.10}$$

The $\mathbf{T_{1.5}}$ is the absolute temperature at the midpoint in the second time step. The next calculation involves

$$N_2 = N_1 \exp\left(-\mathbf{k_{1.5}}\Delta t\right) \tag{6.11}$$

to obtain the microbial survivors (N_2) at the end of the second time step. Note that the initial population (N_1) is the value obtained from the previous time step, and the rate constant ($\mathbf{k}_{1.5}$) is obtained for the midpoint in the second time step.

The same three steps are repeated in an iterative manner as the process continues. The computations are continued until the process is complete, as indicated by the time when the number of survivors reaches the target microbial population. When the computations use a target microbial population, the required process time is obtained. If the process time is the established endpoint for computations, the reduction in microbial population becomes the outcome from the time-step computations.

Example 6.1

A food product has received a process described by the following temperature–time profile:

Time (s)	Temperature (°C)
0	90
1	104
2	111
3	119
4	127
5	131
6	135
7	138
8	139
9	140
10	140
11	140
12	140
13	127
14	114
15	110
16	106
17	102
18	98
19	95
20	92

(*Continued*)

Example 6.1 (Continued)

Estimate the reduction in population of a microbial population with the following kinetic constants: $\mathbf{k}_{121} = 2.094/\text{min}$ and $E_A = 282.5\,\text{kJ/mole}$ as a result of this process. The initial population is 1×10^8 per kg of product.

Given:

1. The temperature–time profile for the process is provided.
2. The kinetic parameters for the microbial population are $\mathbf{k}_{121} = 2.094/\text{min}$, $E_A = 282.5\,\text{kJ/min}$.
3. The initial population of the microbial population is No $= 1 \times 10^8$ per kg.

Approach:

1. The reduction in microbial population for each 1-second increment of the process is computed.
2. The computation for each increment is accomplished using the appropriate rate constant (**k**) magnitude for the product temperature at that time during the process.
3. The temperature for a given time increment is the mean of the temperatures at the beginning and end of the increment.
4. The rate constant (**k**) for the appropriate time increment is computed using Eq. (6.7) and the magnitude at the reference temperature.
5. The change in microbial population for the first time increment is computed by using Eq. (6.8).
6. The changes in microbial population for later time increments are accomplished using the same equations at different temperatures.

Solution:

1. Given that the temperatures at the beginning and end of the first 1-second time increment are 90°C and 104°C, the mean is 97°C.
2. The rate constant for the first time increment at 97°C is obtained using Eq. (6.7), as follows:

$$k_{97} = k_{121} \exp\{-(E_A/R_g)\,[(1/394) - (1/370)]\}$$
$$k_{97} = (2.094) \exp\{-(282{,}500/8.31441)\,[1.646 \times 10^{-4}]\}$$
$$k_{97} = (2.094) \exp\{-5.584\} = 0.00779/\text{min} = 0.00013/\text{s}$$

This rate constant will be used for the first time increment because it has been computed at the mean temperature (97°C) for that time period.

3. Using Eq. (6.8), the microbial population after the first time increment becomes

$$N_1 = (10^8)\ \exp\,[-(0.00013)(1)]$$
$$N_1 = 1 \times 10^8 \text{ per kg}$$

indicating that the impact of the thermal process during the first 1-second increment is not sufficient to cause a detectable change in the microbial population.

4. For the second time increment, the mean temperature is 107.5°C, from 104°C to 111°C. The rate constant is

$$k_{107.5} = (2.094)\ \exp\,\{-(282{,}500/8.31441)\,[(1/394) - (1/380.5)]\}$$
$$k_{107.5} = 0.0982/\text{min} = 0.00164/\text{s}$$

5. Using the rate constant for the second time increment and Eq. (6.8), the microbial population at the end of the second time increment can be computed as

$$N_2 = (10^8)\ \exp\,[-(0.00164)\,(1)]$$
$$N_2 = 9.98 \times 10^7 \text{ per kg}$$

6. The computations illustrated for the first two time increments are repeated for each of the 20 time increments for the process. The results of these computations are presented in Table 6.1.
7. The final computation for the process is at a temperature of 93.5°C, which is the mean of the initial and final temperatures for the time increment. The rate constant is

$$k_{93.5} = (2.094)\ \exp\,\{-(282{,}500/8.31441)\,[(1/394) - (\ 1/366.5)]\}$$
$$k_{93.5} = 0.000054/\text{s}$$

(*Continued*)

Example 6.1 (Continued)

Table 6.1 Time-step computations for solution of Example 6.1

Time (s)	Temp. (°C)	Mid T (°K)	k (1/s)	N
0	90			
		370	0.000129847	1.00E + 08
1	104			
		380.5	0.00163669	99823502
2	111			
		388	0.009195921	98909741
3	119			
		396	0.053938834	93715996
4	127			
		402	0.194117009	77180890
5	131			
		406	0.446381202	49391122
6	135			
		409.5	0.912740595	19826711
7	138			
		411.5	1.366105482	5057765
8	139			
		412.5	1.668844248	953210.4
9	140			
		413	1.843842097	150805.9
10	140			
		413	1.843842097	23858.76
11	140			
		413	1.843842097	3774.656
12	140			
		406.5	0.49477843	2301.431
13	127			
		393.5	0.031272434	2230.573
14	114			
		385	0.004647812	2220.23
15	110			
		381	0.001840179	2216.148
16	106			
		377	0.000714386	2214.565
17	102			
		373	0.000271764	2213.963
18	98			
		369.5	0.000114674	2213.71
19	95			
		366.5	5.40222E-05	2213.59
20	92			

8. The change in microbial population for the final increment, when the population at the beginning of the increment is 2213.7 per kg, is as follows:

$$N_{20} = (2213.7)\ \exp\ [-(0.000054)\ (1)]$$
$$N_{20} = 2213.6 \text{ per kg}$$

   The final microbial population is the predicted value for the process described. To reach a lower final population, the length of the process is extended beyond 20 minutes.
9. The example describes the reduction in microbial population using 1-second time intervals. The size of the time interval influences the population magnitudes at each time during the process, and the time interval used should be carefully selected and evaluated.

6.2.2 The General Method

The historical approach to estimating the lethal impact of a preservation process is referred to as the General Method. This approach was first described by Bigelow et al. (1920), and was based on a graphical method for integrating microbial inactivation kinetics using the temperature–time relationship of a thermal process.

The basis for the General Method is the definition of a thermal death time (F). The F-value is the process time required to reduce a microbial population to some acceptable or predetermined level. In most situations, these acceptable microbial population levels are based on safety or spoilage expectations for the product being preserved. For pathogenic spores, the levels are sufficient to prevent food-borne illness among consumers of the product. When the concerns are about microbial populations causing spoilage, the levels are based on the shelf-life expectations for the product. The thermal death time (F) is some multiple of the Decimal Reduction Time (D). The most referenced relationship is

$$F = 12\ D \qquad (6.12)$$

This relationship is specific for the thermal death time for *Cl. botulinum* spores in a low-acid food and suggests a reduction of 12 log cycles from the initial population. The relationship of the thermal death time (F) to temperature is similar to Eq. (2.24) from Chapter 2:

$$\log (F_1/F_2) = (T_2 - T_1)/z \tag{6.13}$$

This indicates that the thermal death time decreases logarithmically as the process temperature increases.

The General Method uses the relationship in Eq. (6.13) to create a parameter referred to as the lethal rate (LR):

$$LR = F_R/F = 10^{(T - T_R)/z} \tag{6.14}$$

where T = product temperature during the thermal process; F = thermal death time at product temperature; F_R = thermal death time at a reference temperature; T_R = reference temperature. Most often, the reference temperature is the temperature of the heating medium used for the process, usually 121°C.

The lethal rate curve for a typical thermal process is illustrated in Figure 6.2. As indicated, the lethal rate (LR) increases with the increase in product temperature. The increase in LR continues until the product temperature begins to decrease, and LR decreases until the process is complete. It should be noted that the magnitude of LR reaches 1.0 when the product temperature reaches the reference

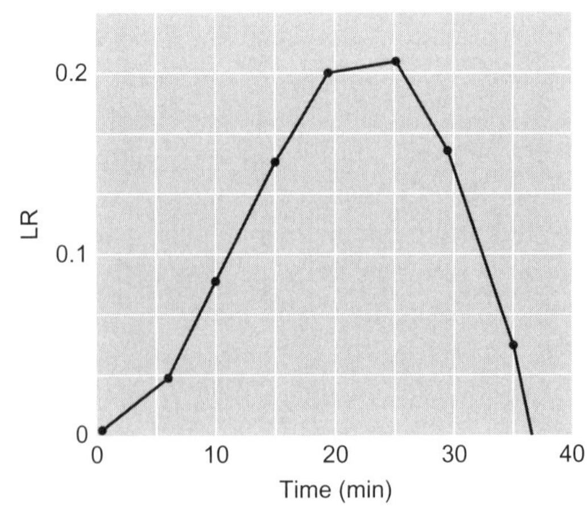

Figure 6.2 A lethal rate curve (from Heldman & Hartel, 1997).

temperature. The area under the lethal rate curve represents the total lethal impact of the process and is expressed as time (F) at the reference temperature (T_R). This approach allows for quantitative evaluation of a process that occurs over a series of temperatures and over the entire duration of the process.

Example 6.2

Use the General Method to evaluate the lethal impact of the thermal process presented in Example 6.1.

Given:

1. A temperature–time profile for a process is provided (Example 6.1).
2. $\mathbf{k_{121}} = 2.094$/min.
3. $E_A = 288{,}500$ kJ/mole.

Approach:

1. The kinetic parameters presented in the previous example are converted to parameters consistent with the General Method.
2. The lethal rates for each temperature and time are computed, and the lethal rate curve is created.
3. The area under the lethal rate curve is determined, and the thermal death time is used to compute the decrease in microbial population during the process.

Solution:

1. The kinetic parameters can be converted in the following manner.
   The rate constant is $\mathbf{k_{121}} = 2.094$/min and can be converted to D_{121} using Eq. (2.22):

$$D_{121} = 2.303/2.094 = 1.1\,\text{min}$$

2. The Activation Energy Constant (E_A) can be converted to the Thermal Resistance Constant (z) by using Eq. (2.27) in the following manner:

$$z = 2.303\,(8.31441)(130 + 273)^2/282{,}500 = 11°\text{C}$$

(Continued)

Example 6.2 (Continued)

3. The lethal rates are computed for each time interval using Eq. (6.13). The temperature used in the computation is at the midpoint. For example, the temperature for the first time interval is 97°C, and

$$LR = 10^{(97-121)/11} = 0.00658$$

Table 6.2 The lethal rate table for Example 6.2

Time (s)	Temp. (°C)	Mid T (°K)	Time (s)	LR
0	90			
		370	0.5	0.006579
1	104			
		380.5	1.5	0.059255
2	111			
		388	2.5	0.284804
3	119			
		396	3.5	1.519911
4	127			
		402	4.5	5.336699
5	131			
		406	5.5	12.32847
6	135			
		409.5	6.5	25.65021
7	138			
		411.5	7.5	38.98604
8	139			
		412.5	8.5	48.06381
9	140			
		413	9.5	53.36699
10	140			
		413	10.5	53.36699
11	140			
		413	11.5	53.36699
12	140			
		406.5	12.5	13.68875
13	127			
		393.5	13.5	0.900628
14	114			
		385	14.5	0.151991
15	110			

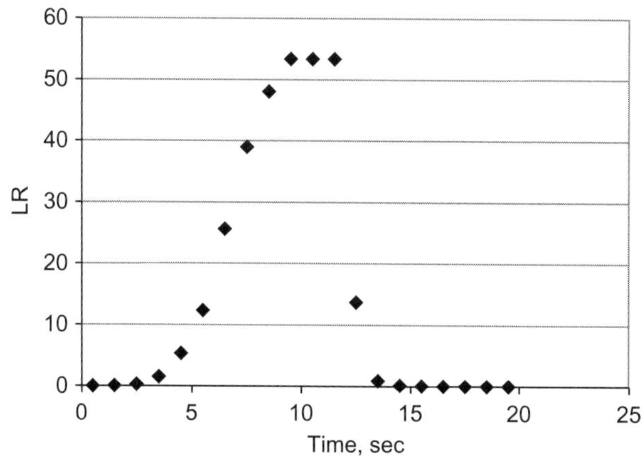

Figure 6.3 Lethal rate curve for Example 6.2.

For the fourth time interval, when the product temperature is 123°C, the lethal rate becomes

$$LR = 10^{(123-121)/11} = 1.52$$

Finally, when the product temperature has reached the maximum value of 144°C, the lethal rate becomes

$$LR = 10^{(140-121)/11} = 53.367$$

The lethal rates for all time intervals are presented in Table 6.2.

The lethal rate curve for the process is presented in Figure 6.3. The lethal rate magnitudes are below 1.0 as long as the product temperature is below the reference temperature, and above 1.0 when the product temperatures are above the reference temperature.

4. The area under the lethal rate curve in Figure 6.3 is 307 sec. This is the magnitude if the lethal impact of the process expressed at 121°C, or

$$F_{121} = 307\ \text{s} = 5.01\ \text{min}$$

(Continued)

Example 6.2 (Continued)

Because Table 6.2 presents the lethal rates at all equally spaced time intervals, the same process lethality magnitude (307.2 sec) can be obtained by obtaining the sum of the lethal rates at each of the 20 time intervals.

5. The reduction in the microbial population is obtained using the survivor curve equation (Eq. 2.21), as follows:

$$N = (10^8)\,10^{(-5.01/1.1)} = 2789 \text{ per kg of product}$$

The predicted population after the process is nearly the same as obtained using the approach in Example 6.1. This difference is small when considering the roundoff associated with the many computations presented in the spreadsheet and is much smaller than the errors in measurement of microbial populations.

6.2.3 Mathematical methods

The time-step approach and the General Method do not provide a direct approach to predicting the operator time for a preservation process. For many types of thermal processes (specifically commercial sterilization), this limitation is significant, and alternative approaches have been developed. One of the more widely referenced approaches is the Formula Method developed by Ball (1923). This approach is based on the equation of the heating curve, as discussed in Chapter 5:

$$\log\,[T_M - T] = -t/f_h + \log\,[\,j_c(T_M - T_o)] \qquad (6.15)$$

where f_h = heating rate constant; j_c = heating lag constant at center of object or product structure.

Although this expression is only valid at the high end (upper 30%) of the temperature range ($T_M - T_o$), the majority of the lethal impact occurs at the higher temperatures. For typical applications, Eq. (6.15) is used at the slowest heating location within a product mass, usually at the geometric center of a container. If the product temperature at the slowest heating location is T_B, the temperature gradient (as compared to the heating medium) at the time when product cooling begins is

$$g = T_M - T_B \tag{6.16}$$

The expression and the parameter (g) define the process time. When Eq. (6.16) is incorporated into Eq. (6.15), the heating rate equation becomes

$$\log(g) = -t_B/f_h + \log[j_c(T_M - T_o)] \tag{6.17}$$

By solving for time (t_B)

$$t_B = f_h \log[j_c(T_M - T_o)/g] \tag{6.18}$$

Equation (6.18) becomes an expression for the process time. This process time (t_B) is equal to the operator time in Eq. (6.1), when the come-up time (t_{cut}) is zero. Given knowledge of the heating rate parameters (f_h, j_c) for the food product, the time required for the slowest heating location to reach the defined final temperature (T_B) has been established. Eq. (6.18) indicates that the process time (t_B) is directly proportional to the heating rate parameters (f_h, j_c) and the temperature difference between the heating medium and the initial product temperature ($T_M - T_o$), but inversely proportional to temperature difference ($T_M - T_B$) at the end of the process. Ball's (1923) Formula Method uses an array of lethal rate curves for typical conditions during thermal preservation processes to compute an array of thermal death times (U). These thermal death times, U, are related to thermal death times at a reference temperature (T_R) by

$$U = F_R 10^{[(T_R - T_M)/z]} \tag{6.19}$$

where the thermal death time (F_R) is the value at the reference temperature (T_R). The relationships between thermal death times and the heating curve characteristics are incorporated into curves of f_h/U versus log g, as illustrated in Figure 6.4.

The relationships in Figure 6.4 are based on specific inputs to lethal rate computations:

- The relationships are specific for microbial populations with z = 10°C (or 18°F). Similar relationships have been developed for

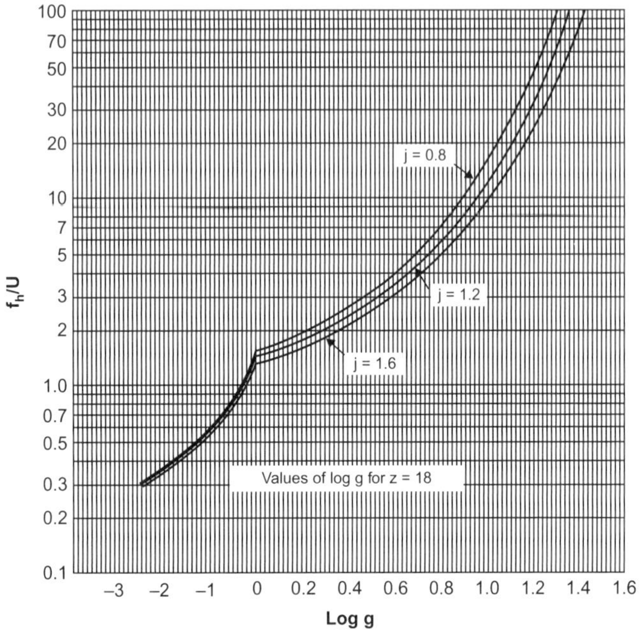

Figure 6.4 The f_h/U versus log g relationship for z = 10°C (z = 18 F) (from Teixeira, 2007).

other values of z, as illustrated by Tables A.6.1 to A.6.6 in the appendices.

- The relationships are dependent on a temperature difference between the heating medium and cooling medium ($T_M - T_{CM}$) of 100°C. Similar relationships can be developed for other conditions.
- The relationships include lethality for the cooling portion of the process, based on four different characteristics of the cooling curve as determined by the cooling lag constant (j_{cc}). These relationships also assume that the cooling rate constant (f_c) is equal to the heating rate constant (f_h).

These assumptions may be viewed as limitations to the approach, but the limitations must be evaluated in contrast to the convenience of estimating the process time.

As suggested, the mathematical method provides a direct approach to predicting the operator time for a thermal preservation process. The procedure begins with a target thermal death time based on the desired reduction in microbial population. Equation (6.19) is used to convert the target thermal death time to the thermal death time (U) at the heating medium temperature. The ratio (f_h/U) is used as an input to Figure 6.4 to obtain log *g*. Finally, Eq. (6.18) is used to compute the process time (t_B).

Example 6.3

A thermal preservation process is being established for a conduction-heating food in a container. The heating rate parameters at the slowest heating location in the container are f_h = 20 min, j_c = 1.5, and j_{cc} = 1.2. The heating medium temperature is 116°C, and the initial product temperature is 34°C. The target thermal death time (F_{121}) for the process is 10 minutes for a microbial population that has a *z* = 10°C.

Given:

1. Heating rate constant is f_h = 20 min.
2. Heating lag constant is j_c = 1.5.
3. Cooling lag constant is j_{cc} = 1.2.
4. Heating medium temperature is T_M = 116°C.
5. Initial product temperature is T_o = 34°C.
6. Thermal Resistance Constant is z = 10°C.
7. The target thermal death time is F_{121} = 10 min.

Approach:

1. Using the target thermal death time and Eq. (6.1), the thermal death time (U) is computed.
2. The thermal death time and the heating rate constant (f_h) are used as inputs to Figure. 6.4 to obtain log g.
3. After computing g, Eq. (6.18) is used to determine the process time (t_B).

Solution:

1. Equation (6.19) is used to compute the thermal death time (U):

$$U = 10 \times 10^{[(121-116)/10]}$$
$$U = 31.6 \text{ min}$$

2. Use the heating rate constant (f_h) and the value of U:

$$f_h/U = 20/31.6 = 0.632$$

(Continued)

Example 6.3 (Continued)

then Figure. 6.4 is used to obtain

$$\log g = -0.8$$

and

$$g = 0.1585$$

3. Note that $g = 0.1585°F$ or $0.088°C$.
4. Equation (6.18) is used as follows:

$$t_B = 20 \log [1.5 (116 - 34)/0.088]$$
$$t_B = 62.9 \text{ min}$$

A process time for the conditions is obtained in an efficient manner. The ease of computation should be evaluated in terms of the limitations to the relationships used in the illustration.

Obviously, the mathematical method is limited to predicting the process time for thermal preservation processes. The application of a similar approach requires the development of appropriate relationships and illustrations for alternative preservation processes. Each alternative process needs to be analyzed to determine the potential for use of a mathematical approach for prediction of a process time. For example, similar relationships can be developed for Ultra-High Pressure (UHP) processes, based on knowledge of the rate of pressure application to the product.

6.3 Process design targets

The purpose of the preservation process is to reduce the microbial population to some acceptable level. The establishment of the acceptable level is based on a variety of factors, and may be dependent on the product. The following factors are used to determine the endpoint for the process:

- Food safety, with specific attention to microbial pathogens in the product.

- Product spoilage, based on microbial populations within the product that limit the product's shelf life.
- Quality attributes, when the preservation process is limited by changes in a defined quality attribute.

When dealing with microbial food safety and product spoilage, the basis for establishing the endpoint for a process is Eq. (2.11) and expressed in terms of microbial populations:

$$N = N_o \exp(-\mathbf{k}t) \tag{6.20}$$

In this equation, $\mathbf{k}$ = the rate constant to describe the reduction on the microbial population.

This expression applies to a first-order survivor curve, which can be used to illustrate the approach to establishing the acceptable process. By solving for the time as

$$t = \mathrm{Ln}\,[N_o/N]/\mathbf{k} \tag{6.21}$$

the relationship between the reduction in microbial population (N_o/N) and time (t) is more evident. Obviously, a longer process time is required to accomplish a larger reduction in the microbial population. The parameter that is most difficult to define is the microbial population at the end of the process (N). When considering spoilage microorganisms or microbial pathogens, the endpoint is even more difficult to express. An examination of the survivor curve equation clearly illustrates that a zero endpoint does not exist. Under these circumstances, the endpoint must be expressed as a probability of survival for the target population. Often, the probability is expressed in terms of risk of spoilage, or the probability of a surviving pathogen in some stated number of processed food containers.

To evaluate the probabilities of survivors, the initial microbial populations in each container and the number of containers must be considered. An expression of the final population as a probability of survivors, based on the quantity of the product in a container, provides insight about the endpoint for the process. The concept of probability becomes more evident when considering the total number of containers being processed. The initial population per container and the total number of containers are both considered in an expression of spoilage probability:

$$1/r_c = N_o/10^{F/D} \tag{6.22}$$

where r_c = number of containers in the preservation process; N_o = initial microbial population per container; F = process time at the specified conditions of the process; D = Decimal Reduction Time.

This expression is used to determine the process time (F) needed to ensure that no more than one container with spoilage exists among some specified total number of containers (r_c) being processed. The alternative expression is

$$1/r_c = N_o/\exp[\mathbf{k}\,F] \quad (6.23)$$

Example 6.4

A thermal preservation process is being used for 10 million containers of a food product. The population of spoilage microorganisms is 10^3 per container and is described by a D_{110} of 5 minutes. Determine the time at 110°C needed to ensure that no more than 1 spoiled container exists for the number being processed.

Given:

1. The number of containers (r_c) = 10 million.
2. The initial population (N_o) = 1000 per container.
3. The Decimal Reduction Time (D) = 5 minutes.

Approach:

1. Use Eq. (6.22) to determine the number of containers processed with one spoiled container.

Solution:

1. Use Eq. (6.22):

$$1/10^7 = 10^3/10^{F/5}$$

and

$$10^{F/5} = 10^{10}$$

2. Then

$$F/5 = 10$$

and

$$F = 50 \text{ min}$$

3. The analysis indicates that a thermal process of 50 min at 110°C is sufficient to ensure no more than 1 spoilage container in 10 million processed. This result can be expressed as a spoilage probability of 10^{-10}.

When considering the potential spoilage of a food, the probability magnitude can be derived in terms of economic impact. In many situations, the survivors of the process cause spoilage within the time frame during distribution, allowing for detection and removal of the specific containers (with spoiled product) prior to the product reaching the consumer. The management decision involves the reduction of the economic impacts of product spoilage as much as possible while minimizing the impact of the process on the quality attributes of the product.

The consideration of risk associated with food safety is much more sensitive than with product spoilage. The management challenge is to establish the probability of risk associated with the survival of a microbial pathogen. The goal of the process is to eliminate any threat of food-borne illness due to consumption of the food product. To achieve this ultimate goal, the food processing industry has relied on a variety of secondary measures to prevent microbial food safety concerns. A typical guideline probability is one survivor per million containers processed, although this risk is not communicated to consumers. This guideline is acceptable due to additional measures and protocols. The best examples are thermal processes and protocols used to eliminate food-borne illness due to *Cl. botulinum* toxin in shelf-stable canned foods. The secondary control for these products is an established period of product storage at typical storage and distribution temperatures. Any survivors of the thermal process are likely to grow in the product during storage, causing the production of by-products from the metabolism associated with the increase in microbial population. For *Cl. botulinum*, the by-product is gas production resulting in swelling of the product container. By eliminating containers with demonstrated symptoms of swelling at the end of the established storage period, the probability for food-borne illness due to pathogens surviving the preservation process can be reduced to zero. Given these circumstances, the process conditions become a function of company policies and requirements of regulatory agencies.

Preservation processes may be established after considering the impact on quality attributes of the food product. Product quality attributes may influence the process in several ways. For some products, the process may be based entirely on the established process needed to create the desired attributes in the final product. This happens when the process is needed to create the desired product quality attributes, and the process exceeds the conditions needed for reaching the probabilities for product spoilage and/or product safety.

Example 6.5

A canned potato product is being thermally processed, and the process is based on achieving the desired texture (hardness) of the potato. The kinetic parameters associated with texture include a rate constant of 0.055/min at 100°C and an Activation Energy Constant of 100 kJ/mole. If the desired texture is an 80% reduction on the hardness of the potato, determine the process.

Given:

1. The product is potato.
2. The rate constant (**k**) for potato hardness = 0.055/min.
3. The Activation Energy Constant (E_A) = 100 kJ/mole.
4. The target texture = 80% of initial magnitude.

Approach:

1. Use Eq. (2.16) from Chapter 2 to determine the rate constant at 121°C.
2. Compute the process time at 121°C using Eq. (2.12).

Solution:

1. Using Eq. (2.16) and the input conditions,

$$k_{121} = (0.055) \exp \{(100{,}000/8.31441)\,[(1/373) - (1/394)]\}$$
$$k_{121} = 0.307/\text{min}$$

2. The process time is determined by using Eq. (2.12) and recognizing that the 80% reduction in hardness is a retention of 0.2:

$$0.2 = \exp [0.307\ t]$$
$$\mathrm{Ln}\ (0.2) = 0.307\ t$$
$$t = 5.24\ \mathrm{min}$$

This indicates that a process time of 5.24 min at 121 °C is needed to achieve the desired texture of the product.

The more typical impact of product quality attributes on the preservation process occurs when the process has negative impact on the quality attributes. For these types of situations, adjustments in the preservation process may be considered in an effort to reduce the negative impacts on quality. These considerations are most evident when dealing with decisions on spoilage probabilities and the potential for accepting a slightly higher rate of spoilage, while improving the retention of quality attributes in the product. Development of processes based on these considerations is feasible by using kinetic parameters for the target microbial populations and for the most sensitive product quality attribute. Ultimately, the process can be designed to minimize the spoilage probability, while maximizing the retention of a sensitive product quality attribute.

6.4 Integrated impacts of preservation processes

The focus of processes presented and analyzed up to this point has been on the microbial populations at a specific location within the product mass or structure. An inherent assumption in all process design is that microbial populations are uniformly and homogeneously distributed within the product mass and structure. This assumption can occur in many preservation processes but is difficult to ensure when the process is applied uniformly to all food product in a container. When the process occurs after the product is placed in a container, the preservation process must be based on a specific location where the intensity of the process is minimum. The importance of this concept is most evident with thermal processing of conduction-heating food products, where the slowest heating location is at the geometric center of the product within the container. This is a core concept when ensuring microbial safety and monitoring rates of product spoilage from preservation processes.

Unless the mechanisms associated with the preservation process can be uniformly applied throughout the product mass or structure, the impact of the process will be greater at all locations within the product structure or mass than at the target location. For the thermal process, the impact of the preservation process is more intense at all locations other than the slowest heating location. The analysis of these impacts is most important when considering the influence of the process on product quality attributes that are sensitive to the mechanism of the preservation process.

When considering preservation processes to be discussed in this book, the thermal processes are the most obvious processes to be considered when evaluating the integrated impact of the process on the entire product structure or mass. The application of UHP to a product structure or mass seems to be instantaneous and uniform, although there is limited evidence to suggest that uniformity of the impact may be a function of product composition and structural components. Most other alternative preservation processes depend on uniformity of application within the product mass or structure, but the development and application of all processes should consider the potential for lack of uniformity.

The integrated impact of a preservation process can be evaluated by using a cross-section of a cylindrical package as an example. Figure 6.5 shows the cross-section divided into five cylinders of equal thickness in the radial direction.

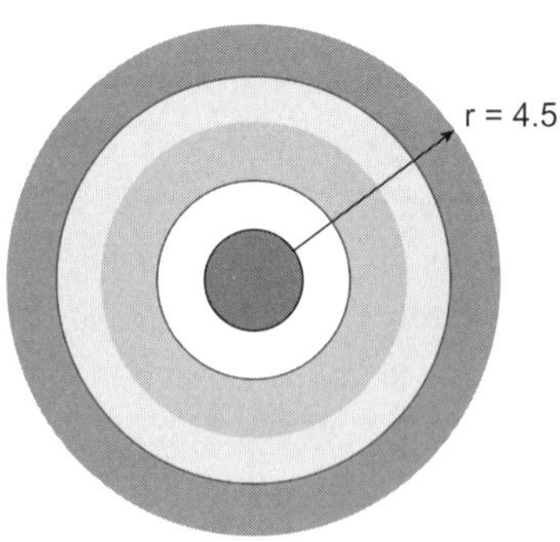

Figure 6.5 The cross-section of a cylindrical container.

The diameter of the cylinder has five units, with the center section having a radius of one-half unit. Each of the outer four cylindrical sections has a thickness of one unit in the radial direction. For this analysis, each section has a one-unit depth along the height of the cylinder. Assuming the cylinder is a conduction-heating food, and the preservation process is thermal, the cylindrical section of product near the surface will increase in temperature most rapidly. With the same assumptions, the product in the core cylinder will be the slowest heating location. The following analysis demonstrates the integrated impact of the process on the cross-section of product illustrated in Figure 6.5. The preservation process is designed to ensure that the slowest heating location receives the process needed for microbial safety or acceptable spoilage rate.

For a thermal process designed to ensure that the slowest heating location receives a specified reduction in microbial population, the cylindrical section near the surface and the heating medium will receive a process much greater than at the slowest heating location. Likewise, cylindrical sections between the center and the surface will receive process intensities less than at the surface but greater than at the slowest heating location. The impact of the thermal process on heat-sensitive food quality attributes is similar to the impact on microbial populations, with more significant losses at the surface and less dramatic impacts at the slowest heating location. The most appropriate expression for the impact of the process on the product quality is a weighted average over the cross-section of the cylinder. These types of analyses provide an opportunity to evaluate the influence of a preservation process on product quality attributes and the tools needed to optimize the process by minimizing the quality loss while meeting the target reduction in a microbial population.

To evaluate the influence of location on heat-sensitive product quality attributes, the temperature distribution within the product structure during the process must be predicted. The prediction of temperature distribution is accomplished by using the concepts introduced in Chapter 5 (Section 5.2.7), but with emphasis on the mathematical functions needed to account for location with the product structure. For a conduction-heating product, the influence of location is expressed in terms of the following expressions.

a. For an infinite slab geometry,

$$j = 1.27324 \cos(\pi\ x/d_c) \tag{6.24}$$

where x = distance from the center of the slab geometry; d_c = half-thickness of the slab.

b. For an infinite cylinder,

$$j = 1.60218\ J_o\ [(\mathbf{R_1})(r/R)] \tag{6.25}$$

where $\mathbf{R_1}$ = the argument of the zero-order Bessel function; r = radial distance from the geometric center of the cylinder; R = the radius of the cylinder.

c. For a sphere,

$$j = 0.63662\ (R/r) \sin(\pi\ r/R) \tag{6.26}$$

where r = the distance from the center of the sphere; R = the radius of the sphere.

Note that $j_c = 2.0$ when $r = 0$.

All expressions (Eq. 6.24, 6.25, 6.26) are based on a condition of negligible resistance to heat transfer at the surface of the product. Pflug et al. (1966) has provided the expressions and charts needed to account for Biot numbers at less than 100. The values needed to evaluate the Bessel functions are presented in Table 6.3.

Using the preceding relationships and those from Chapter 5, the temperature distribution histories within the food product can be predicted. More specifically,

$$\log(T_M - T) = -t/f_h + \log[\,j_c(T_M - T_o)] \tag{6.27}$$

and

$$f_h = 0.934\ d_c/\alpha \text{ for an infinite slab} \tag{6.28}$$

$$f_h = 0.399\ R/\alpha \text{ for an infinite cylinder} \tag{6.29}$$

$$f_h = 0.233\ R/\alpha \text{ for a sphere} \tag{6.30}$$

Table 6.3 Values of zero-order Bessel functions as a function of the argument $\{J_o[(R_1)(r/R)]\}$

Argument	Function	Argument	Function
0	1.0000	1.3	0.6201
0.1	0.9975	1.4	0.5669
0.2	0.9900	1.5	0.5118
0.3	0.9776	1.6	0.4554
0.4	0.9604	1.7	0.3980
0.5	0.9385	1.8	0.3400
0.6	0.9210	1.9	0.2818
0.7	0.8812	2.0	0.2239
0.9	0.8463	2.1	0.1666
1.0	0.8075	2.2	0.1104
1.1	0.7652	2.3	0.0555
1.2	0.6711	2.4	0.0025
		2.5	−0.0484

Source: Adapted from Schneider (1955).

Example 6.6

Estimate the retention of a heat-sensitive product quality attribute during a thermal process for a specified microbial population when the heating medium temperature is 125°C and the initial product temperature is 70°C. The product is in a cylindrical container and has thermophysical properties as follows: density (ρ) = 1000 kg/m^3, thermal conductivity (k) = 0.45 W/m K, and specific heat (c_p) = 3.5 kJ/kg K. The initial microbial population is 10 per gram, and the survivor curve is described by a first-order rate constant (**k**) of 0.1/min at 121°C and an Activation Energy Constant (E_A) of 250 kJ/mole. The heat-sensitive quality attribute is described by first-order rate constant (**k**) of 0.15/min at 121°C and Activation Energy Constant (E_A) of 75 kJ/mole. The product container is a cylinder with a diameter of 9 cm and height of 18 cm, and the cooling medium temperature is 35°C. The thermal process is designed to ensure a spoilage rate of no more than one container per one million processed.

Given:

1. Product density is ρ = 1000 kg/m^3.
2. Product thermal conductivity is k = 0.45 W/m K.

(*Continued*)

Example 6.6 (Continued)

3. Product specific heat is $c_p = 3.5\,kJ/kg\,K$.
4. Initial microbial population is $N_o = 10/g$.
5. Rate constant for microbial population is **k** = 0.5/min.
6. Activation Fnergy Constant for microbial population is E_A = 250 kJ/mole.
7. Rate constant for quality attribute is **k** = 0.15/min.
8. Activation Energy Constant for quality attribute is E_A = 75 kJ/mole.
9. Container diameter is 9 cm.
10. Container height is 18 cm.
11. Spoilage rate is 1 per 10^6 containers.
12. Heating medium temperature is $T_M = 125\,°C$.
13. Initial product temperature is $T_o = 70\,°C$.
14. Cooling medium temperature is $T_{CM} = 35\,°C$.

Approach:
1. The process time to ensure that the desired spoilage rate is achieved is computed.
2. The retention of the heat-sensitive quality attribute is predicted based on the process time needed for the target spoilage rate.

Solution:
1. Based on the spoilage rate, the process time (F) at 121 °C is

$$1/10^6 = N_o/10^{F/D}$$

   where N_o = initial population per container; D = Decimal Reduction Time, D.
2. The population per container must be computed based on the volume of the container and the population per gram:

$$\text{Volume of container} = (\pi\,R^2)\,(H) = [\pi\,(4.5)^2]\,[18]$$
$$= 1145.11\,cm^3$$
$$V = 1.145 \times 10^{-3} m^3$$

Given the density of the product, the mass of product in a container is

$$\text{Mass} = [1000\ \text{kg/m}^3]\,[1.145 \times 10^{-3}\text{m}^3] = 1.145\ \text{kg}$$

Then, the population becomes

$$\begin{aligned}\text{Population per container} &= (1.145\ \text{kg})\,(10/\text{g})\,(1000\ \text{g/kg}) \\ &= 1.145 \times 10^4\end{aligned}$$

3. The D-value is determined from the first-order rate constant:

$$D = 2.303/0.5 = 4.606\ \text{min}$$

4. The process time (F) becomes

$$\begin{aligned}1/10^6 &= 1.145 \times 10^4/10^{F/4.606} \\ 10^{F/4.606} &= 1.145 \times 10^{10} \\ F &= 46.33\ \text{min}\end{aligned}$$

where F is the time at 121°C required to reduce the microbial population at the geometric center of each container.

5. The process time (U) at 125°C is computed by using Eq. (6.19), and the z-value is determined by using Eq. (2.27):

$$\begin{aligned}z &= 2.303\ R_g\mathbf{T}^2/E_A = \\ &2.303\ (8.31441)\ (121 + 273)^2/250{,}000 = 12.2\ \ C\end{aligned}$$

Then

$$U = (46.33)\ 10^{[(121-125)/11.9]} = 21.37\ \text{min}$$

6. The operator process time can be determined by using Eq. (6.18) and Figure 6.3. To use Figure 6.3, the magnitude of f_h must be established, using Eq. (6.29) and the assumption that the container is an infinite cylinder:

$$f_h = 0.399\ R/\alpha$$

(Continued)

Example 6.6 (Continued)

where

$$\alpha = k/\rho\, c_p = 0.45/(1000)\,(3500) = 1.2857 \times 10^{-7}\,m^2/s$$

then

$$f_h = 0.399\,(4.5/100)^2/1.2857 \times 10^{-7} = 6.2843 \times 10^3\,s$$
$$= 104.74\ min$$

7. Given the preceding values,

$$f_h/U = 104.74/21.37 = 4.9$$

8. Given the j-value of 1.60218 at the geometric center of the cylinder and the preceding value of f_h/U and z = 12.2°C (22°F), Table A.6.6 indicates that g = 7.89°F (4.38°C).
9. Using Eq. (6.18), the operator time for the process becomes

$$t_B = (104.74)\log[(1.60218)\,(125 - 70)/4.38]$$
$$= 136.5\ min$$

10. To evaluate the impacts of the thermal process on the heat-sensitive component in each region, the j-values for each region must be determined using Eq. (6.25) and Table 6.3. For the region with midpoint radius of 0.25 cm (Region 1),

$$j = 1.60218\,J_o\,[R_1(0.25/4.5)] = 1.60218\,J_o[2.4048\,(0.056)]$$
$$= 1.60218\,J_o\,[0.1336] = 1.5942$$

For the increment with a midpoint radius of 1 cm (Region 2),

$$j = 1.60218\,J_o\,[2.4048\,(1/4.5)] = 1.60218\,[0.93248]$$
$$= 1.494$$

For the increment with a midpoint radius of 2 cm (Region 3),

$$j = 1.60218\, J_o[2.4048\,(2/4.5)] = 1.60218\,[0.7784] = 1.247$$

For the increment with a midpoint radius of 3 cm (Region 4),

$$j = 1.60218\, J_o\,[2.4048\,(3/4.5)] = 1.60218\,[0.45] = 0.721$$

For the increment with a midpoint radius of 4 cm (Region 5),

$$j = 1.60218\, J_o\,[2.4048\,(4/4.5)] = 1.60218\,[0.3735] = 0.2331$$

11. These values of j have been used with Eq. (6.14) to compute the temperature within each of the five regions during the process. These computations are presented in the spreadsheet in Table A.6.7.
12. The impact of the process on retention of the heat-sensitive component can be evaluated by computing the retention in different regions of the container in the radial direction. The evaluation is accomplished by dividing the container into 1 cm increments in the radial direction from the center. By reference to Figure 6.3, the center portion has a 0.5 cm radius, and the evaluation is completed for a 1 cm height (Region 1):

$$\text{Volume} = \pi\,(0.5)^2(1) = 0.7854\ \text{cm}^3 = 7.854 \times 10^{-7}\,\text{m}^3$$
$$\text{Mass} = [7.854 \times 10^{-7}]\,[1000\ \text{kg/m}^3] = 7.854 \times 10^{-4}\,\text{kg}$$

The volume and mass of the incremental region with a radius of 1.5 cm (Region 2) is

$$\text{Volume} = \pi\,(1.5)^2(1) - 0.7854 = 6.2832\ \text{cm}^3 = 6.283 \times 10^{-6}\ \text{m}^3$$
$$\text{Mass} = [6.283 \times 10^{-6}]\,(1000\ \text{kg/m}^3) = 6.283 \times 10^{-3}\ \text{kg}$$

(*Continued*)

Example 6.6 (Continued)

The volume and mass of the incremental region with a radius of 2.5 cm (Region 3) is

$$\begin{aligned}\text{Volume} &= \pi\,(2.5)^2(1) - 6.283 - 0.7854 = 12.566\ \text{cm}^3\\ &= 12.566 \times 10^{-6}\ \text{m}^3\\ \text{Mass} &= 12.566 \times 10^{-3}\ \text{kg}\end{aligned}$$

The volume and mass of the incremental region with a radius of 3.5 cm (Region 4) is

$$\begin{aligned}\text{Volume} &= \pi(3.5)^2(1) - 12.566 - 6.283 - 0.7854\\ &= 18.85\ \text{cm}^3 = 18.85 \times 10^{-6}\ \text{m}^3\\ \text{Mass} &= 18.85 \times 10^{-3}\ \text{kg}\end{aligned}$$

The volume and mass of the incremental region with a radius of 4.5 cm (Region 5) is

$$\begin{aligned}\text{Volume} &= \pi(4.5)^2(1) - 18.85 - 12.566 - 6.283 - 0.7854\\ &= 25.133\ \text{cm}^3 = 25.133 \times 10^{-6}\ \text{m}^3\\ \text{Mass} &= 25.133 \times 10^{-3}\ \text{kg}\end{aligned}$$

The total mass of the 1 cm high region is 63.62×10^{-3} kg.

13. The spreadsheet in Table A.6.7 indicates that the retention and losses of the heat-sensitive attribute at the completion of the process is as follows:

Region 1	41.9%	58.1%
Region 2	17.6	82.4
Region 3	17.9	82.1
Region 4	8.6	91.4
Region 5	0.5	99.5

(Continued)

14. Using the mass of product in each region, the weighted average of the loss of heat-sensitive product attribute can be estimated:

Region 1	58.1 $(7.854 \times 10^{-4}/63.62 \times 10^{-3})$ =	0.717
Region 2	82.4 $(6.283 \times 10^{-3}/63.62 \times 10^{-3})$ =	8.138
Region 3	82.1 $(12.566 \times 10^{-3}/63.62 \times 10^{-3})$ =	16.216
Region 4	91.4 $(18.85 \times 10^{-3}/63.62 \times 10^{-3})$ =	27.081
Region 5	99.5 $(25.133 \times 10^{-3}/63.62 \times 10^{-3})$ =	39.307
Mass Average Loss of Quality Attribute =		91.64%

These also apply as long as conditions at the product surface have negligible resistance to heat transfer. These expressions provide the basis for predicting the impact of the thermal process as a function of location within the product structure. Given knowledge of these distributions, the relative impacts on microbial populations and heat-sensitive quality attributes within the product can be evaluated. In addition, the opportunity to begin optimization of the process becomes more evident.

The computations in Example 6.6 reveal several insights about the impacts of a thermal process on the product:

- A process designed to ensure a target rate of microbial spoilage delivers a process that exceeds the target for most of the product in the container. This is typical for thermal processes for conduction heating/cooling foods, when the target location for the reduction in microbial population is at the geometric center of the container.
- The regions of the container near the surface in close contact with the heating/cooling medium are impacted more significantly by the process, both in terms of the reduction in microbial population and the heat-sensitive product quality attribute. These regions of a cylindrical container have larger volume and mass, so the impact of the process on the outer regions is greater than on the product in the interior regions of the container.
- During a period of the process when cooling of the product begins, the product temperature continues to increase for a short period of time. This product temperature profile is most dramatic at the center of the container, and these high temperatures contribute to the negative impact of the process on the heat-sensitive product quality attributes.

- The regions of maximum retention of heat-sensitive quality attribute are near the geometric center of a cylindrical container. Quality retention within regions between the center and the surface is nearly as good as at the center due to the lower temperatures at the onset of cooling.

The influence of the kinetic parameters for both microbial populations and quality attributes on optimization of product quality retention can be explored using the relationships demonstrated in the spreadsheet in Table A.6.7. These relationships are investigated in more detail in Chapter 8.

6.5 Design of a microwave process

There has been a continuing interest in preservation processes based on applying electromagnetic energy to food product. Although microwave heating has become a routine method for rapid warming of individual food entries prior to consumption, the applications for food preservation have been limited. The primary limitation has been the inability to establish the process time, due to the uncertainties in locating the slowest heating location within the product structure. The incentives associated with applications of the microwave process include (a) the rapid heating of the product, (b) the uniformity of product temperature increase throughout the product structure during the process, and (c) the opportunity to improve retention of quality attributes by applying the high-temperature, short-time concept. Although the monitoring of the slowest heating location during microwave heating must be addressed, the rapid and uniform heating throughout the product structure provides incentive for exploring process design.

Example 6.7

A preservation process is being developed using microwave energy for potato pieces with dimensions of 10 cm × 10 cm × 5 cm. The process is being designed to reduce the initial population of 1000 *L. monocytogenese* by 12 log cycles. The thermal inactivation of microbial population is described by a first-order rate constant of 0.55/min at 60°C and an Activation Energy Constant of 386 kJ/mole. The retention of quality attribute is used to compare the microwave process to a traditional hot water process. The degradation of folic acid is described by a rate constant of 0.068/min at 80°C, and an Activation Energy Constant of 80 kJ/mole. Microwave heating is accomplished using a system with an electric field intensity of 12 V/cm and frequency of 2450 MHz. The properties of the potato include a relative dielectric constant of 64, a loss tangent of 0.23, and a density of 900 kg/m^3. The thermal process uses hot water at 65°C with a forced convection coefficient of 5000 W/m^2°C to heat the product from an initial temperature of 20°C. Cooling for both processes occurs using 20°C air with a forced convection coefficient of 50 W/m^2°C. The thermophysical properties of potato are thermal conductivity of 0.554 W/mC and specific heat of 3.6 kJ/kg°C. Determine the process times required for the traditional thermal process and for the microwave process, and estimate the retention of folic acid for both processes.

Given:

1. Product is potato with 10 cm by 10 cm by 5 cm dimensions.
2. Reduce population of *L. monocytogenes* by 12 log cycles.
3. Rate constant for microbial population is **k** = 0.55/min at 60°C.
4. Activation Energy Constant for microbial population is E_A = 386 kJ/mole.
5. Rate constant for folic acid is **k** = 0.068/min at 80°C.
6. Activation Energy Constant for folic acid is E_A = 80 kJ/mole.
7. Electric field intensity is E_V = 12 V/cm.
8. Frequency of microwave is f = 2450 Hz.
9. Relative dielectric constant is ε' = 64.
10. Loss tangent is $\tan \delta = 0.23$.
11. Potato density is ρ = 900 kg/m^3.
12. Heating medium is water: T_M = 65°C.
13. Surface heat transfer coefficient for heating medium is h = 5000 W/m^2 K.
14. Product initial temperature is T_o = 20°C.
15. Product cooling is air: T_{CM} = 20°C.
16. Convective heat transfer coefficient for cooling medium is h = 50 W/m^2 K.
17. Thermal conductivity of potato is k = 0.554 W/m K.

(Continued)

Example 6.7 (Continued)

18. Specific heat of potato is $c_p = 3.6\,kJ/kg\,K$.
19. Initial microbial population is $N_o = 1000$.

Approach:

1. Estimate the process time for hot water heating based on a 12 log cycle reduction in the pathogen, and determine the retention of thiamine for the traditional thermal process.
2. Estimate the process time for the microwave heating process and the retention of thiamine for the microwave heating process.
3. Compare the retention of thiamine for the two processes, based on same target reduction in the microbial pathogen.

Solution:

1. Heating the product for the traditional thermal process by using Eq. (5.19) of Chapter 5 to estimate the temperature at the center of the potato piece is

$$\log[T - T_M] = -t/f_h + \log[j_c(T_o - T_M)]$$

where f_h is determined by using Figure 5.4, after computing the Biot number from

$$N_{Bi} = hd_c/k$$

with

$$f\alpha/d_c^2$$

and

$$\alpha = k/\rho\, c_p$$

and j_c is obtained by using Figure 5.5, and the Biot number. For example, the temperature at the center of the potato during the eighth time interval of 5 minutes after 40 minutes of heating is

$$\log[T - 65] = -40/f_h - \log[j_c(20 - 65)]$$

2. To determine the appropriate values for f_h and j_c, the Biot number must be computed as

$$N_{Bi} = (5000)(5/100)/0.554 = 451 \text{ (for 10 cm dimensions)}$$

and

$$N_{Bi} = (5000)(2.5/100)/0.554 = 250.5 \text{ (for 5 cm dimension)}$$

Given these Biot numbers, Figure 5.4, and

$$\alpha = (0.554)/(900)(3600) = 1.7 \times 10^{-7} \, m^3/s$$

then

$$f_h(1.7 \times 10^{-7})/(0.05)^2 = 0.94$$
$$f_h = 13{,}823.5 \text{ s} = 230.4 \text{ min} = 3.84 \text{ hr (for the 10 cm dimensions)}$$
$$f_h = 3455.9 \text{ s} = 57.6 \text{ min} = 0.96 \text{ hr (for the 5 cm dimension)}$$

Using Eq. (5.22),

$$1/f_h = 1/230.4 + 1/230.4 + 1/57.6$$

$f_h = 38.4$ min for the shape of the potato piece.

Using Figure 5.5 and $N_{Bi} = 451$ or 250.5,

$j_c = 1.27$ for all three dimensions of the potato.

And,

$$j_c = 1.27 \times 1.27 \times 1.27 = 2.048$$

then

$$\log[65.5 - T] = -40/38.4 + \log[2.048(65.5 - 20)]$$

and

$T = 57°C$ for temperature at center of potato piece at 40 min of heating.

(Continued)

Example 6.7 (Continued)

3. The reduction in microbial population is estimated by Eq. (6.7), as applied at time intervals of 5 minutes during the heating portion of the process:

$$N_i = N_{i\text{-}5} \exp(-\mathbf{k}\Delta t)$$

where $N_{i\text{-}5}$ = microbial population at end of time interval; N_i = microbial population at beginning of time interval; Δt = time interval; **k** = first-order rate constant rate for microbial population, based on temperature at the midpoint in time interval, and computed using Eq. (6.3).

4. For the eighth time interval, the rate constant for the microbial population is

$$\mathbf{k} = (0.55)\exp\{-(386{,}000/8.314)[(1/57 + 273) - (1/60 + 273)]\}$$
$$\mathbf{k} = 0.157/\text{min at } 57^\circ\text{C}$$

5. Then, for initial microbial population at the beginning of the eighth time interval (see Table A.6.8), $N_{i\text{-}5} = 770$:

$$N_i = (770)\exp[(0.157)(5)]$$
$$N_i = 351$$

6. As indicated in Table A.6.8, the preceding computations have been completed for each time interval during the process, until the microbial population is reduced to 2.69×10^{-8}; or a 12 log cycle reduction from the initial population of 1000.
7. To evaluate the impact of the thermal process on the heat-sensitive component, the mass average temperature with the product during the process is used. These temperatures are predicted using a mass average lag coefficient from Figure 5.6 of Chapter 5:

$$j_m = 0.82 \text{ (for } N_{Bi} = 451 \text{ and } 250.5)$$

Then

$$j_m = 0.82 \times 0.82 \times 0.82 = 0.551$$

Using this lag coefficient, the temperature change for the eighth time interval becomes

$$\log(65.5 - 39) = -40/38.4 + \log[(0.551)(65.5 - 20)]$$
$$T = 63.2°C$$

This is the mass average temperature for the product after 40 minutes of heating. This temperature can be used to estimate the retention of the heat-sensitive product component.

8. The rate constant for the heat-sensitive component is estimated at the preceding temperature:

$$k = (0.068)\exp\{(80{,}000/8.314)[(1/63.2 + 273) - (1/80 + 273)]\}$$
$$k = 0.0082/\text{min}$$

The thiamin retention during the eighth time interval is

$$C = (83.9)\exp[-(0.0082)(5)]$$
$$C = 80.5\%$$

9. The temperatures during cooling at the potato center are estimated by using Eq. (5.19), with appropriate constants for cooling conditions. The Biot number during cooling is

$$N_{Bi} = (50)(0.05)/0.554 = 4.5 \text{ (for the 10 cm dimensions)}$$

or

$$N_{Bi} = (50)(0.025)/0.554 = 2.26 \text{ (for the 5 cm dimension)}$$

10. Given the Biot numbers, and the properties of the potato,

$$f_c(1.7 \times 10^{-7})/(0.05)^2 = 1.06$$
$$f_c = 15{,}588 \text{ s} = 259.8 \text{ min}$$
$$= 4.33 \text{ hr (for the 10 cm dimensions)}$$

and

$$f_c(1.7 \times 10^{-7})/(0.025)^2 = 1.98$$
$$f_c = 7279 \text{ s} = 121.3 \text{ min}$$
$$= 2.02 \text{ hr (for the 5 cm dimension)}$$

(*Continued*)

Example 6.7 (Continued)

and

$$1/f_c = 1/259.8 + 1/259.8 + 1/121.3$$
$$f_c = 62.7 \text{ min}$$

Using Figure 5.5 and the $N_{Bi} = 4.5$ and 2.26,

$$j_c = 1.235 \text{ (for 10 cm dimensions)}$$
$$j_c = 1.19 \text{ (for the 5 cm dimension)}$$

Then

$$j_c = 1.235 \times 1.235 \times 1.19 = 1.815$$

11. For evaluating retention of the heat-sensitive component during cooling, the mass average lag coefficient is used and estimated from the Biot number, using Figure 5.6:

$$j_m = 0.92 \text{ (for } N_{Bi} = 4.5)$$
$$j_m = 0.96 \text{ (for } N_{Bi} = 2.26)$$

Then

$$j_m = 0.92 \times 0.92 \times 0.96 = 0.813$$

As indicated by Table A.6.8, after completion of all steps in the computation, the retention of folic acid after the thermal process is 69.1%.

12. The microwave heating process requires conversion of electromagnetic energy to thermal energy, and the estimation the rate of temperature increase, as provided by Eq. (5.27):

$$dT/dt = (55.61 \times 10^{-14})(12)^2\ (2450 \times 10^6)(64)(0.219)/ (900)(3880)(5 \times 10^{-4})$$
$$dT/dt = 1.375 \times 10^{-3}\,^{\circ}\text{C/s per cm}^3$$

For the volume of the potato piece, 10 × 10 × 5 or 500 cm^3,

$$dT/dt = 0.788°C/s$$

13. Tables 6.4 and A.6.8 shows the temperatures at 2-second intervals during the heating of the potato. For each 2-second interval, the reduction microbial population has been computed. For example, for the 2-second interval from 47 to 49 seconds, the change in population is

$$N_i = N_{i-2} \exp(-\mathbf{k}\,\Delta t)$$

where

$$\mathbf{k} = (0.55) \exp\{-(386{,}000/8.31441)[(1/57.8 + 273) - (1/60 + 273)]\}$$
$$\mathbf{k} = 0.217/\text{min}$$

and then

$$N_i = (985.4) \exp[-(0.217)(2/60)] = 974.3$$

14. The microwave heating portion of the process is continued until the reductions in microbial population during cooling are sufficient to reach the overall 12 log cycle reduction or 1×10^{-9} population at the end of the process. As indicated by Tables 6.4 and A.6.8, this is accomplished by heating until 50.8 seconds. Although cooling begins at this time, the change in temperature at the slowest cooling point remains at 62°C until 20 minutes of cooling. As expected, the reductions in microbial population are significant during the initial portions of cooling.
15. The retention of the heat-sensitive component is computed at time intervals during the entire process, using the same expressions as used for the traditional heating and cooling process. During cooling, the mass average temperature is used in the estimation of the retention of folic acid. For the entire process, the retention is 96.8%, as compared 69.1% for the traditional process.

(Continued)

Example 6.7 (Continued)

Table 6.4 Illustration of changes in microbial population and quality retention during a microwave process

Time (s)	Temp (°C)		k (1/min)	N		k (1/min)	Retention %
0	20			1000			100
2	21.57493	18.18662	6.95E-09	1000	6.157335	0.000144	99.99952
4	23.14986	17.34845	1.61E-08	1000	5.983622	0.000171	99.99895
6	24.72479	16.51915	3.68E-08	1000	5.811746	0.000203	99.99827
8	26.29972	15.69858	8.37E-08	1000	5.641679	0.000241	99.99747
10	27.87465	14.88659	1.88E-07	1000	5.473392	0.000285	99.99652
12	29.44958	14.08307	4.21E-07	1000	5.306858	0.000337	99.99539
14	31.02451	13.28787	9.32E-07	999.9999	5.14205	0.000398	99.99407
16	32.59944	12.50086	2.05E-06	999.9999	4.97894	0.000468	99.99251
18	34.17437	11.72193	4.46E-06	999.9997	4.817503	0.00055	99.99067
20	35.7493	10.95094	9.65E-06	999.9994	4.657713	0.000645	99.98852
22	37.32423	10.18778	2.07E-05	999.9987	4.499544	0.000756	99.986
24	38.89916	9.43232	4.41E-05	999.9972	4.342973	0.000884	99.98306
26	40.47409	8.684456	9.31E-05	999.9941	4.187975	0.001032	99.97962
28	42.04902	7.944069	0.000195	999.9876	4.034527	0.001203	99.97561
30	43.62395	7.211047	0.000406	999.9741	3.882606	0.001401	99.97094
32	45.19888	6.485282	0.000839	999.9461	3.732188	0.001628	99.96552

34	46.77381	5.766666	0.001722	999.8888	3.583252	0.001889	99.95922
36	48.34874	5.055093	0.003507	999.7719	3.435776	0.00219	99.95193
38	49.92367	4.350462	0.007095	999.5354	3.289738	0.002534	99.94348
40	51.4986	3.65267	0.014257	999.0605	3.145118	0.002928	99.93373
42	53.07353	2.961619	0.028454	998.1134	3.001895	0.003379	99.92247
44	54.64847	2.277211	0.056413	996.2383	2.860049	0.003894	99.9095
46	56.2234	1.599351	0.111115	992.5552	2.71956	0.004481	99.89458
48	57.79833	0.927946	0.217451	985.3868	2.580409	0.005151	99.87743
50	59.37326	0.262904	0.422849	974.338	2.442576	0.005912	99.86169
50.8	60.00323	−0.00135	0.550744	970.7672	2.387808	0.006244	99.85753
2.5	62	−0.83237	1.264323	41.32611	2.215576	0.007418	98.02512
5	62	−0.83237	1.264323	0.074265	3.555685	0.001942	98.89251
10	62	−0.83237	1.264323	0.000133	3.988297	0.00126	98.27138
15	62	−0.83237	1.264323	2.4E-07	4.357949	0.000871	97.84448
20	59.30481	0.291675	0.410856	3.07E-08	4.672494	0.000636	97.53396
25	52.71153	3.119867	0.02429	2.72E-08	4.939205	0.000487	97.2968
30	47.22426	5.562434	0.002112	2.69E-08	5.16468	0.000389	97.10793
35	42.65746	7.660015	0.000259	2.69E-08	5.354812	0.000321	96.95204
40	38.85673	9.452572	4.32E-05	2.69E-08	5.514799	0.000274	96.8194

The solution to Example 6.7 illustrates several concepts related to thermal processes, the influence of heating and cooling rates on the reductions in microbial populations, and the retention of a heat-sensitive component of a food product. During microwave heating of the product, the temperature increases rapidly and reaches 62°C in 50.8 seconds, but the reduction in microbial population is relative small during this time period. A significant portion of the reduction in microbial population occurs during the initial phase of product cooling. Nearly 20 minutes elapses before the temperature at the slowest cooling location (geometric center) of the product decreases below 62°C. During this 20-minute period, the majority of the reduction in microbial population occurs at the slowest heating location. A similar pattern occurs when considering the retention of the heat-sensitive component of the product. Because the change in retention of the heat-sensitive component is based on the mass average temperature of the product, which temperature decreases more rapidly than the geometric center, the concentration of the component decreases slowly during product cooling.

During heating of the product by the traditional heating process, the temperature increase is much slower at the slowest heating location. For the process being considered, 53 minutes is required for the slowest heating location to reach 61.6°C. During heating, the reduction of microbial population is from 1000 to less than 1. The majority of the target reduction in microbial population occurs during the first 20 minutes of cooling before the temperature at the slowest cooling location decreases below 61.6°C. The influence of the process on the heat-sensitive product component is evaluated at the mass average temperature throughout the process. More significant impact of the process on the heat-sensitive component occurs during heating and during the initial phases of cooling.

When considering the concept of process time, the difference of the microwave heating and traditional heating is significant. As illustrated by the example, the process time for the microwave system is 50.8 seconds, while the process time for traditional heating is 53 minutes. These times reflect the heating time required to reach the temperature needed to accomplish the reductions in microbial population during the entire process: time for an increase in temperature, the holding time (initial phase of cooling), and the time period while the temperature decreases.

6.6 Design of an ohmic heating process

The application of ohmic heating to the product is similar to the microwave system. An alternating electric current is passed through the product to create the internal generation of thermal energy. The temperature increase within the product is rapid and uniform. Key factors associated with the process are the electrical properties of product being heated. These properties include the electrical conductivity at a reference temperature and a coefficient to compensate for the influence of temperature on the electrical conductivity. These properties have been measured by Palaniappan and Sastry (1991) and discussed by Singh and Heldman (2009).

The concept of ohmic heating using the food product as a resistance in an electrical circuit was presented in Section 5.1.5 of Chapter 5. An estimate of thermal energy created during ohmic heating is evident in the following expression:

$$q = E\,I = I^2 R_e \qquad (6.31)$$

Based on Eq. (5.6) and (5.7), the electrical conductivity in food products and similar materials becomes

$$\sigma_E = L/A\,R_e \qquad (6.32)$$

where L = distance between electrodes of the ohmic heating system; A = cross-sectional area of product between electrodes; R_e = electrical resistance.

By combining Eq. (6.31) and (6.32),

$$q = E^2 A\,\sigma_E/L \qquad (6.33)$$

When this expression is set equal to the thermal capacity, as in Eq. (5.24), of the product, the equation is

$$E^2 A \sigma_E/L = \rho c_p V\,dT/dt \qquad (6.34)$$

or

$$dT/dt = E^2 \sigma_E/L\,\rho\,c_p \qquad (6.35)$$

Using this expression, the rate of temperature increase in a food product can be estimated for a given set of ohmic heating conditions. A more specific expression for the temperature at the exit from a tubular heat exchanger using ohmic heating was presented as Eq. (5.22). The expression applies to the flow of product through the heat exchanger and considers the thermal energy loss at the surface of the tubular heat exchanger. It assumes the properties of the product flowing through the tube are uniform throughout the heat exchanger.

The application of ohmic heating has occurred in aseptic processing systems. When the product being heated contains food particles in a carrier fluid, rapid heating is difficult in a continuous system. For these applications, ohmic systems provide uniform and rapid heating of the particles and the carrier fluid. Normal heat exchangers provide rapid heating of the carrier liquid, but the transfer of heat to the interior structure of the food particles is relatively slow. To ensure the target reduction of the microbial population within the particles, the carrier fluid is exposed to more significant process than required.

Example 6.8

An aseptic processing system is being designed for a vegetable soup containing pieces of potato. The product is flowing through a tubular heat exchanger with a 4.75-cm diameter at a rate of 1.874 m^3/hr. The inactivation constants of spoilage microorganism include a rate constant of 1.38/s at 121°C and Activation Energy Constant of 300 kJ/mole. Product heating is to be accomplished using an 0.94-m-long ohmic heater with a voltage of 5000 volts over the length of the heat exchanger, followed by a holding tube of length to be determined. The properties of potato include a density of 1000 kg/m^3, specific heat of 3.88 kJ/kg K, and electrical conductivity of 0.32 S/m at 25°C, with temperature coefficient of 0.035°C. Determine the length of the ohmic heating system and holding tube required to achieve a 12 log cycle reduction in the population of the spoilage microorganism from an initial population of 1000 per container, when product enters the system at 80°C, and is cooled to 60°C following the process. Cooling of the product is accomplished using a counter-current tubular heat exchanger with cold water entering the heat exchanger at 20°C and leaving at 100°C, with an overall heat transfer coefficient of 2000 w/m^2K.

Given:

1. Product flow rate js m = 1.874 m^3/hr.
2. Heat exchanger diameter is d = 4.75 cm = 0.0475 m.
3. Rate constant for microbial population is **k** = 1.38/s.
4. Activation Energy Constant for microbial population is E_A = 300 kJ/mole.
5. Ohmic heat system length is L = 0.94 m.
6. Voltage gradient is E = 5000 volts.
7. Potato density is ρ = 1000 kg/m^3.
8. Potato specific heat is c_p = 3.88 kJ/kg K.
9. Electrical conductivity is σ_E = 0.32 S/m at 25°C.
10. Electrical conductivity temperature coefficient = 0.035/°C.
11. Target reduction in microbial population = 12 log cycle.
12. Initial microbial population is N_o = 1000 per container.
13. Initial product temperature is T_o = 80°C.
14. Cooling medium temperature increases from 20°C to 100°C.
15. Overall heat transfer coefficient is U = 2000 W/m^2K.

Approach:

1. The length of holding tube needed to reduce the microbial population is determined by estimating the reduction of microbial population in the ohmic heater, followed by estimating the length of holding tube required to accomplish the target reduction in the microbial population.

Solution:

1. The rate of heating in the ohmic heater is computed using Eq. (6.35), as follows:

$$dT/dt = (5000)^2\ [0.32 + 0.035(100)]/(1.175)(1000)(3880) = 20.93°C/s$$

2. Given the volumetric flow rate of 1.874 m^3/hr and a tube diameter of 4.75 cm, the mean velocity is

$$u = 1.874/(\pi/4)(4.75/100)^2 = 1057.5\ m/hr = 0.29375\ m/s$$

3. Because flow in the heater is laminar, the maximum velocity is

$$u_{max} = 0.29375/0.5 = 0.5875\ m/s$$

to obtain and estimate the velocity of the product at the center of the heat exchanger tube.

(*Continued*)

Example 6.8 (Continued)

4. Because the length of the heater is 1.175 m, the residence time for a particle moving through the ohmic heater at the maximum velocity becomes

$$\text{Residence Time} = 1.175/0.5875 = 2.0 \text{ s}$$

5. Using the spreadsheet in Table A.6.9, the temperature at 0.5-second intervals is computed. At the exit from the ohmic heating system, the product temperature is 121.9°C. At each time interval, the reduction in the microbial population is computed. The rate constant for the fourth time interval is

$$\mathbf{k} = (1.38/\text{s}) \exp\{(300{,}000/8.31441)[(1/121.9 + 273) - (1/121 + 273)]\}$$
$$\mathbf{k} = 1.7/\text{s}$$

6. The reduction in microbial population for the fourth time interval is

$$N = (927) \exp[-(1.7)(0.5)]$$
$$N = 397$$

7. The product cooling is accomplished in a counter-current heat exchanger using cold water. The length of the counter-current tubular heat exchanger is determined by the thermal energy balance:

$$(0.52)(3880)(121.9 - 60) = (2000)[\pi(4.75/100)L][(40 - 21.9)/\text{Ln}(40/21.9)]$$

where

$$m = 0.52 \text{ kg/s, based on } 1.874 \text{ m}^3/\text{hr}$$
$$L = 13.9 \text{ m}$$

8. To estimate the reduction in microbial population during cooling, the product temperature is based on the maximum velocity, so the residence time is

$$\text{Time} = 13.9/0.5875 = 23.7 \text{ sec}$$

and

$$\text{Rate} = 61.9/23.7 = 2.61°\text{C/s}$$

9. The spreadsheet in Table A.6.9 illustrates that after 2.0 seconds in the ohmic heater, a residence time of 14 seconds is required in the holding tube. Holding is followed by 23.7 seconds in the heat exchanger for cooling. The length of the holding tube is 8.225 m.

The solution in the preceding example illustrates the effectiveness of ohmic heating to increase the product temperature rapidly and uniformly. These types of continuous systems provide very good efficiency, with high capacity, and contribute to retention of temperature-sensitive product quality.

6.7 Design of ultra-high pressure processes

The use of ultra-high pressure (UHP) for food preservation has been demonstrated as an effective process for many types of food products. For many products, the process can be accomplished without elevating product temperature and without the recognized impacts of thermal processes on product quality attributes. The primary process design parameter is the holding time, or the time the product must be held at a given pressure to accomplish the desired reduction in microbial population. Although the majority of the reduction in the microbial population occurs during the holding period, the portions of the overall process accomplished during the increase in pressure and during the pressure drop should be considered.

The application of UHP processes to the food product should consider a variety of factors. More specifically, applying pressure to the product results in an increase in temperature. The magnitude of the temperature increase is a significant function of product composition and a lesser function of initial temperature. A typical magnitude is 3°C per 100 MPa of pressure change. The application of UHP does result in a decrease in product volume. For some products, this change in volume may result in an irreversible impact on product structure. Although the impact varies with the product, a typical change in volume is 15% for a pressure increase of 600 MPa.

The process design for UHP can be similar to a thermal process and uses the appropriate kinetic parameters for the microbial population being considered. The key expressions needed for the design include the appropriate survivor curve equation and the equation to consider the influence of pressure on the rate of microbial survival. Equations (2.11) or (2.23) of Chapter 2 are the most likely expressions to describe the reduction in microbial population at a constant pressure. The most appropriate expression to describe the influence of pressure on microbial survival rate is Eq. (2.20). The use of these expressions during an interval step analysis of the process will lead to the desired result.

Example 6.9

A UHP process is being designed for inactivation of *E. coli 0157: H7* in hamburger. The initial product temperature is 25°C, and the mass average temperature of the product will increase 3°C for each 100 MPa increase in product pressure. Determine the holding time required at 400 MPa to decease the population of the microbial pathogen by 6 log cycles. The UHP system requires 4 minutes to increase the pressure to 400 MPa and allows 1 minute to reduce the pressure after the holding period. Evaluate the impact of the process on the thiamine in the product.

Given:

1. For *E. coli 0157:H7*, $\mathbf{k_R} = 0.768$/min at 400 MPa and 50°C, $\Delta V = -4.5 \times 10^{-5}\,m^3$/mole.
2. Time to reach 400 MPa = 4 minutes.
3. Time to reduce pressure = 1 minute.
4. Target reduction of microbial population is 6 log cycles, assuming the initial population of 1000 per container of product.
5. Initial product temperature is $T_o = 25$°C.
6. Because temperature increase due to pressure is 12°C (3°C per 100 MPa), a final temperature of 37°C is assumed to have no impact on microbial inactivation.
7. For thiamine, $\mathbf{k_R} = 0.0025$/minutes at 98°C, and $E_A = 113\,kJ$/mole.

Approach:

1. The reduction in microbial population due to elevated pressure is computed at time intervals during the process, beginning with the 4 minutes of pressure increase to 40 MPa.

2. The reductions in microbial populations continue during the holding period, until approaching the desired 6 log cycle reduction, or a probability of 10^{-3} survivors.
3. The reduction in population during the pressure reduction is computed, and the holding period is adjusted.
4. After the holding time is established, the influence of temperature on thiamine retention is computed using the same time intervals used for computing the reductions in microbial population.

Solution:

1. The reduction in microbial population is computed at 0.5-minute time intervals through the process. During each time interval, the change in population is estimated using the first-order survivor equation. The rate constant is computed using Eq. (2.20), as follows for 25 MPa:

$$\mathbf{k} = (0.768)\exp\{[-4.5 \times 10^{-5}/(8.31441)(25.75 + 273)][(25 - 400)(10^{6})]$$

$$\mathbf{k} = 0.00086/\text{min at 25 MPa and 25.75 C}$$

2. Given the rate constant for 25 MPa, the reduction in microbial population for the first time interval becomes

$$N = (1000)\exp[-(0.00086)(0.5)]$$

$$N = 999.57$$

3. The computations for each half-minute time interval are repeated until the pressure has increased 400 MPa at 4 minutes, as illustrated in Table 6.5. The same computations are continued during the holding period, while the rate constant remains constant at the pressure of 400 MPa.
4. As illustrated in Table 6.5, some portion of the reduction in microbial population occurs during the pressure release at the end of the holding period. Based on the time-step computations, by extending the holding period ends to 21 minutes, the desired reduction in microbial population to 1×10^{-3} is achieved.
5. Given the computations, the holding time of 17 minutes (21 min − 4 min) is needed to accomplish the process.
6. The retention of this quality attribute during the process can be estimated using the appropriate reference rate constant

(Continued)

Example 6.9 (Continued)

Table 6.5 Influence of UHP on microbial population and quality retention

Time (min)	Pressure (Mpa)		k (1/min)	N	Temp (°C)		k (1/min)	Quality (%)
0	0	−7.26518	0.000537	1000	25	8.974311	3.17E-07	100
0.25	25	−6.794	0.000861	999.5698	25.75	8.859811	3.55E-07	99.99998
0.75	75	−5.85872	0.002193	998.4746	27.25	8.632527	4.46E-07	99.99996
1.25	125	−4.93274	0.005535	995.7153	28.75	8.407503	5.58E-07	99.99993
1.75	175	−4.01591	0.013844	988.8466	30.25	8.184704	6.97E-07	99.9999
2.25	225	−3.10811	0.034318	972.0236	31.75	7.9641	8.69E-07	99.99985
2.75	275	−2.20921	0.084317	931.8963	33.25	7.745656	1.08E-06	99.9998
3.25	325	−1.31906	0.205352	840.9612	34.75	7.529341	1.34E-06	99.99973
3.75	375	−0.43756	0.495831	656.308	36.25	7.315125	1.66E-06	99.99965
4.25	400	0	0.768	447.032	37	7.208795	1.85E-06	99.99956
4.75	400	0	0.768	304.4875	37	7.208795	1.85E-06	99.99946
5.25	400	0	0.768	207.396	37	7.208795	1.85E-06	99.99937
5.75	400	0	0.768	141.264	37	7.208795	1.85E-06	99.99928
6.25	400	0	0.768	96.21932	37	7.208795	1.85E-06	99.99919
6.75	400	0	0.768	65.538	37	7.208795	1.85E-06	99.99909
7.25	400	0	0.768	44.63999	37	7.208795	1.85E-06	99.999
7.75	400	0	0.768	30.4057	37	7.208795	1.85E-06	99.99891
8.25	400	0	0.768	20.71028	37	7.208795	1.85E-06	99.99882
8.75	400	0	0.768	14.10642	37	7.208795	1.85E-06	99.99872

9.25	400	0	0.768	9.608327	37	7.208795	1.85E-06	99.99863
9.75	400	0	0.768	6.544534	37	7.208795	1.85E-06	99.99854
10.25	400	0	0.768	4.457688	37	7.208795	1.85E-06	99.99845
10.75	400	0	0.768	3.036271	37	7.208795	1.85E-06	99.99835
11.25	400	0	0.768	2.0681	37	7.208795	1.85E-06	99.99826
11.75	400	0	0.768	1.408648	37	7.208795	1.85E-06	99.99817
12.25	400	0	0.768	0.959474	37	7.208795	1.85E-06	99.99808
12.75	400	0	0.768	0.653528	37	7.208795	1.85E-06	99.99798
13.25	400	0	0.768	0.445138	37	7.208795	1.85E-06	99.99789
13.75	400	0	0.768	0.303198	37	7.208795	1.85E-06	99.9978
14.25	400	0	0.768	0.206518	37	7.208795	1.85E-06	99.99771
14.75	400	0	0.768	0.140666	37	7.208795	1.85E-06	99.99761
15.25	400	0	0.768	0.095812	37	7.208795	1.85E-06	99.99752
15.75	400	0	0.768	0.06526	37	7.208795	1.85E-06	99.99743
16.25	400	0	0.768	0.044451	37	7.208795	1.85E-06	99.99734
16.75	400	0	0.768	0.030277	37	7.208795	1.85E-06	99.99724
17.25	400	0	0.768	0.020623	37	7.208795	1.85E-06	99.99715
17.75	400	0	0.768	0.014047	37	7.208795	1.85E-06	99.99706
18.25	400	0	0.768	0.009568	37	7.208795	1.85E-06	99.99697
18.75	400	0	0.768	0.006517	37	7.208795	1.85E-06	99.99687
19.25	400	0	0.768	0.004439	37	7.208795	1.85E-06	99.99678
19.75	400	0	0.768	0.003023	37	7.208795	1.85E-06	99.99669
20.25	400	0	0.768	0.002059	37	7.208795	1.85E-06	99.9966
20.75	400	0	0.768	0.001403	37	7.208795	1.85E-06	99.9965
21.25	300	−1.78045	0.129456	0.001315	31	8.07413	7.79E-07	99.99647
21.75	100	−5.44888	0.003303	0.001313	25	8.974311	3.17E-07	99.99645

Example 6.9 (Continued)

and activation energy coefficient for thiamin, as illustrated in Table 6.5. As a result of the modest temperature increase during the UHP process, the retention is over 99.99%. A thermal process to accomplish the same reduction in microbial population results in much lower retention of the heat-sensitive product component.

As illustrated by Example 6.9, design of UHP processes are relatively easy to visualize. Most of the reduction in microbial population occurs during the holding period at a constant pressure. This situation changes if the product temperature entering the process is higher, and the additional elevation in product temperature due to the pressure increase results in a thermal contribution to the overall process. This type of process is analyzed in detail later in this chapter. The contribution of the pressure decrease at the end of the process is relatively small. Recent publications by Noma et al. (2002) and Ramaswamy, Riahi & Idziac (2003) suggest that microbial inactivation may be increased by rapid reductions in pressure. The contributions of this factor can be incorporated by modifying the rate constant during the pressure reduction portion of the process design.

6.8 Design of pulsed-electric-field processes

As discussed in Chapter 3, the exposure of a microbial population to a pulsed electric field (PEF) provides a rapid reduction in the population. The typical design of a system for exposure of a food product to PEF requires the flow of a liquid food through a rectangular channel while the electric field is maintained across the channel. The two important process design parameters are the length of the channel and the product flow rate through the channel. A third significant process design parameter is the intensity of the electric field. In general, exposure time to accomplish reductions in microbial population is short, but due to the small dimensions of the flow channel, the product flow rates and system capacities may be limited.

A typical system includes a flow channel with a 0.5-cm gap between the electrodes maintaining the electric field intensity. This dimension must be relatively small to ensure uniform exposure of the product to the electric field. The width of the channel is usually 2 cm, and the length is 20 cm. Both of these dimensions may be increased by modest amounts. The number of channels incorporated into the system determines the overall capacity. Although the increase in product temperature during exposure to the PEF is relatively small, the system includes cooling to maintain uniform product temperature.

Example 6.10

A PEF system is being designed to accomplish a 6 log cycle reduction in *E. coli* population in a liquid food product. Determine the length of channel required for a capacity of 400 kg/min.

Given:

1. A 6 log cycle reduction in *E. coli* population is required, assuming an initial population of 1000 per kg.
2. The process is designed for a product flow rate of 500 kg/min.
3. The kinetic parameters for *E. coli* in a phosphate buffer include a first-order rate constant of 0.00853/microsec at an electric field intensity 20 kV/cm, and an electric field intensity coefficient of 41 kV/cm.
4. The cell for product flow has a 0.5-cm gap and 2-cm width.

Approach:

1. The kinetic parameters for *E. coli* is used to determine the required exposure time to accomplish the desired reduction in microbial population
2. The length of cell needed for the target flow rate is determined using the exposure time.

Solution:

1. Using the first-order survivor curve equation of

$$N = N_o \exp[-\mathbf{k}t]$$

(*Continued*)

Example 6.10 (Continued)

or

$$10^{-3} = 10^{3} \exp[-(0.00853)\, t]$$

and

$$t = 1620 \text{ microsec at } 20 \text{ kV/cm}$$

2. The velocity within the cell can be determined from the target flow rate:

$$m = \rho A u$$

where A = the cross-section area of the cell or (0.5)(2) = $1\,cm^2 = 1 \times 10^{-4} m^2$.
Then

$$u = (500 \text{ kg/min})/(1000)(1 \times 10^{-4}) = 5000 \text{ m/min}$$
$$= 83.33 \text{ m/s}$$

3. To determine the maximum velocity, the Reynolds number for flow in the cell must be calculated with

$$N_{Re} = \rho\, d\, u/\mu$$

where d = diameter of a channel for a rectangular cross-section = 4 (area of cross-section)/perimeter,
or

$$d = 4\,[(0.5)(2)]/[2(0.5) + 2(2)] = 0.8 \text{ cm}$$

then

$$N_{Re} = (1000)(0.8/100)(83.33)/(0.0015)$$
$$= 444{,}427$$

4. Because liquid flow in the cell is turbulent, the maximum velocity is

$$u_{max} = 83.33/0.8 = 104 \text{ m/s}$$

5. Using the required exposure time and the maximum velocity on the cell, the length of the cell becomes

$$L = [(104 \text{ m/s})/(1 \times 10^{6} \text{ microsec/s})]\,(1620 \text{ microsec})$$
$$L = 0.169 \text{ m} = 16.9 \text{ cm}$$

6. The dimensions of the cell needed for the process have a 0.5-cm gap, with a 2-cm width and 16.9-cm length.
7. The electric field intensity coefficient can be used to evaluate the process at a lower intensity. For example, if the electric field intensity is decreased to 15 kV/cm,

$$\log[D/270] = (20 - 15)/41$$
$$D = 357.5 \text{ microsec or } \mathbf{k} = 0.00644/\text{microsec}$$

8. The time of product exposure to the PEF becomes

$$10^{-3} = 10^{3} \exp[-(0.00644)\,t]$$
$$t = 2145 \text{ microsec}$$

9. For this process time, the length of the cell becomes

$$L = [(104)/(1 \times 10^{6})]\,(2145) = 0.223 \text{ m} = 22.3 \text{ cm}$$

10. As illustrated, the length of the cell is increased from 16.9 to 22.3 cm to accommodate a decrease in electric field intensity from 20 to 15 kV/cm.

The PEF process is influenced by temperature. Investigations by Pagan et al. (1998), Alvarez et al. (2000, 2003), and Evrendilek and Zhang (2003) have quantified the impact of temperature and demonstrated that a modest increase in temperature increases the effectiveness of the PEF process. In many reports, the influence of temperature has been expressed as an activation energy coefficient at a constant electric field intensity. Based on results from the Hulsheger et al. (1980) research, a more specific model for PEF has been proposed:

$$N/N_o = [t/t_c]^{-[(E-E_C)/K]} \tag{6.36}$$

where t_t = threshold treatment time; E_c = threshold electric field intensity; K = temperature coefficient function.

Hulsheger et al. (1980) have published the coefficients for Eq. (6.36) for several different microorganisms of interest in food preservation.

Example 6.11

A PEF process is being designed for reduction of the microbial population of *S. enteritis* in liquid egg whites by 8 log cycles from an initial population of 10,000 per kg. Determine the initial product temperature needed to process the product at a rate of 600 kg/min in a PEF cell with a 0.5-cm gap, a 2-cm width, and a 5-cm length.

Given:

1. Product flow rate is m = 600 kg/min.
2. Kinetic parameters for *S. enteritis* (from Table 3.1 of Chapter 3) include the first-order rate constant k = 0.0303/microsec at 30 kV/cm and 20°C, and an Activation Energy Constant E_A = 40 kJ/mole.
3. Microbial population is to be reduced from 10,000/kg to 10^{-6}.
4. The properties of liquid egg white include $\rho = 1000\ kg/m^3$ and $\mu = 0.002$ Pa s.

Approach:

1. The dimensions of the PEF cell and the flow rate are used to establish the exposure time in the cell.
2. The exposure time is used to the estimate the rate constant required for the process.
3. The temperature can be determined from the activation energy relationship.

Solution:

1. Given the flow rate of 600 kg/min, the mean velocity in the cell becomes

$$u = (600)/[(0.5)(2)(1\times 10^{-4})(1000)] = 6000\ m/min = 100\ m/s$$

2. To determine flow characteristics within the cell, the Reynolds Number must be computed:

$$N_{Re} = [(1000)(0.8/100)(100)]/0.002 = 400{,}000 \text{ to confirm turbulent flow}$$

3. Then, the maximum velocity of product within the cell is

$$u = (100)/(0.8) = 125\ m/s$$

4. Based on the length of the cell,

$$\text{Process Time} = (5\ cm/100\ cm/m)/(125\ m/s) = 0.0004\ s = 400\ \text{microsec}$$

5. Using the first-order survivor curve expression and the process time,

$$10^{-6} = 10^{4} \exp[-\mathbf{k}(400)]$$
$$\mathbf{k} = 0.0576/\text{microsec}$$

6. Because the first-order rate constant must be 0.0576/microsec, the temperature can be determined by using the Arrhenius expression:

$$(0.0576) = (0.0303) \exp\{-(40{,}000/8.31441)[(1/\mathbf{T}) - (1/20 + 273)]\}$$
$$\mathbf{T} = 304.9^{\circ}\text{K} = 31.9^{\circ}\text{C}$$

7. The results indicate that the process can be accomplished at the target flow rate by increasing the temperature from 20°C to 32°C.

Example 6.12

A PEF process is being designed for reduction of the *E. coli* population in a new food product from 100 to 10^{-4} per kg. The kinetic parameters are: E_c = 8.3 kV/cm, t_t = 18 μs, and K = 7.875 [1 − 0.04(T − 15)]. Determine the length of PEF cell needed when the product entering the system is 25°C at a flow rate of 100 kg/min. The dimensions of the cell include a 0.5-cm gap and a 2-cm width, and an electrical field intensity of 20 kV/cm will be used. Product properties are density of 990 kg/m^3 and viscosity of 0.065 Pa s.

Given:

1. Initial population of *E. coli*, N_o = 100 per kg, is to be reduced to 10^{-4}.
2. Kinetic parameters are the following:

$$E_c = 8.3 \text{ kV/cm}$$
$$t_t = 18\ \mu\text{s}$$
$$K = 7.875[1 - 0.04(T - 15)]$$

3. Initial product temperature is T_o = 25°C.
4. PEF cell dimensions are 0.5-cm gap and 2-cm width.

(*Continued*)

Example 6.12 (Continued)

5. Product physical properties are $\rho = 990\,kg/m^3$, and $\mu = 0.065$ Pa s.

Approach:

1. The model (Eq. 6.36) is used to compute the process time needed for the process.
2. The product flow rate is used to compute the maximum velocity through the cell.
3. The process time and the maximum velocity are used to compute the length of the PEF cell required.

Solution:

1. Using the temperature-dependent kinetic parameter, the coefficient can be estimated:

$$K = 7.875[1 - 0.04\,(25 - 15)]$$
$$K = 4.725 \text{ kV/cm at 25 C}$$

2. Using Eq. (6.36),

$$\log[10^{-4}/10^2] = -[(20 - 8.3)/4.725]\log(t/18)$$
$$-6/-2.476 = \log(t/18)$$
$$264.9 = t/18$$
$$t = 0.004768 \text{ s}$$

3. The maximum product velocity is computed from the flow rate, as follows:

$$u = 100/(990)[(0.5)(2)/1 \times 10^4] = 1010.1 \text{ m/min}$$
$$= 16.835 \text{ m/s}$$

4. The Reynolds number is

$$N_{Re} = (990)(0.8/100)(16.835)/(0.065) = 2051$$

indicating that the flow through the cell is laminar, and

$$u_{max} = 16.385/0.5 = 32.77 \text{ m/s}$$

5. Given the product velocity and the process time, the length of the PEF cell becomes

$$L = (32.77)(0.004768) = 0.156 \text{ m} = 15.6 \text{ cm}$$

6. This seems to be a reasonable length of cell for the conditions presented. Note that the process is very dependent on temperature, and small changes in product temperature have significant influence on dimensions of the cell.

6.9 Design of combined processes

Current trends consider the use of combinations of two or more process technologies for product preservation. Many process combinations decrease the probability of safety concerns, while improving the product quality attributes. One process combination receiving consideration is pressure-assisted thermal processing (Min et al., 2010; Nguyen et al., 2010; Somerville & Balasubramaniam, 2009). This process uses the rise in temperature during the application of pressure to accomplish the thermal process in a more efficient manner. The advantages of this approach are most obvious with conduction-heating food products, where rates of heat transfer within the product structure are very slow. As was illustrated in Section 5.5 of Chapter 5, the application of pressure causes the increase in temperature to occur uniformly throughout the product structure.

During a pressure-assisted thermal process (PATP), the initial portion of the process is accomplished by traditional heating. The product is preheated before being placed in a vessel for application of pressure, and the pressure increase is used to elevate the temperature to the final temperature needed for the thermal process. As soon as the target reduction in microbial population is achieved, the pressure is released, and the product temperature decreases to the same level as before the application of pressure. The final step of the process is cooling to the final temperature. This type of process should increase the retention of product quality attributes by the rapid and uniform temperature increase and the similar rapid decrease following the holding time at elevated pressure. The process design is accomplished without considering any influence of elevated pressure on the reduction in microbial population.

Example 6.13

A PATP is being designed for a conduction-heating food product in a pouch with dimensions of 15 cm × 15 cm × 3 cm. The product is preheated to 80°C, before the pressure is increased to 750 MPa to reduce the population of *Cl. botulinum 62A*. The

(*Continued*)

Example 6.13 (Continued)

pressure increase is accomplished at a rate of 300 MPa per minute, and the pressure is released in 1.5 minutes. The temperature elevation due to pressure increase is 5°C per 100 MPa. Determine the holding time, at 750 MPa, required to reduce the microbial population by 12 log cycles, and compare the process with a thermal process designed to provide the same reduction in microbial population. The thermal process uses steam at 130°C with forced convection coefficient of 5000 W/m^2K to heat the product from an initial temperature of 80°C. Cooling of the product uses water at 30°C and a forced convection heat transfer coefficient of 500 W/m^2K. The thermophysical properties of the product include a thermal conductivity of 0.554 W/mK, specific heat of 3.6 kJ/kgK, and density of 900 kg/m^3. Compare the retention of a product quality attribute, with a rate constant of 0.01/min at 121°C and an Activation Energy Constant of 75 kJ/mole, for the two processes.

Given:

1. Product package has dimensions of 10 cm × 10 cm × 5 cm.
2. The population of *Cl. botulium 62A* must be reduced by 12 log cycles.
3. Rate constant for microbial population is **k** = 1/min at 110°C.
4. Activation Energy Constant for microbial population is E_A = 240 kJ/mole.
5. Rate constant for quality attribute is **k** = 0.01/min at 121°C.
6. Activation Energy Constant for quality attribute is E_A = 75 kJ/mole.
7. Pressure is increased to 750 MPa in 2.5 min.
8. Pressure is released in 1.5 min.
9. Product temperature increases 5°C for each 100 MPa.
10. Product density is $\rho = 900\ kg/m^3$.
11. Heating medium is water: T_M = 130°C.
12. Surface heat transfer coefficient is h = 5000 W/m^2 K during heating.
13. Initial temperature of product is T_o = 80°C.
14. Cooling medium is water: T_{CM} = 30°C.
15. Convective heat transfer coefficient is h = 500 W/m^2 K for cooling.

16. Thermal conductivity of potato is $k = 0.554\,W/m\,K$.
17. Specific heat of potato is $c_p = 3.6\,kJ/kg\,K$.
18. Initial microbial population is $N_o = 1000$.

Approach:

1. The process time for the PATP is determined by evaluating the reductions in microbial population at 0.5-min intervals throughout the process, including the increase in pressure, during holding and during the release of pressure.
2. All reductions in microbial population are due to the temperature increase associated with the pressure change.
3. The holding time for the PATP process is established by the time needed to accomplish the target reduction in the microbial population.
4. The traditional thermal process is established by the heating time needed to accomplish the target reduction in the microbial population, while including the contributions during production cooling.
5. The influence of both processes on retention of the heat-sensitive product quality attribute is evaluated by summation of retention at each time interval throughout the processes.

Solution:

1. The computations for reductions in microbial populations during the PATP are presented in the spreadsheet in Table A.6.10.a.
2. A typical reduction in microbial population for the time period between 3 and 3.5 minutes is

$$\mathbf{k} = (1.0)\exp\{-(240000/8.314)[(1/390.5) - (1/383)]$$
$$\mathbf{k} = 4.2525/\text{min}$$

and

$$N = (63)\exp[-(4.2525)(0.5)]$$
$$N = 7.5$$

3. As illustrated, the product temperature reaches 117.5°C at 750 MPa, and holding time required at that pressure is 5.5 minutes.
4. The retention of quality attribute is computed at each 0.5 time interval, and accumulated over the entire process. The retention for the process was 94.4%.

(*Continued*)

Example 6.13 (Continued)

5. The thermal process requires a process time of 30 minutes to accomplish the target reduction in microbial population, including a significant portion during the cooling portion of the process (see Table A.6.10.b).
6. The retention of the heat-sensitive quality attribute is computed for each interval on a cumulative basis at the mass average temperature of the product during heating and cooling. The computed retention is 61.3%.

Example 6.13 illustrates the improvement in quality retention achieved by using PATP. In addition, PATP can be completed in a shorter time period than the traditional thermal process. Whereas the thermal process required nearly 50 minutes, the PATP required less than 10 minutes. It is important to emphasis that the illustration is a process for a pathogenic spore population (*Cl. botulinum 62A*), and the process should be considered for populations of spoilage microorganisms.

List of symbols

A = area, m^2
α = thermal diffusivity; m^2/s
C = concentration or intensity of product quality attribute
c_p = specific heat; kJ/kg K
D = Decimal Reduction Time, min
d_c = characteristic dimension, m
E = electric field strength, V/m
E_c = threshold electric field intensity; kV/cm
E_V = voltage intensity, V
E_A = Activation Energy Constant, kJ/mole
ε' = relative dielectric constant
F = thermal death time, min
F_R = thermal death time at reference temperature, min
f = frequency of microwave energy, Hz
f_h = heating rate constant, s or min
f_c = cooling rate constant, s or min
g = temperature difference between heating medium and product slowest heating location, and end of process, C

h = convective heat transfer coefficient, W/m^2 K
I = electric current, amp
J_o = Bessel function
j_c = heating lag constant at geometric center
j_m = mass average heating lag constant
j_{cc} = cooling lag constant at geometric center
K = temperature coefficient from Eq. (6.36)
k = thermal conductivity, W/m K
k = rate constant, 1/s
$\mathbf{k_o}$ = reference rate constant, 1/s
L = length, m
LR = lethal rate
m = mass flow rate, kg/s
μ = viscosity; Pa s
N = microbial population
N_o = initial microbial population
N_{Bi} = Biot number
N_{Re} = Reynolds number
P = power dissipation, W/cm^3
q = thermal energy transfer, W
R = radius, m
R_g = gas constant; kJ/kg K
R_e = electrical resistance, ohms
R_1 = argument of Bessel function
r = radial direction
r_c = number of containers in a process
ρ = density, kg/m^3
σ_E = electrical conductivity
T = temperature, °C
T_{CM} = cooling medium temperature, °C
T_M = medium temperature, °C
T_o = initial temperature, °C
T_R = reference temperature, °C
T = absolute temperature, K
t = time, s
t_B = process time, min
t_p = operator process time, min
t_{cut} = come-up time for heating system, min
t_r = time when heating system reaches temperature, min
t_c = time when cooling is started, min
t_t = threshold time, s
tan δ = loss tangent

U = thermal death time at process temperature, min
u = mean velocity, m/s
u_{max} = maximum velocity, m/s
V = volume, m^3
ΔV = activation volume; m^3/mole
z = Thermal Resistance Constant, C

Bibliography

Alvarez, I., Pagan, R., Condon, S., & Raso, J. (2003). The influence of process parameters for the inactivation of *L. monocytogenes* by pulsed electric fields. *International Journal of Food Microbiology*, *87*, 87–95.

Alvarez, I., Raso, J., Palop, A., & Sala, F. J. (2000). Influence of factors on the inactivation of *Salmonella senftenberg* by pulsed electric field. *International Journal of Food Microbiology*, *85*, 143–146.

Ball, C. O. (1923). Thermal process time for canned foods. Part 1. *Bulletin of the National Research Council*, *7*(37), 76.

Bigelow, W. D., Bohart, G. S., Richardson, A. C., & Ball, C. O. (1920). Heat penetration in processing canned foods. *National Canners Association Bulletin*, *No. 16L*.

Calderon-Miranda, M. L., Barbosa-Canovas, G. V., & Swanson, B. G. (1999a). Inactivation of *Listeria innocua* in liquid whole egg by pulsed electric fields and nisin. *International Journal of Food Microbiology*, *51*, 7–17.

Calderon-Miranda, M. L., Barbosa-Canovas, G. V., & Swanson, B. G. (1999b). Inactivation of *Listeria innocua* in skimmed milk by pulsed electric fields and nisin. *International Journal of Food Microbiology*, *51*, 19–30.

Cserhalmi, Z., Vadac, I., Beczner, J., & Czukor, B. (2002). Inactivation of *Saccharomyces cerevisiae* and *Bacillus cereus* by pulsed electric field technology. *Innovative Food Science and Emerging Technology*, *3*(1), 41–45.

Evrendilek, G. A., & Zhang, Q. H. (2003). Effects of ph, temperature and pre-pulsed electric field treatment on pulsed electric field and heat inactivation of *Escherichia coli* 0157:H7. *Journal of Food Protection*, *66*, 755–759.

Heinz, V., Phillips, S. T., Zenker, M., & Knorr, D. (1999). Inactivation of *Bacillus subtilis* by high intensity pulsed electric fields. *Food Biotechnology*, *13*, 155–168.

Heldman, D. R., & Hartel, R. W. (1997). *Principles of food processing*. New York: Chapman & Hall, Inc.

Hulsheger, H., & Nieman , E. G. (1980). Lethal effect of high voltage pulsed on *E. coli* K12. *Radiation and Environmental Biophysics*, 18(4), 281–288.

Martin-Belloso, O., Vega-Mercado, H., Qin, B. L., Chang, E. J., Barbosa-Canovas, G. V., & Swanson, B. G. (1997). Inactivation of *Escherichia coli* suspended in liquid egg using pulsed electric fields. *Journal of Food Processing and Preservation*, *21*(3), 193–208.

Martin-Belloso, O., Qin, B. L., Chang, E. J., Barbosa-Canovas, G. V., & Swanson, B. G. (1997). Inactivation of *Escherichia coli* in skim milk by high intensity pulsed electric fields. *Journal of Food Process Engineering*, *20*, 317–336.

Min, S., Sastry, S. K., & Balasubramaniam, V. M. (2010). Compressibility and density of select liquid and solid foods under pressures up to 700 MPa. *Journal of Food Process Engineering*, *96*(4), 568–574.

Nguyen, L. T., Balasubramaniam, V. M., & Sastry, S. K. (in press). Determination of in-situ thermal conductivity, thermal diffusivity, volumetric specific heat and isobaric specific heat of selected foods under pressure. *International Journal of Food Properties*.

Noma, S., Shimoda, M., & Hayakawa, I. (2002). Inactivation of vegetative bacteria by rapid decompression. *Journal of Food Science*, *67*(9), 3408–3411.

Pagan, R., Esplugas, S., Gongora-Nieto, M. M., Barbosa-Canovas, G. V., & Swanson, B. G. (1998). Inactivation of *Bacillus subtilus* spores using high intensity pulsed electric fields in combination with other food conservation technologies. *Food Science and Technology International*, *4*(1), 33–44.

Palaniappan, S., & Sastry, S. K. (1991). Electrical conductivity of selected solid during ohmic heating. *Journal of Food Process Engineering*, *14*, 221–236.

Pflug, I. J., Blaisdell, J. L., & Kopelman, I. J. (1965). Developing temperature-time curves for objects that can be approximated by a sphere, infinite plate or infinite cylinder. *ASHRAE Transaction*, *71*(1), 238–248.

Ramaswamy, H. S., Riahi, E., & Idziak, E. (2003). High-pressure destruction kinetics of *E. coli* (29055) in apple juice. *Journal of Food Science*, *68*(5), 1750–1756.

Schneider, P. J. (1955). *Conduction heat transfer*. Reading, MA: Addition-Wesley Pub. Co., Inc.

Singh, R. P., & Heldman, D. R. (2009). *Introduction to food engineering* (4th ed.). San Diego, CA: Elsevier Academic Press.

Somerville, J., & Balasubramaniam, V. M. (2009). Pressure-assisted thermal sterilization of low-acid shelf-stable foods. *Resource: Engineering and Technology for a Sustainable World*, *16*(7), 14–17.

Teixeira, A. A. (2007). Thermal processing of canned foods. In D. R. Heldman & D. B. Lund (Eds.), *Handbook of food engineering* (2nd ed., pp. 745–797). Boca Raton, FL: CRC Press, Inc.

Zhang, Q. H., Monsalve-Gonzales, A., Barbosa-Canovas, G. V., & Swanson, B. G. (1994). Inactivation of *E. coli* and *S. cerevisiae* by pulsed electric fields under controlled temperature conditions. *Transactions of the ASAE. American Society of Agricultural Engineers*, *37*(2), 581–587.

Zhang, Q., Qin, B. L., Barbosa-Canovas, G. V., & Swanson, B. G. (1995). Inactivation of *Escherichia coli* for food pasteurization by high-strength pulsed electric fields. *Journal of Food Processing and Preservation*, *19*, 103–118.

7 Process Validation and Evaluation

7.1. Process validation for microbial inactivation 219
7.2. Alternative approaches to validation 229
7.3. Process validation for alternative process technologies 237
List of symbols 243
Bibliography 243

The process design models presented in the previous chapters provide a sound and efficient basis for establishing and evaluating processes. The input parameters for these models are based on experimental measurements, so the conditions during these experiments may not duplicate the conditions during an actual product manufacturing operation. To ensure that the outcomes from the process design are acceptable, the process must be validated under conditions similar to manufacturing-scale operations. This chapter describes and discusses the steps to validate the processes obtained from process design.

Validation is very important. The primary purpose of many preservation processes is to ensure microbiological safety of the food product. As indicated during the development of process design, the microbial population of concern is a pathogen, and the process is designed to reduce the risk of a food safety hazard to some negligible level. To ensure that the process design is appropriate to prevent the manufacturing of product causing a food-borne illness outbreak, the process design must be validated. The conditions of the validation should consider as many of the process variables as possible.

Food Preservation Process Design
ISBN: 978-0-12-372486-1

There are four factors that emphasize the need for process validation:

1. **The acceptability of kinetic constants.** The kinetic constants used as inputs to the process design models are usually based on experiments conducted under controlled laboratory conditions. Most often, these experiments are conducted using pure cultures of a given microbial population, and the substrates may not have been food product similar to the potential application. Both of these situations represent variations from the commercial application. The microbial population of concern will not exist as a pure culture in the product being processed, and the substrate used in the experiment is unlikely to match the composition of the product.
2. **Physical properties of product.** The process design models require physical properties of the product as inputs to predict the changes in intensity of the agent being used to reduce the microbial population. Physical properties data for food products have been assembled in handbooks and similar references, but the composition of products available for experimental measurements may not be the same as the products being manufactured. The use of models to predict physical properties based on product composition tends to overcome many of these differences. Other properties and coefficients needed for predicting agent intensity may or may not adequately duplicate the conditions during commercial manufacturing operations.
3. **Shape and size of product container or package.** Most models for predicting transport phenomenon require simplifying the product and/or package geometry. The process design must be validated to ensure that deviations in shape and size do not significantly impact the process.
4. **Quality considerations.** When models incorporate impacts of the process on product quality attributes, the kinetic constants for the attributes have been determined under controlled experimental conditions. A process validation should include evaluation of quality attributes for the product manufactured using the proposed process to ensure that attributes predicted by the model are comparable to products from commercial operations.

7.1 Process validation for microbial inactivation

Because the primary purpose of a preservation process is the reduction of the microbial population, the validation must involve evaluation of the microbial survivors from the proposed process. Validations should be conducted in both pilot-scale and commercial-scale manufacturing conditions. Some of the procedures to be considered for the two scales of operation will be similar. One of the significant challenges in both scales of validation is the quantitative determination of microbial survivors. Because the proposed process design ensures appropriate levels of safety, the number of survivors will be very small and must be expressed in terms of probabilities.

Two approaches to evaluating the results for process validation have been identified. One approach is based on an expression from Halvorson and Zeigler (1932), which was demonstrated by Stumbo, Murphy, and Cochran (1950). The expression is

$$s = \mathbf{n} \, \text{Ln} \, [n/q] \tag{7.1}$$

where s = the most probable number of survivors; **n** = the total number of containers processed; q = the number of containers with no evidence of survivors.

The Halvorson/Zeigler (H-Z) expression can be used in a variety of situations when evaluating outcomes from food preservation processes. These applications are described in the following sections.

7.1.1 Laboratory evaluation of experiment results quantify parameters describing the microbial survivor curve

Although significant portions of the survivor curve are created by measurements of microbial populations at times when populations are one or higher, the ultimate objective of the process is to reduce the microbial population to a probability of survivors.

These experiments are referred to "endpoint" measurements. The measurement of the kinetic parameters must include times when the process is expected to ensure probabilities of survival similar to those expected under commercial operations. In the laboratory, the medium containing the microorganism can be divided into an established number of containers before the proposed procedure is applied.

Example 7.1

An experiment is being conducted to validate the survivor curve parameters for a pathogenic spore. Initial populations of 25,600 spores were placed in each of six test tubes, and a process of 7 minutes at 121°C has been applied. The results of the experiment indicate that four of the six tubes contained survivors. Assuming the survivor curve can be described by a first-order model, estimate the first-order rate constant for the microbial population.

Given:

1. A total of six tubes are used.
2. Four tubes have microbial survivors; two tubes are negative.
3. Process time is 7 minutes at 121°C.
4. The initial microbial population was 25,600 spores per tube.

Approach:

1. Use the Halvorson-Zeigler expression to compute the most probable number of survivors.
2. Estimate the first-order rate constant for the microbial spores.

Solution:

1. Using the Halvorson-Zeigler equation,

$$s = (6)\ \mathrm{Ln}\ [6/2] = 6.59$$

2. The first-order survivor expression provides

$$N = N_o \exp [-k\ t]$$

where

$$N_o = 6\ (25{,}600) = 153{,}600$$

then

$$6.59 = (153{,}600)\ \exp\left[-k\ (7)\right]$$

and

$$k = -10.056/-7 = 1.437/\text{min}$$

As indicated by this example, the number of survivors are expected to be less than one in each of the six tubes. The number of tubes in the experiment needs to be adjusted as the temperature or time of the process is changed. An alternative is to increase or decrease the initial population in each tube.

7.1.2 Pilot-scale verification of preservation processes

After the kinetic parameters for the microbial population have been established and verified in the laboratory, the recommended process needs to be validated under conditions more similar to commercial operations. The pilot-scale operation may be the first opportunity to evaluate the response of the microbial population in the food product. In addition, the process can be evaluated in the same container to be used for commercial operations. Although these factors introduce new variables, the process has been designed to ensure that the target reduction in microbial population is accomplished at the location within the container where intensity of the process is minimum. The Halvorson-Ziegler expression can be used to determine the probability of survivors after a pilot-scale process by determining the total numbers of containers with evidence of microbial growth after applying the proposed process. Because the typical process is designed to ensure a very small number of microbial survivors, the Halvorson-Ziegler expression is needed to estimate the probability of survivors.

Example 7.2

A preservation process for a new food product is being validated in a pilot-scale operation. The survivor curve for the microbial population of concern is non-log-linear and described by a rate constant (k) of 2.0 per minute and an index parameter (**n**) of 0.83. Based on these parameters, a process time of 25 minutes has been recommended. The pilot scale inoculated-pack experiment involves the production of 1,000 containers, with an initial population of 1000 per container. The results of the experiment indicate that two containers had evidence of microbial growth. Do these results validate the recommended process?

Given:

1. Survivor curve parameters for the microbial population include $k = 2/\text{min}$ and $n = 0.83$.
2. The number of containers (**n**) is 1000.
3. The initial population (N_o) in each container is 1000.
4. Two containers had evidence of microbial activity after the process.

Approach:

1. The Halvorson-Ziegler expression is used to determine the most probable number of survivors from the pilot-scale operation.
2. The number of survivors and the equation of the survivor curve are used to compute the apparent process time and validation of the process.

Solution:

1. Using the Halvorson-Ziegler expression, where q is 998,

$$s = (1000)\,\text{Ln}\,[1000/998]$$
$$s = 2.002$$

2. Using the survivor equation or Eq. (2.24) from Chapter 2,

$$2.002 = (1000)(1000)\exp\,[-2\,(\text{time})]^{0.83}$$
$$\text{time } t_p = 11.12 \text{ min}$$

3. Based on the results of the pilot-scale operation, the process time of 25 minutes was not validated. These results indicate that the initial population and/or the rate constant were higher than reported.

The results of Example 7.2 illustrate the challenges associated with verifying a process. In this situation, the recommended process was not valid, as indicated by the actual process time based on the actual number of survivors. There are several potential explanations for the results, including the influence of the product substrate on the magnitude of the kinetic parameters of the microbial population or an error in measurement of the initial population. A second explanation is the inadequacy of the process at a defined location within the product structure or container. In addition, the example illustrates that the sample size in this experiment should be much larger to ensure that survivors are detected. The next step in validation is to complete another pilot-scale experiment, using a larger rate constant, a higher initial population and a larger number of containers.

7.1.3 Commercial-scale validation of processes

After the process is validated at pilot scale, the process is monitored at commercial scale. The outcomes from commercial-scale preservation operations are large numbers of containers, and a given number of containers are usually identified as part of the same production lot. As a production lot is placed in storage, these containers are monitored for evidence of microbial survivors. If the results of these monitoring procedures are within acceptable limits, this is additional evidence of process validation. Alternatively, any evidence of microbial survivors requires follow-up and additional evaluation to determine the need for process adjustments.

Example 7.3

The output from a commercial food manufacturing operation for a shelf-stable food includes 10 million containers. The kinetic parameters for the spoilage microorganism are a rate constant (k) of 3 per minute at 121°C, and an activation energy constant (E_A) of 300 kJ/mole. The process time at the slowest heating location within each container is 6 minutes at 121°C and is based on an initial microbial population of 100 per container. During storage, two containers with evidence of microbial growth were discovered. Demonstrate that these results confirm the adequacy of the process.

(Continued)

Example 7.3 (Continued)

Given:

1. The number of containers (n) in the sample is 10 million.
2. The kinetic parameters for the spoilage microorganism are $k = 3/\text{min}$ at 121 °C, and $E_A = 300\,\text{kJ/mole}$.
3. The initial population of spoilage microorganism (N_o) is 100/container.
4. There are two containers (q) with evidence of survivors.
5. The process time is 6 minutes.

Approach:

1. The Halvorson-Ziegler expression is used to estimate the most probable number of survivors.
2. The survivor curve equation is used to compute the effective process time to be compared to the actual process time.

Solution:

1. Using the Halvorson-Ziegler expression,

$$s = (10^7)\ \text{Ln}\ [10^7/9{,}999{,}998]$$
$$s = 2$$

2. The survivor curve equation provides

$$2 = (10^7)(100)\ \exp\ [-3\ (\text{time})]$$
$$\text{time} = 6.68\ \text{min}$$

3. The results of the analysis indicate that the effective process time is more than the actual process time. The results suggest that the assumptions used during the original process development are not appropriate. The most likely parameter in the process development is initial population per container. This assumption can be evaluated:

$$2 = (10^7)(N)\ \exp\ [-(3)(6)]$$
$$N = 13$$

4. The results indicate that the initial population of spoilage microorganism was 13 per container. This observation can be evaluated, and the process time can be adjusted to achieve the probability of 1 container in 10 million:

$$1 = (10^7)(13) \exp[-(3)(\text{time})]$$

$$\text{time } t_p = 5.9 \text{ min}$$

5. The results indicate that the process time can be reduced from 6 minutes to 5.9 minutes, based on the results of the observations of survivors during storage of the processed product.

The analysis in Example 7.3 provides considerable insight into process development. The most probable number of survivors from a preservation process indicates the actual effectiveness of the process and becomes a component of quality assurance procedures. The outcomes can be used in continuous process evaluation and adjustment to meet the expectations of the manufacturing operations.

Schmidt (1954) proposed an alternative approach to evaluating outcomes from preservation processes. The proposed approach uses normal probability graph paper (Figure 7.1) to determine the process time when the probability of survivors is 50%. In this application, the Halvorson-Ziegler expression for the most probable number of survivors becomes

$$s = n \text{ Ln}[n/q] = n \text{ Ln}[2] = 0.693n \qquad (7.2)$$

This quantity for the most probable number is an input to the appropriate survivor curve equation, along with the time obtained from the normal probability graph paper.

The results for an experiment with the same number of containers exposed to the preservation process at different process

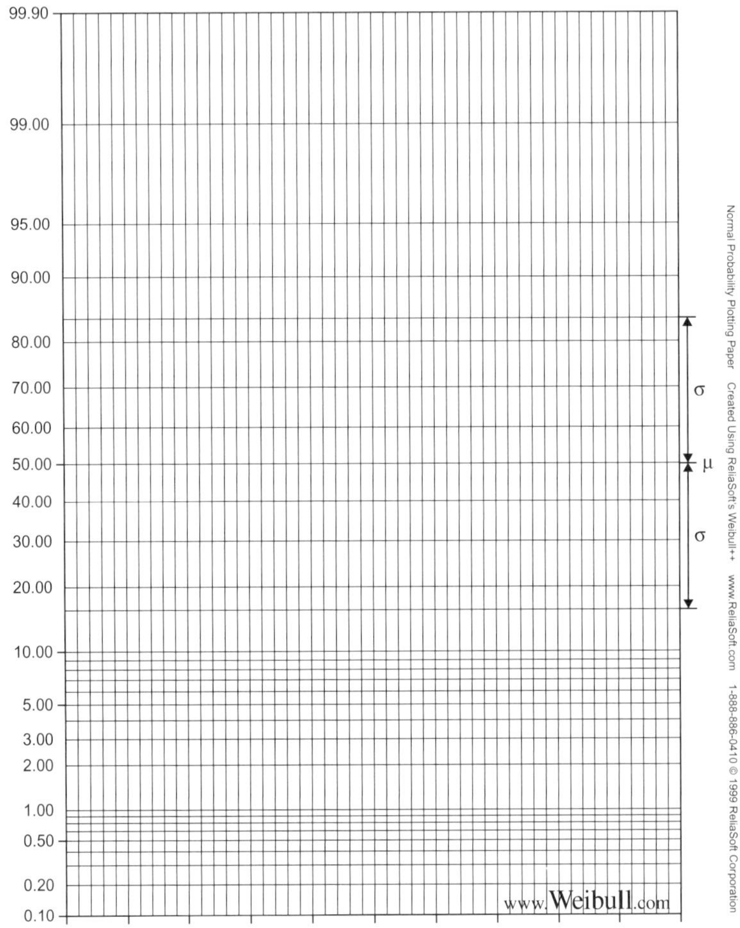

Figure 7.1 Normal probability paper.

times are the inputs to the analysis. The number of containers with survivors is used to compute the probability of survivors, or the number of containers with survivors divided by the total number of containers. These probability values are located on the probability axis, while the corresponding times are found on the horizontal axis, and each point is plotted on the normal probability paper. This approach provides additional statistics about the variability of the results from survivor data. The most useful outcome is the standard deviation on the process time needed to establish a 0.5 probability of survivors.

Example 7.4

A pilot-scale experiment has been used to establish an effective rate constant for pathogen survivors from a new process. The experiment included a total of 500 containers with initial populations of 1000 per container at five different process times. The results of the experiment were as follows:

Time (minutes)	Number of Containers With Survivors
10	500
12	385
14	215
16	95
18	1

Determine the first-order rate constant for the pathogen survivor curve. Use the normal probability paper approach to estimate the mean rate constant and the standard deviation on the rate constant.

Given:

1. The number of containers at five different process times are 500.
2. The initial population is 1000 per container.
3. The survivor curve equation is first order.
4. The number of containers with survivors at each process time is provided.

Approach:

1. The fraction number of survivors for each process time and the normal probability paper are used to estimate the time for 50% survivors.
2. The survivor curve equation is used to compute the first-order rate constant.

Solution:

1. The survivor curve data are plotted on normal probability paper as illustrated in Figure 7.2. The results indicate that the mean process time is 14.25 minutes, and the standard deviation is 2.3 minutes.
2. Based on these outcomes, the first-order rate constant is

$$N = N_o \exp[-k\,t]$$
$$0.693\,(500) = (1000)(500) \exp[-k\,(14.25)]$$
$$k = 0.511/\text{min}$$

(*Continued*)

Example 7.4 (Continued)

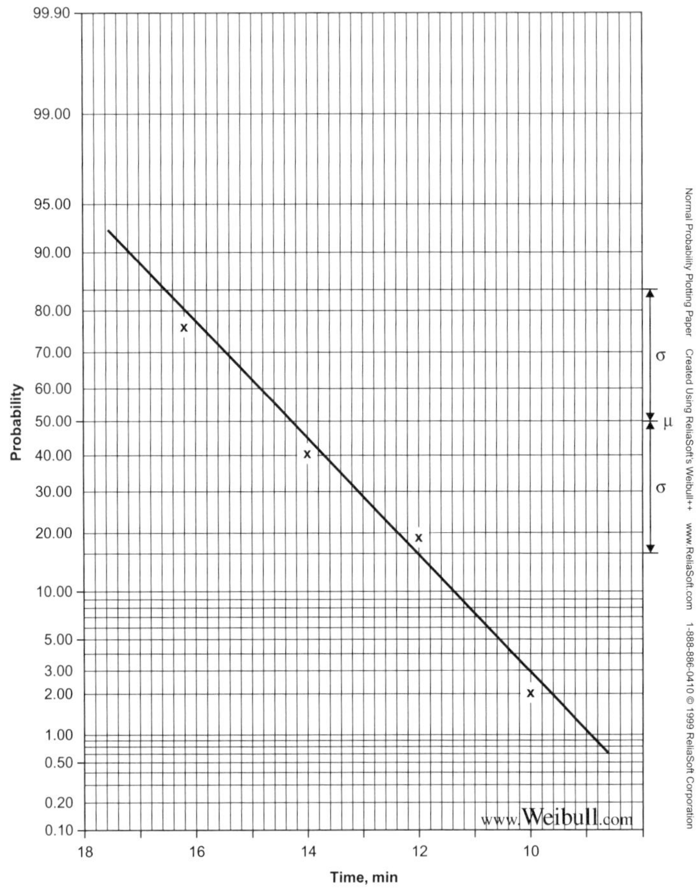

Figure 7.2 Normal probability plot for Example 7.4.

3. The standard deviation on the rate constant is

$$0.693\,(500) = (1000)(500)\exp\left[-k\,(16.55)\right]$$
$$k_{-SD} = 0.44/\text{min}$$
$$0.693\,(500) = (1000)(500)\exp\left[-k\,(11.95)\right]$$
$$k_{+SD} = 0.609/\text{min}$$

and then

$$k = 0.511/\text{min, and a range } 0.44 \text{ to } 0.609/\text{min}$$

The results in Example 7.4 illustrate an approach to using most probable number data to evaluate a preservation process. In addition, the Schmidt approach provides additional insight on the process by providing statistical parameters for use in process development.

7.2 Alternative approaches to validation

Several alternatives to validating preservation processes have evolved and should be considered. All processes must ensure that the risk of pathogen survival is negligible. Validating processes designed for pathogens normally involves using the pathogenic microorganism during the validation experiments. All steps associated with handling a pathogen in a laboratory and/or pilot facility require extra levels of precaution, and the frequency of these activities should be minimized. Certainly, the incorporation of a pathogen into the environment of a commercial operation should be avoided.

Traditionally, significant attention has been given to identifying surrogate microorganisms. In applications associated with process validation for inactivation of pathogen populations in food products, the ideal surrogate microorganism responds to process conditions in the same manner as the microbial pathogen. More specifically, the surrogate has the same survivor curve as the pathogen when incorporated into the food product and the preservation process is applied. Many surrogates used to validate thermal processes for foods have been identified and used in validation of verifying thermal processes for shelf-stable food products.

The most visible example of a surrogate microorganism in preservation processes is a spore-bearing mesophilic anaerobe referred to as PA 3679. This microorganism has an almost identical response to a thermal process as the most resistant spores of *Cl. botulinum*. Due to these similarities, PA 3679 has become the surrogate microorganism for evaluating thermal processes used for shelf-stable foods. Other surrogate microorganisms include the spores of *Bacillus sterothermophilusm, Cl. sporogenes,* and *Bacillus coagalans*. These surrogates have been used most recently to verify thermal processes for liquid foods carrying particulates through a continuous system of heat exchangers and holding tube. The obvious key to identifying the optimum surrogate is the microorganism with the same response to the preservation process, without being pathogenic.

Example 7.5

The surrogate (PA 3679) is being used for pilot-scale validation of a thermal process. An acceptable risk of survivors in 1 container per million processed has been established, and 1000 containers will be used in the pilot plant experiment. Determine the initial population per container needed to ensure that the most probable number method can be used for detecting survivors when the process time is 5 minutes.

Given:

1. The kinetic parameters for PA 3679 include a k = 1.55/minute at 121°C and an activation energy constant of 250 kJ/mole.
2. The proposed thermal process is intended to ensure survivors in less than 1 container per 1 million processed.
3. The pilot-scale experiment includes 1000 containers, and a process time of 5 minutes at 121°C.

Approach:

1. The Halvorson-Ziegler expression is used to estimate the most probable number of survivors in 1000 containers to be processed in the pilot plant.
2. The first-order survivor curve equation is used to estimate the initial population.

Solution:

1. Based on the Halvorson-Ziegler expression and allowing for 1 container with evidence of survivors in the 1000 containers processed:

$$s = (1000) \text{ Ln } [1000/999] = 1.0005$$

2. Using the first-order survivor curve equation,

$$N = N_o \exp[-k\,t]$$
$$1.0005 = N_o \exp[-1.55\,(5)]$$
$$N_o = 2323$$

3. The result indicates that an initial population of more than 2323 per container is needed to ensure that the process time can be validated.

Several observations about the conditions described in Example 7.5 are important. If the process is conducted with the product in the containers, the process of 5 minutes must be applied at the slowest heating location within the container. Assuming the initial population of 2323 is uniformly distributed within the container, the initial population is not concentrated at the slowest heating location. This suggests that the initial population could be lower and still be detected with the 1000 containers in the pilot-scale experiment. If the 5-minute process is a lethal process delivered to the slowest heating location, the operator process time is much longer depending on the heating and cooling characteristics of the product in the container.

A variety of different Time-Temperature-Integrators (TTIs) have evolved for use in validating preservation processes. In general, TTIs are materials or reactions that integrate the influence of temperature over time to simulate the impact of a process. These sensors include the following:

- **A reaction occurring within the product that is sensitive to the preservation process, with products of the reactions that can be evaluated or easily measured after the process.** This approach may be the ideal because it does not require any additions to or modification of the product. Obviously, the approach requires significant insights about the product and the response of product components to the process.
- **A biosensor designed to be incorporated into the product to detect the impact of the preservation process.** These types of TTIs may be food grade and compatible with the product, and may or may not require recovery to evaluate the impact of the process. Alternatively, the biosensor may be designed to be recovered from the product after the process.

- **A sensor containing a specific reaction to integrate the temperature–time impact of the preservation process.** In most cases, these types of TTIs are not incorporated into the product and need to be carefully positioned to allow appropriate interpretation of the outcomes.

Example 7.6

A biosensor TTI is being used to validate a thermal process for a new formulation of a food being processed in a container. The TTI is positioned at the slowest heating location and is used as a surrogate for a microbial pathogen found in an ingredient in the product. The kinetic parameters of the pathogen include a rate constant of 2 per minute at 121°C and an activation energy constant of 300 kJ/mole. If the activation energy constant of the TTI is 280 kJ/mole, select the rate constant (at 121°C) for TTI needed to simulate the pathogen.

Given:

1. Microbial pathogen has $k = 2/\text{min}$ at 121°C and $E_A = 300$ kJ/mole.
2. Proposed surrogate TTI has $E_A = 280$ kJ/mole.

Approach:

1. The influence of a preservation process on a microbial population is expressed by the first-order survivor curve equation:

$$N = N_o \exp[-k\,t]$$

2. Solve for the product of the rate constant (k) and time (t):

$$k\,t = \text{Ln}\,[N_o/N]$$

3. Because the rate constant is a function of temperature and the activation energy constant (E_A) describes the influence of temperature, the relationship of rate constants with temperature is described by

$$k = k_o \exp[-E_A/R_g\mathbf{T}]$$

4. Then, the equivalent processes for different activation energy constants become

$$k_{o1} \exp[-E_{A1}/R_g\mathbf{T}]\, t = k_{o2} \exp[-E_{A2}/R_g\mathbf{T}]\, t$$

5. The preceding expression can be used to determine the rate constant (k) for the surrogate TTI at 121°C.

Solution:

1. Using the expression presented in the Approach section of this example,

$$\begin{aligned} k_{o1} &= 2 \exp[-300{,}000/(8.31441)(394)]/ \\ &\quad \exp[-280{,}000/(8.31441)(394)] \\ k_{o1} &= 2 \exp[-20{,}000/(8.31441)(394)] \\ k_{o1} &= 4.46 \times 10^{-3}/\text{min} \end{aligned}$$

2. The rate constant (k) at 121°C for the surrogate TTI needs to be 4.46×10^{-3}/min for the TTI to provide an equivalent simulation of the microbial pathogen response to the preservation process.

There are many opportunities for using TTIs in validating the processes for food preservation. The previous example illustrates one of these opportunities. Note that the TTI respond to the process at the specific location of the sensor. For thermal processes, this location is always the slowest heating location, so the location of the TTI is very important to interpreting the outcomes from the sensor.

Another alternative approach to validation of a preservation process involves evaluating the impact of the process on a product quality attribute. As indicated in Chapter 4, the concentration or intensity of many product quality attributes changes in a defined manner during the process and may be quantified by appropriate kinetic parameters. This approach to process validation is not as direct as those previously described but provides a more flexible approach for many products. Following are the specific steps associated with this approach:

1. Identify a product quality attribute that can be easily measured and is sensitive to the preservation process. Color and texture are examples of attributes that are sensitive to preservation processes, and the kinetic parameters to describe the process impact may be available in the literature. Alternatively, a heat-sensitive product component (vitamin or nutrient) may be selected.
2. A container or sample of the product is collected after the process is completed, and an analysis of the selected attribute is completed. If the product is in a container, note that the impact of the process varies with location within the container. The more practical approach is to mix the contents of the container until the concentration of the attribute is uniform and homogeneous. The analysis provides the mass average retention of the attribute in the product after the process.
3. The measured mass average concentration of the selected quality attribute is compared to the predicted mass average concentration based on the process time used to reduce the target microbial population by the desired amount.

Example 7.7

The chlorophyll content of green beans is being used to validate the thermal process for the product. The thermal process is designed to reduce the population of a spoilage microorganism by sufficient amounts to achieve a spoilage probability of 1 spoiled container per 1 million processed. Measurement of chlorophyll in green beans from 10 of the containers after processing indicated an average retention of 39.25%. The kinetic parameters for the spoilage microorganism are the rate constant of 3/minute at 121°C with an activation energy constant of 250 kJ/mole. The kinetic parameters for chlorophyll include a rate constant of 0.076/minute at 100°C and activation energy constant of 38.5 kJ/mole. The initial population of the spoilage microorganism is 1000 per container.

Given:

1. For the spoilage microorganism, $k = 3/\text{min}$ at 121°C and $E_A = 250\,\text{kJ/mole}$.

2. Initial population of spoilage microorganism is $N_o = 1000$ per container.
3. For chlorophyll, $k = 0.076/\text{min}$ at 100°C and $E_A = 38.5\,\text{kJ/mole}$

Approach:

1. The process time needed to achieve the desired reduction in the population of the spoilage microorganism is determined.
2. The impact of the thermal process on the mass average concentration of chlorophyll is computed.
3. The computed retention of chlorophyll is compared to the measured concentration.

Solution:

1. The process time is computed by using the spoilage probability equation:

$$1 = (1000)(10^6)\exp[-3\,(\text{time})]$$
$$\text{time } t_p = 6.9 \text{ min at } 121^\circ\text{C}$$

2. To evaluate the impact of the process on the chlorophyll, the rate constant for chlorophyll must be expressed at 121°C, as follows:

$$k = (0.076)\exp\{-(38500/8.31441)\,[(1/394) - (1/373)]\}$$
$$k = 0.147/\text{min at } 121^\circ\text{C}$$

3. To evaluate the impact of the process on the chlorophyll, the influence on the mass average contents in the content must be considered.
4. Based on the rate constant at 121°C, the retention of chlorophyll in the green beans is

$$C = 100\exp[-0.147\,(6.9)]$$
$$C = 0.362 \text{ or } 36.2\% \text{ of the initial concentration}$$

5. Because the measurements indicated a higher retention, the process cannot be validated. A higher process time is required to ensure the desired rate of spoilage.

As indicated by the previous example, most preservation processes impact the quality attributes of the food product. More specifically, the influence of the process can be expressed quantitatively as long as the appropriate kinetic parameters to describe the quality attribute have been measured. The impact of the process on a quality attribute is different from the influence on the microbial population. In general, the focus of currently available kinetic data is on attributes that are impacted in a negative manner. Historically, information has been prepared on defining the influence as a part of investigations to reduce or minimize the influence of the process on food quality attributes. Although these data are useful, kinetic parameters to provide a better understanding of the potential positive impacts of the preservation process on quality attributes are valuable as well.

Interpreting the impacts on quality attributes in process analysis is important as well. When considering thermal processes, the general influence of time and temperature on a microbial population is similar to the influence on a quality attribute. Those portions of the product where the process has a higher intensity cause a greater reduction on microbial population, so there is a more significant loss of a heat-sensitive quality attribute. The interpretation of these analyses is most important when the process is conducted after the product is placed in a container. As has been demonstrated in several examples, the design of a preservation process is most sensitive at the slowest heating location within the container. The target reduction in the microbial population must be accomplished at the slowest heating location, even though the overall impact on the product is much greater than required. The corresponding impact of the same process on retention of the quality attributes has a similar distribution within the container, but the important output is the mass average retention of the attribute. This distinction must be recognized when evaluating the influence of the preservation process on the product quality.

When the thermal process does not occur while the product is in a container, the differences in impact of the process on the microbial population and the product quality attribute are much smaller. For these situations, the intensity of the process is still dictated by the target reduction in microbial population. If these processes are designed to meet the unique distribution of temperatures within the product during the process, the distribution must be considered when evaluating the impact of the process on quality attributes.

7.3 Process validation for alternative process technologies

The validation of preservation processes when using alternative technologies requires special considerations. Due to the potential differences in mechanisms of microbial inactivation, many of the approaches need to be modified or changed, as compared to the approach followed for traditional thermal processes. The following discussion of process validation focuses on several of the alternative technologies being considered for preservation of foods.

7.3.1 Microwave processes

The use of microwave energy for preservation of foods is receiving significant attention. As indicated earlier, microwave heating is a well-known process, but the more intense application of the process needed for food preservation has not been explored in depth. Because the reductions in microbial populations are caused by elevating the product temperature, the same surrogate microorganisms used for traditional thermal processes can be used during process validations. The differences from the traditional thermal processes are more obvious during processing of the product in a container, where the heating of the slowest heating location depends on conduction heat transfer from the product surface to a center location. During the microwave process, a more uniform distribution of temperatures occurs during the heating of the product. Although this observation can be viewed as positive when considering quality retention, it does introduce a unique challenge for the process validation. Assumptions accepted for estimating process time must be evaluated as part of the validation steps. Because slowest heating locations may exist at many locations within the product mass, the surrogate microorganism must be uniformly distributed within the product for the process validation. The primary difference in approach to validation of traditional thermal processes is that the process is based on a mass average population of survivors.

Example 7.8

A preservation process using microwave energy is being validated using PA 3679 as the surrogate microbial population. An initial population of 1000 per container has been established for a pilot-scale experiment with 3000 containers. The kinetic parameters for the surrogate include a rate constant of 1.55/min at 121°C and an activation energy constant of 250 kJ/mole. The results of the experiment conducted by using a process time of 5 minutes provided 12 containers with indication of survivors. Use the results of the experiment to estimate the process time needed at 115°C to ensure a risk of no more than 1 survivor per 1 million containers to be processed.

Given:

1. Initial population of the surrogate microorganism is $N_o = 1000$ per container.
2. Number of containers in the experiment is 3000.
3. Kinetic constant for surrogate is $k = 1.55$/min at 121°C and $E_A = 250$ kJ/mole.
4. Results indicate 12 containers with survivors.

Approach:

1. Use the Halvorson-Zeigler equation to estimate the probability of survivors.
2. Use the first-order survivor equation to estimate the effective rate constant.
3. Use the effective rate constant to propose the process time needed to provide a survivor probability of one per million.

Solution:

1. Determine the most probable number of survivors, using the Halvorson-Zeigler equation:

$$s = (3000)\ \mathrm{Ln}\ [3000/2988] = 12.024$$

2. Use the survivor curve equation to estimate the effective process time:

$$12.024 = (1000)(3000)\ \exp\ [-k\ (5)]$$

$$k = 2.485/\mathrm{min\ at\ } 110°\mathrm{C}$$

3. To establish the process time at 115°C, the rate constant must be converted to 115°C:

$$k = (2.485)\ \exp\{-(250{,}000/8.31441)[(1/383) - (1/388)]\}$$
$$k = 0.9035/\text{min at } 115°\text{C}$$

4. Using the rate constant at 115°C, and the one per one million survivor probability, the process time can be estimated:

$$10^{-6} = (1000)(10^{6})\ \exp\,[-(0.9035)\ (\text{time})]$$
$$\text{time } t_p = 38.23 \text{ min}$$

The previous example illustrates the approach to estimating the process time using outcomes from a pilot-scale experiment. The process time obtained using the best-available kinetic date for the surrogate microorganism did not provide a sufficient reduction in the microbial population. By using an effective rate constant based on survivors from the pilot-scale experiment, a more appropriate process time was determined that can be used for commercial-scale operations.

7.3.2 Ohmic heating processes

The preservation process for food products heated in an ohmic heating system is the result of the temperature–time profile within the product. Because the benefits of ohmic heating include more uniform temperature distribution during heating, the validation of a preservation process can be accomplished in the same manner as microwave heating. The specific applications for ohmic heating systems differ from microwave process applications in several ways. One of the specific applications of ohmic heating is continual online heating where the process is applied to food particles in a carrier liquid. For these applications, the surrogate microorganism must be uniformly distributed within the food particle to ensure that the outcomes from the pilot-scale experiment can be interpreted in a valid manner. In these applications, the validation is intended to confirm the flow rate of product through the heating and holding steps of the preservation process, when the ohmic heating system is being used to ensure that the microbial populations within the food particles are reduced to the target levels.

7.3.3 Ultra-high pressure processes (UHP)

Preservation processes based on use of UHP create several challenges to process validation. Because the primary mechanism of inactivation is not thermal, the use of surrogate microorganisms requires careful consideration. At this time, there is limited information on the response of specific microbial population to UHP, including, more specifically, the response of food-borne pathogens to UHP. In addition, little attention has been given to identifying nonpathogens with responses to UHP similar to those of the pathogens of concern.

The impact of the UHP process on the microbial population in a food is expected to be uniform throughout the product structure. As with thermal processes where temperature distributions are expected to be uniform within the product structure, the validation of the process must be based on the mass average of the surviving population. The situation becomes more complex when attempting to consider the impact of temperature increase due to elevated pressures. Because any thermal contribution to the overall reduction of the microbial population during the UHP process is a function of the initial product temperature, the initial product temperature must be considered during process design and validation.

Example 7.9

A UHP process is being used for preservation of a new product. The process design is based on the first-order rate constant for the spoilage microorganism, $k = 1.25/\text{min}$ at 550 MPa, and an initial population of 100 spoilage microorganisms per container. The first commercial process run of 400,000 containers, using a process time of 12 minutes, had 270 containers with evidence of spoilage. At that point, the process time was increased to 15 minutes for the second production of 250,000 containers, and there were 128 containers with surviving spoilage microorganisms. The third production run of 550,000 containers used a process time of 20 minutes, and the number of spoiled containers was reduced to 72. For the fourth production run of 750,000 containers, the process time was increased to 25 minutes. The number of containers with evidence of spoilage was 15. Recommend a process time, based on the statistical variability of the information provided and the objective of an acceptable spoilage rate of 5 containers per 1 million processed.

Given:

1. First-order rate constant for spoilage microorganism is $k = 1.25/\text{min}$ at 550 MPa.
2. Initial population is $N_o = 100$ per container.
3. Spoilage rates and process times were

$t_p = 12\,\text{min}$	400,000 containers	270 spoiled containers
13	250,000	128
14	550,000	72
15	750,000	15

4. Accepted spoilage rate is five containers per one million

Approach:

1. The most probable number of survivors is determined for each of the four commercial runs.
2. Based on the most probable number of survivors, the process times, and the first-order rate constant, the initial population associate with each commercial run is established.
3. If there is significant variability among the initial populations, the highest value is used to establish the recommended process time.

Solution:

1. Using the Halvorson-Zeigler equation, the most probable number of survivors for each of the four commercial runs can be determined:

$$s = (400{,}000)\ \text{Ln}\,(400{,}000/399{,}730)$$

$$s = 270 \text{ for process time} = 12 \text{ min}$$

$$s = 128 \text{ for process time} = 13 \text{ min}$$

$$s = 72 \text{ for process time} = 14 \text{ min}$$

$$s = 15 \text{ for process time} = 15 \text{ min}$$

(Continued)

Example 7.9 (Continued)

2. The initial population for each of the production runs is estimated by using the first-order survivor curve equation:

$$270 = (400{,}000)\, N_o \exp\left[-1.25\,(12)\right]$$
$$N_o = 2207 \text{ for process time of 12 min}$$
$$N_o = 5842 \text{ for process time of 13 min}$$
$$N_o = 5213 \text{ for process time of 14 min}$$
$$N_o = 2780 \text{ for process time of 15 min}$$

3. Because the initial population of the spoilage microorganism is significantly higher than 100, an initial population of 5850 is used to establish the final process time:

$$1 = (1{,}000{,}000)(5850) \exp\left[-(1.25)(\text{time})\right]$$

time t_p = 18 min for the ultrahigh pressure process at 550 MPa

4. The initial population of microorganism is very important, and the quality of the raw material used for the product needs to be monitored continuously to ensure that initial populations are not exceeding the values used in establishing the process time.

7.3.4 Pulsed-electric-field processes (PEF)

The validation of preservation processes using PEF processes requires considerations similar to processes based on other alternative technologies. Because the mechanism of inactivation for microbial populations is unique, the selection of surrogate microorganisms requires careful consideration. Limited information and kinetic data for surrogate microorganisms are available at this time. Most applications of PEF are continuous processes for liquid products, and the impact of the field is uniform throughout the product. As with most continuous processes, residence times are dictated by the velocity profiles within the flow stream during application of the process. An additional consideration is the elevation of product temperature during the PEF process. In a manner similar to the

UHP process, the initial product temperature must be incorporated into the process design and the process validation.

Several other preservation processes based on alternative technologies are evolving and may find applications in food preservation in the future. Each alternative technology is unique and requires specific considerations when developing a validation protocol. In most cases, selecting surrogate microorganisms requires special consideration. In addition, the uniformity of the process impact within the product structure should be evaluated and must be established. Finally, the potential influence of the alternative technology on product temperature or other properties should be established and considered in both process design and validation.

List of symbols

C = concentration or intensity of product quality attribute
E_A = activation energy constant, kJ/mole
j_{cc} = cooling lag constant at geometric center
k = rate constant, 1/s
k_o = reference rate constant, 1/s
N = microbial population
N_o = initial microbial population
n = number of containers
n = index parameter for non-log linear survivor curve
q = number of containers with evidence of survivors
R_g = gas constant, kJ/kg K
s = most probable number of survivors
T = temperature, °C
T_R = reference temperature, °C
T = absolute temperature, K
t = time, s
t_p = process time, s

Bibliography

Ball, C. O., & Olson, F. C. W. (1957). *Sterilization in food technology*. New York: McGraw-Hill.

Halvorson, H. O., & Zeigler, N. R. (1932). Application of statistics in bacteriology. *Journal of Bacteriology*, *25*, 101.

Heldman, D. R., & Hartel, R. W. (1997). *Principles of food processing*. New York: Chapman & Hall, Inc.

Noma, S., Shimoda, M., & Hayakawa, I. (2002). Inactivation of vegetative bacteria by rapid decompression. *Journal of Food Science*, *67*(9), 3408–3411.

Pflug, I. J., Blaisdell, J. L., & Kopelman, I. J. (1965). Developing temperature-time curves for objects that can be approximated by a sphere, infinite plate or infinite cylinder. *ASHRAE Transactions*, *71*(1), 238–248.

Pflug, I. J., & Esselen, W. B. (1963). Food processing by heat sterilization. In J. L. Heid & M. A. Joslyn, (Eds.), *Food processing operations* (Vol. 2, pp. 410–479). Westport, CT: The AVI Pub Co., Inc.

Schmidt, C. F. (1954). Thermal resistance of microorganisms. In G. F. Reddish (Ed.), *Antiseptics, disinfectants, fungicides and sterilization* (pp. 720–759). Philadelphia Lea & Febiger.

Singh, R. P., & Heldman, D. R. (2009). *Introduction to food engineering* (4th ed.). San Diego, CA: Elsevier Academic Press.

Somerville, J., & Balasubramaniam, V. M. (2009). Pressure-assisted thermal sterilization of low-acid shelf-stable foods. *Resource: Engineering & Technology for a Sustainable World*, *16*(7), 14–17.

Stumbo, C. R. (1965). *Thermobacteriology in food processing*. New York: Academic Press.

Stumbo, C. R., Murphy, J. R., & Cochran, J. (1950). Nature of thermal death time curves for P.A. 3679 and *Clostridium botulinum*. *Food Technology*, *4*, 321.

Teixeira, A. A. (2007). Thermal processing of canned foods. In D. R. Heldman & D. B. Lund (Eds.), *Handbook of food engineering* (2nd ed., pp. 745–797). Boca Raton, FL: CRC Press, Inc.

8 Optimization of Preservation Processes

8.1 The HTST concept 246
8.2 Applications to nonliquid foods 249
List of symbols 263
Bibliography. 265

Although the primary purpose of a preservation process is to ensure food product safety or acceptable levels of product spoilage, the impact of the process on product quality attributes has received increasing attention. Due to the different mechanisms associated with various processes, the impacts on product quality will be different, but these changes should be considered and minimized during the design of the preservation process. In general, a more intense preservation process is expected to be more detrimental to the quality attributes of the food product. This is most evident with thermal processes, when high temperatures accelerate the losses of temperature-sensitive food components. All processes must be carefully evaluated using the kinetics parameters of the target microbial populations as well as similar parameters of the temperature-sensitive product quality attributes. Opportunities for optimizing these processes become evident due to the different magnitudes of the kinetic parameters. In this chapter, process optimization refers to defining the process needed to ensure product safety or acceptable spoilage rate, while providing a maximum retention of a product quality attribute. The approaches presented and discussed illustrate the opportunities for optimum processes. These discussions include background and illustrations of specific processes and products.

Food Preservation Process Design
ISBN: 978-0-12-372486-1

8.1 The HTST concept

The opportunities for process optimization are most evident when considering the impact of processes accomplished at higher temperatures for shorter times. These opportunities were first revealed during the development of continuous processing of liquid foods, where controlling the time that a product is held at a specific temperature is relatively easy. As indicated by Figure 8.1, the retention of a temperature-sensitive food component is improved when the preservation process is accomplished at a higher temperature for a shorter time.

The key to the relationships presented in Figure 8.1, is the recognition that the magnitude of the kinetic parameters favor improved retention of quality attributes, while ensuring the target reduction of the microbial population. The solid curve in Figure. 8.1 represents a series of time-temperature processes required to accomplish a

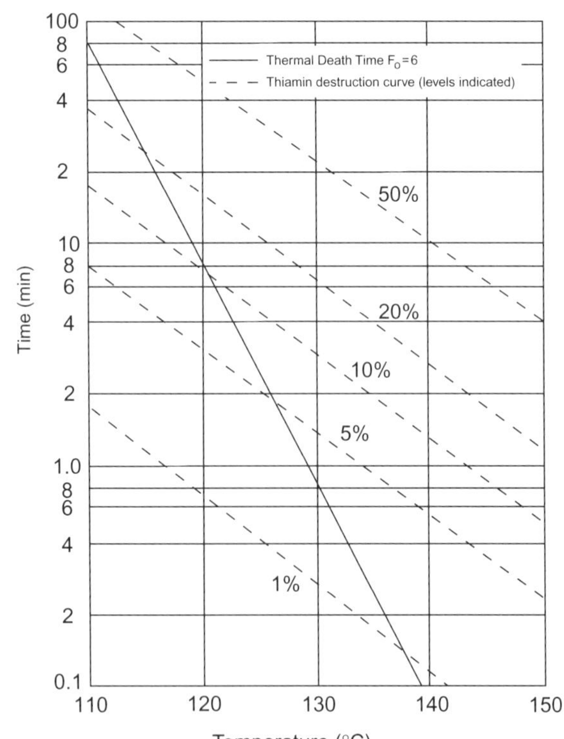

Figure 8.1 Quality improvement through high-temperature short-time processes (from Lewis & Heppel, 2000).

target reduction on microbial population. Each dash-curve represents a series of time-temperature relationships and defines a given retention of the temperature-sensitive product component. The other dash-curves represent different magnitudes of retention for the same product component. Based on these relationships, the process design needed to improve retention of the quality attribute can be identified, while achieving the reduction of the microbial population needed for product safety.

The relationships used to create the curves in Figure 8.1 are based on kinetic models presented and discussed in Chapter 2 and the kinetic models and parameters illustrated in Chapters 3 and 4. As presented in Table 4.1, the relationships between the kinetic parameters for product quality attributes and the parameters for microbial pathogens or spoilage microorganisms favor the design of processes using higher temperatures at shorter times. High-temperature short-time (HTST) processes can be accomplished with excellent control when designing continuous thermal processes for liquid food products. For these types of thermal processes, the temperature increase to the process temperature is very rapid, and cooling of the product may be very rapid as well. Given these conditions, the process temperature is the temperature after heating, and the process time is the holding time at the process temperature. For these types of processes, optimizing the process is limited by equipment or process control constraints. Most often, these constraints are control of product temperature for very short periods of time.

Example 8.1

A new liquid product is processed to ensure shelf stability by using a HTST process. The rate constant for the spoilage microorganism is 0.3/min at 100°C, and the activation energy constant is 250 kJ/mole. The most temperature-sensitive quality attribute has a rate constant of 0.1/min at 120°C and an activation energy constant of 100 kJ/mole. The process is based on an acceptable spoilage rate of 1 container per 10 million processed and packaged, and the initial population is 1000 per container. Determine the improvements in quality retention to be achieved by using a process temperature of 115°C as compared to 100°C.

(*Continued*)

Example 8.1 (Continued)

Given:

1. The rate constant for the microbial population is $\mathbf{k} = 0.3$/min at 100°C.
2. The activation energy for the microbial population is $E_A = 250$ kJ/mole.
3. The rate constant for the product quality attribute is $\mathbf{k} = 0.1$/min at 120°C.
4. The activation energy for the product quality attribute is $E_A = 100$ kJ/mole.
5. The process temperatures are 100°C and 115°C.
6. Acceptable spoilage rate is 1 container per 10 million.
7. The initial microbial population is $N_o = 1000$ per container.

Approach:

1. Process times are established for each process temperature, using the target spoilage rate.
2. The influence of both processes on product quality attributes are well determined and compared.

Solution:

1. Given the initial population of 1000 per container and the spoilage rate of 1 per 10 million:

$$10^{-7}/10^{3} = \exp\,[-0.3\,(\text{time})]$$
$$\text{time } t_p = 76.75 \text{ min at } 100°\text{C}$$

2. To determine the retention of the quality attribute at 100°C, the rate constant at 100°C must be computed:

$$\text{Ln}\,(\mathbf{k}) = -(100{,}000/8.31441)\,[(1/373) - (1/393)] + \text{Ln}\,(0.1)$$
$$= -1.64095 - 2.3026 = -3.9435$$

$$\mathbf{k} = 0.0194/\text{min at } 100°\text{C}$$

3. The retention of the product quality attribute is

$$\text{Retention} = 100 \exp\,[-0.0194\,(76.75)] = 0.226$$
$$\text{Retention} = 22.6\% \text{ at } 100°\text{C}$$

4. To determine the process time at 115°C, the rate constant for the microbial population at 115°C is computed:

$$\begin{aligned}\text{Ln}\,(\mathbf{k}) &= -(250{,}000/8.31441)\,[(1/388) - (1/373)] + \text{Ln}\,(0.3)\\ &= 3.116 - 1.204 = 1.9124\end{aligned}$$

$$k = 6.77/\text{min at } 115°\text{C}$$

5. The process time at 115°C becomes

$$\begin{aligned}10^{-7}/10^{3} &= \exp\,[-6.77\,(\text{time})]\\ \text{time } t_p &= 3.4 \text{ min at } 115°\text{C}\end{aligned}$$

6. To compute the quality retention at 115°C, the rate constant is computed:

$$\begin{aligned}\text{Ln}\,(\mathbf{k}) &= -(100{,}000/8.31441)\,[(1/388) - (1/393)] + \text{Ln}\,(0.1)\\ &= -0.3944 - 2.3026 = -2.697\end{aligned}$$

$$\mathbf{k} = 0.0674/\text{min at } 115°\text{C}$$

7. The retention of quality is

$$\begin{aligned}\text{Retention} &= 100 \exp\,[-0.0674\,(3.40)]\\ &= 79.5\% \text{ at } 115°\text{C}\end{aligned}$$

8. The results indicate that the retention of the temperature-sensitive quality attribute is 22.6% at 100°C but improves to 79.5% at 115°C. These results are typical of HTST processes.

8.2 Applications to nonliquid foods

The evaluation of quality retention in nonliquid foods requires an analysis of thermal energy transfer within the product structure. As illustrated in Chapter 6, the retention of quality attributes varies with location within the product structure. The outer regions of the product are in close proximity to the heating medium, so the temperature in these regions increases more rapidly than regions near the geometric center of the product structure. In addition, the outer regions will be at the elevated temperature for a longer period of time. As a result of the conditions, lower retention of quality attributes occur within these regions, as compared to the regions of the product at greater distance from the heating medium. As illustrated in Example 6.6, the overall retention of the heat-sensitive quality attribute is the mass average of the retention for the various regions of the product.

The challenge for optimization of thermal processes of nonliquid foods occurs within any product with thermal energy transfer by conduction. These products include high viscosity or consistency products in a continuous heat exchanger, as defined by the regions of product flow with temperature gradients. When these temperature gradients exist for significant periods of time during the process, the impact on product quality attributes becomes more evident. The most obvious situations occur within the product holding tube due to the velocity distribution within the product during flow through a tube or pipe. These types of temperature gradients and the corresponding variability in holding times require careful attention during process design. Although process design is always based on the slowest moving product particle, negative impacts on product quality attributes can occur when portions of the product are held for longer times than required.

When a thermal process is applied with the product in the container or package, the temperature distribution within the product structure may be significant. In these situations, operator process time is established by the time needed to ensure that the microbial population at the slowest heating location is reduced to some target level. All other regions of the product are exposed to higher temperatures and longer times, and more intense processes than required for the target reduction in microbial population. It follows that the distribution of quality retention within the container will be significant as well. In general, quality retention will be greater within the internal regions of the product, as compared to those portions of the product located near the container surface in direct contact with the heating medium.

Teixeria, Dixon, Zahradnik, and Zinsmeinster (1969) illustrated the concept of optimization for thermal processes for conduction-heating food products. The results of this investigation demonstrated that a

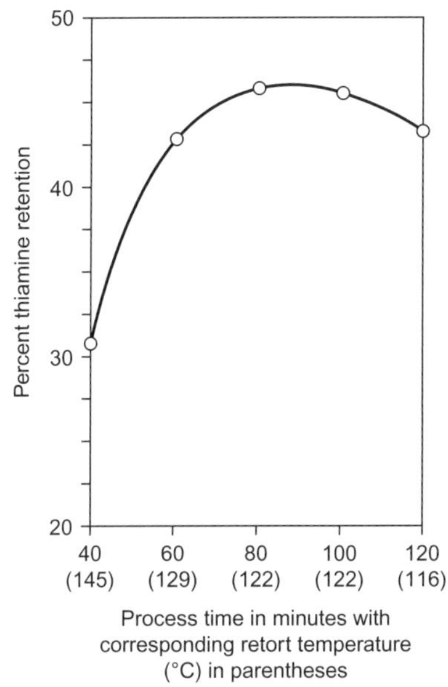

Figure 8.2 Illustration of optimization of thiamine retention in a conduction-heating food product (from Teixeira et al., 1969).

temperature–time process could be identified for maximum retention of a quality attribute. As illustrated by Figure 8.2, the retention of thiamine in a conduction-heating product during thermal processing in a can is maximized at a defined temperature–time process.

These results are obtained by comparing a series of processes where the combination of heating medium temperature and operator process time provide the same reduction in a defined microbial population at the geometric center of the can. The retention of thiamine is maximum for a process of 122°C for 80 minutes, as compared to equivalent processes at higher temperatures for shorter times, or lower temperatures for longer times. The thiamine retention is based on mass or volume-average within the product container and considers the lack of retention in the outer regions of the container (near the heating medium) as compared to the higher retention in the regions at greater distance from the surface. Because a greater proportion of the product mass is located in the outer regions of the container, the influence of quality retention in these locations on overall retention is important and influence the magnitudes of the optimum process.

Example 8.2

Determine the optimum process, in terms of quality retention, for a thermal process being used to ensure a spoilage rate of no greater than 1 container per 10 million processed. The initial product temperature is 70°C, and the product has the following thermophysical properties: density (ρ) = 1000 kg/m^3, thermal conductivity (k) = 0.45 W/m K, and specific heat (c_p) = 3.5 kJ/kg K. The initial microbial population is 1000 per container, and the survivor curve is described by a first-order rate constant (**k**) of 2.1/min at 121°C, and an activation energy constant (E) of 240 kJ/mole. The retention of the heat sensitive quality attribute is described by a first-order rate constant (**k**) of 0.0435/min at 138°C and an activation energy constant (E_A) of 97 kJ/mole. The product container is a cylinder with a diameter of 9 cm and height of 24 cm. The thermal process is accomplished using steam, with a convective heat transfer coefficient of 5000 W/m^2 K, over a range of temperatures and pressures needed to ensure the desired spoilage rate. The cooling medium is 35°C, and the convective heat transfer coefficient is 1500 W/m^2 K.

Given:

1. Spoilage rate is 1 per 10 million containers processed.
2. Initial product temperature is T_o = 70°C.
3. Product density is ρ = 1000 kg/m^3.
4. Product thermal conductivity is k = 0.45 W/m K.
5. Product specific heat is c_p = 3.5 kJ/kg K.
6. Initial microbial population is N_o = 1000 per container.
7. Rate constant for microbial population is **k** = 2.1/min at 121°C.
8. Activation energy constant for microbial population is E_A = 240 kJ/mole.
9. Rate constant for retention of product quality attribute is **k** = 0.0435/min at 138°C.
10. Activation energy constant for retention of product quality attribute is E_A = 97 kJ/mole.
11. Product container is 9 cm in diameter and 16 cm in height.
12. Surface heat transfer coefficient for heating medium is h = 5000 W/m^2 K.
13. Cooling medium temperature is T_c = 35°C.
14. Surface heat transfer coefficient for cooling medium is h = 1500 W/m^2 K.

Approach:

1. The process time required for the target reduction in microbial population is determined at five different temperature levels.
2. The quality retention within various regions of the product structure is computed, and the mass average quality retention for each of the five processes is determined.
3. The process with the maximum retention of the product quality attribute can be selected.

Solution:

1. To compute the temperature change as a function of time, the parameters of the following heating rate equation must be estimated, using Figures 5.4 and 5.5 of Chapter 5:

$$\log(T_M - T) = -t/f_h + \log[j_c(T_M - T_o)]$$

2. The Biot number for the heating medium is

$N_{Bi} = (5000)(4.5/100)/0.45 = 500$ for container diameter
$N_{Bi} = (5000)(12/100)/0.45 = 1333.3$ for the container height

3. Using Figure 5.4 for an infinite cylinder,

$$f\alpha/r^2 = 0.39975 \text{ for an infinite cylinder}$$

where

$$\alpha = k/\rho\, c_p = 0.45/(1000)(3500) = 1.2857 \times 10^{-7}\, m^2/s$$

Then

$$f_h = 0.39975\,(4.5/100)^2/1.2857 \times 10^{-7} = 6296.1\, s$$
$$= 104.94 \text{ min} = 1.75 \text{ hr}$$

and

$$f\alpha/d^2 = 0.935 \text{ for an infinite slab}$$

$$f_h = 0.935\,(0.12/100)^2/1.2857 \times 10^{-7} = 104{,}721.2\, s$$
$$= 1745.4 \text{ min} = 29.1 \text{ hr}$$

(Continued)

Example 8.2 (Continued)

or

$$1/f = 1/1.75 + 1/29.1$$
$$f_h = 1.65 \text{ hr} = 99 \text{ min}$$

4. To obtain the parameter (j_c) at the geometric center of the container using N_{Bi} of 500 and 1333.3, Figure 5.5 provides

$$j_c = 1.602 \times 1.273 = 2.04$$

5. During cooling, the lag parameter (j_{cc}) is obtained as follows:

$$N_{Bi} = (1500)(4.5/100)/0.45 = 150$$
$$N_{Bi} = (1500)(12/100)/0.45 = 400$$

6. The cooling parameter (j_{cc}) is obtained by using Figure 5.5:

$$j_{cc} = 1.6017 \times 1.2732 = 2.0393$$

7. These parameters are used to estimate the temperatures at several locations within the product after increments of time during heating of the product, using the heating rate equation. These computations are illustrated in Table A.8.1.a through Table A.8.1.e.
8. The magnitude of the cooling rate parameter (f_c) at the N_{Bi} of 150 and 400 is obtained from Figure 5.4 as follows:

$$f\alpha/r^2 = 0.404 \text{ for an infinite cylinder}$$

And

$$f\alpha/d^2 = 0.938 \text{ for an infinite slab}$$

then

$$f_c = 0.404\,(4.5/100)^2/1.2857 \times 10^{-7} = 6363 \text{ s}$$
$$= 106.05 \text{ min} = 1.7675 \text{ hr}$$

$$f_c = 0.938(12/100)^2/1.2857 \times 10^{-7} = 105{,}057.16 \text{ s}$$
$$= 1750.95 \text{ min} = 29.2 \text{ hr}$$

and

$$1/f = 1/106.05 + 1/1750.95$$
$$f_c = 100 \text{ min}$$

9. To estimate operator process time, the spoilage probability equation is used to establish the target process for the spoilage microbial population:

$$1/10^7 = 10^3/10^{F/1.097}$$

where

$$D = 2.303/2.1 = 1.097 \text{ min at } 121°C$$

and

$$F = 1.097 \times 10 = 10.97 \text{ min at } 121°C$$

10. The thermal resistance constant is based on the activation energy constant of 240 kJ/mole. Using Eq. (2.27),

$$z = (2.303)(8.3144)(394)^2/240{,}000 = 12.4°C$$

11. For a heating medium temperature of 145°C,

$$U = 10.97 \times 10^{121-145/12.4}$$
$$U = 0.1273 \text{ min}$$

Given:

$$f_h/U = 99/0.1266 = 778$$

Table A.6.4 is used with z = 22 F (12.4°C) and $j_{cc} = 2.04$:

$$g = 55°F \text{ or } 30.6°C$$

(Continued)

Example 8.2 (Continued)

12. The operator process time is computed by using Eq. (6.17):

$$t_b = (99)\log\{[(2.04)(145 - 70)]/30.6\}$$
$$t_b = 69.2\ \text{min}$$

13. This operator process time has been used to determine the initiation of cooling in the spreadsheet in Tables A.8.1.a to A.8.1.e.
14. The operator process time for several other heating medium temperatures have been estimated using the same procedures, as follows:

Temperature (°C)	Operator time (min)
145	69.2
135	82.0
125	101.4
120	113.4
115	144.4

15. The retention of the heat-sensitive quality attribute is computed within the spreadsheet in Tables A.8.1.a to A.8.1.e. The table contains computations for various locations within the product and for different operator process times. For example, the retention of the quality attribute in the center region for a heating medium temperature of 145°C is computed for 2-minute increments during the increase and decline in product temperature. The results are as follows:
    a. The product temperature reaches nearly 116.2°C in the center region by the operator process time of 70 minutes (69.2 min). The retention of quality attribute at this location and time is 87.5%.
    b. At the operator process time of 70 minutes, the cooling portion of the process begins. The temperature is computed by using the cooling rate equation, as follows:

$$\log(T - T_c) = t/f_c + \log[j_{cc}(T_o - T_c)]$$

where: T_c = cooling medium temperature, f_c = cooling rate constant, j_{cc} = cooling lag parameter, T_o = initial product temperature.

As indicated by the spreadsheet, the product temperature continues to increase, and the temperature reaches a temperature of 127°C and does not begin to decrease until 22 minutes after the beginning of the cooling process.

c. At the completion of cooling, the temperature in the center region of the product is 65.3°C, and the product quality retention is 53.5% at that location.

16. To estimate the quality retention at other locations within the product, temperature histories at other locations must be computed. This requires use of the heating rate equation but with the appropriate magnitude of the heating lag constant (j) for a given location. The magnitude of these constants is determined by the steps as illustrated in Example 6.6. Using the heating lag constant of 2.04 and cooling lag constant of 2.093 at the geometric center of the product, the constants at various distances from the center is as follows (for a 1-cm vertical volume):

Location (cm)	Heating lag (j)	Cooling lag (j)
0.25	1.959	1.958
1.0	1.902	1.901
2.0	1.588	1.587
3.0	0.918	0.918
4.0	0.297	0.297

17. When these heating and cooling lag values are entered into the appropriate equations, the temperature histories of the various locations are obtained, as illustrated in the spreadsheet in Tables A.8.1.a to A.8.a.e. In addition, the retention of the quality attribute within each production region is computed. The product temperature at the end of heating (beginning of cooling), and the quality retention when the heating medium is 145°C are as follows:

Location (cm)	Temperature (°C)	Retention (%)
0.25	116.2	53.5
1.0	117.0	51.6
2.0	121.6	50.1
3.0	131.5	45.5
4.0	140.6	8.4

(Continued)

Example 8.2 (Continued)

These temperature differences become less as the heating medium temperature is decreased. In addition, the distribution of quality retention is changed as indicated in the spreadsheet.

18. In Example 6.6, the portions of product mass in the various regions of product were computed. Using these fractions of the total product mass, the quality retention within each region can be determined. When these computations are completed, the following results are obtained for the five heating medium temperatures:

Temperature (°C)	Quality retention (%)
145	32.5
135	38.5
125	45.1
120	49.9
115	48.7

The results in Figure 8.3 demonstrate that the maximum retention of the quality attribute occurs at a heating medium temperature of 120°C. In addition, it is evident that higher medium temperatures and temperatures lower than 120°C result in less retention.

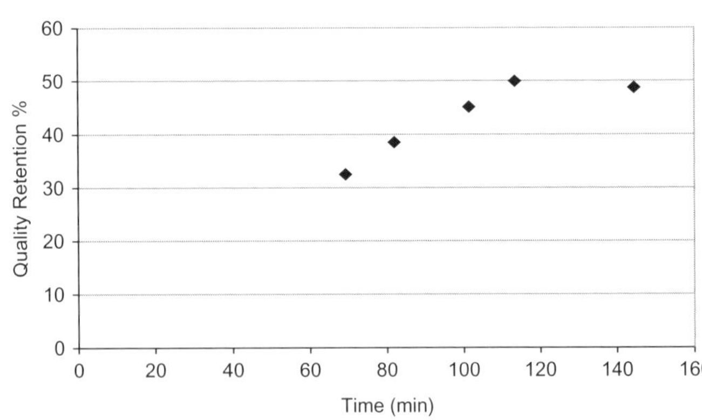

Figure 8.3 Optimization of process for Example 8.2.

The previous example illustrates that several factors impact the retention of quality attributes during a thermal process. For products with thermal energy transfer by conduction, the retention of a quality attribute decreases with location, from the center region to the region product near the product surface. The most significant decrease in retention occurs within the region near the surface where the product is in contact with the heating medium. Because the outer region of a container with a cylindrical geometry has a higher proportion of the product mass, the influence of quality retention in this region is more significant. The distribution of quality retention magnitudes within the container is influenced by heating medium temperature, and the optimization of quality retention becomes feasible. Note that the heating medium temperature influence on quality retention is not independent because each heating medium temperature is part of a thermal process and requires an appropriate operator time to ensure the desired reduction in the target microbial population. Ultimately, the combinations of quality attribute distribution within the container and the proportions of product mass within each region of the container create the opportunity for process optimization.

The opportunity for optimization of thermal processes should exist within any container geometry where a temperature and quality retention distribution history exists within a conduction-heating food product. The magnitudes of quality retention and optimization are influenced by many factors, including the dimensions of the container. The most critical dimension is the distance from the geometric center to the surface of the container in contact with the heating medium. When the radius of the cylindrical container (can) and the vertical distance from center to surface are similar, the analysis must include temperature and quality distributions in both the radial and axial direction. Each combination of container dimension is likely to provide a different optimum process.

Any variation in thermal energy transfer from conduction within the product structure impacts the potential for thermal process optimization. The heating of a liquid food within the container occurs by convection and causes a shift in the slowest heating location within the product structure. In addition, the magnitudes of temperature gradients between the slowest heating location and the container surface are much less than within conduction-heating products. Due to these differences, the distribution of quality retention between the container surface and the slowest heat location is much less than for a conduction-heating product. Because the temperature increase within the product and the temperature decrease during cooling are more rapid due to convection, the optimum processes are expected to be at higher temperatures and shorter times. As the viscosity (or consistency) of the product increases, the impact of convection heating declines. These impacts cause the optimum temperature–time process to shift closer to that of the product with thermal energy transfer by conduction. Any product with solid food particles within a liquid phase carrier becomes more difficult to analyze. If the quality retention is dependent on the temperature and quality retention distributions within the food particles, the focus of the analysis is on heating and cooling characteristics within the particles. More specifically, the analysis of the particle with largest dimensions is handled in a manner similar to a container with a conduction-heating product. These types of analyses lead to quality retention distributions within individual particles, and the results need to be combined with quality retention within the liquid phase as well as within other food particles with different dimensions.

The retention of quality attributes within the food product and optimization of the thermal process are influenced by the magnitude of kinetic parameters for the quality attribute. The magnitudes of both the first-order rate constant (at reference temperature) and the activation energy constant for the quality attribute have direct impact on the retention of the product quality attribute during the thermal process and on the distributions of quality retention within the product structure. The magnitude of the activation energy constant appears to have more significant influence on the retention of quality attribute. Ultimately, the difference in magnitudes of the activation energy constants for the target microbial population and that of the product quality attribute has the most significant influence on the optimization of the thermal process. The larger the difference in magnitude of the two constants, the more evident the optimum process becomes.

The thermophysical properties of the product have a direct impact on the optimization of the thermal process. The density, specific heat, and thermal conductivity of the product structure are among the primary factors influencing the increase in product temperature during the initial part of the process and the temperature decrease during the cooling portion of the process. In addition, these properties influence the temperature distribution within the product and the distribution of quality retention. The key factor influencing these properties is water content. The density, specific heat, and thermal conductivity of high moisture foods are similar to water. Food products with lower water content have lower values of specific heat and thermal conductivity but higher density magnitudes.

An evaluation of the product temperature profile during the thermal process indicates that a very important part of the process occurs during transition from heating to cooling. At interior locations, the product temperature continues to increase after the cooling process begins. This increase in temperature at the product center may continue until portions of the product near the surface are reduced in temperature, and thermal energy transfer from the interior regions begins. The temperature profile during the transition follows a parabolic pattern, with a continuing but more gradual increase, followed by a gradual decrease in temperature, before the predicted cooling curve occurs. The temperatures during the transition from heating to cooling are the highest of the process and have significant impact on the quality retention. The accuracy of the optimization is directly related to the simulation of the temperature during transition.

Optimizing processes based on alternative preservation technologies is an appropriate part of process evaluation. These types of optimization and evaluation are based on knowledge of impacts of preservation technology on quality attributes of the product. This information is limited at this time, and the quantification of the impact in terms of kinetic parameters is not available. Processes based on microwave energy, ohmic heating, and the pressure-assisted thermal process can be evaluated in the same manner as demonstrated, but only the impact on heat-sensitive attributes are considered.

Example 8.3

A preservation process for new highly viscous food product is being developed. The process is designed to reduce the population of *Salmonella* (**k** = 2.5/min at 65°C, E_A = 400 kJ/mole) from 1000 per product container to a probability of survivors of 10^{-6}. The traditional process of a continuous system of heat exchanger and holding at a process temperature (followed by aseptic packaging) causes significant losses to a heat-sensitive quality attribute (**k** = 0.25/min at 95°C; E_A = 80 kJ/mole) due to laminar flow in the holding tube. The alternative is a pressure-assisted thermal process, with the product placed in the

(*Continued*)

Example 8.3 (Continued)

package prior to the process. Compare the quality retention for the two processes, when the initial product temperature is 40°C, and heating and cooling rates in the heat exchanger are 5°C/minute. The pressure process increases the pressure at 100 MPa/minute, the temperature increase is 5°C per 100 MPa, and pressure release is 200 MPa/minute.

Given:

1. Kinetic parameters for *Salmonella* are **k** = 2.5/min at 65°C and E_A = 400 kJ/mole.
2. Kinetic parameters for quality attribute are **k** = 0.35/min at 95°C and E_A = 80 kJ/mole.
3. Initial product temperature is T_o = 40°C.
4. Initial microbial population is N_o = 1000 per package.
5. Target probability of survivors is 10^{-6}.
6. Pressure process increases pressure at 100 MPa/min.
7. Temperature increase due to pressure is 5°C per 100 MPa.
8. Thermal process increases temperature at 5°C/min, and cools product at 5°C/min.
9. Pressure release is 200 MPa/min.

Approach:

1. The holding times for both processes, as required to accomplish the target probability of survivors, are established.
2. Given the process times, the quality retention can be computed using time-step intervals throughout the process.

Solution:

1. The holding time for the traditional thermal process is determined by determining the reduction in microbial population at 1-minute intervals, starting at 40°C and increasing to 60°C in 4 minutes.
2. Beginning at 4 minutes, the product was in the holding tube until 12 minutes, or a holding time of 8 minutes.
3. The product was then cooled at a rate of 5°C/min, until cooled to 40°C at 16 minutes, as illustrated in the spreadsheet in Table A.8.2.
4. At the completion of the process, the probability of survivors is 3.2×10^{-7}, which exceeds the target of 10^{-6}.
5. The pressure-assisted thermal process is computed in the same manner as the traditional thermal process, with the

pressure increase to 400 MPa accomplished in 4 minutes, and the temperature increase to 60°C is accomplished as well.

6. The process has a holding period of 8 minutes, followed by pressure release in 2 minutes.
7. By using time-step computations, the microbial population is reduced to 8×10^{-7}, as illustrated in the spreadsheet in Table A.8.2.
8. The quality retention for both processes has been computed at 1-minute time intervals throughout the processes by using the kinetic parameters for the quality attribute. For the pressured-assisted process, the quality retention is 84.2%, as illustrated in Table A.8.2.
9. For the traditional thermal process, with laminar flow in the holding tube, the mass average velocity is double the maximum velocity. Although the reduction of microbial population must be based on the maximum velocity, the impact of the process on the quality attribute should be based on the average velocity. This approach results in quality retention of 74.1%. In Table A.8.2, the time-step computations during the holding period for the traditional thermal process have been for 2 minutes at each time step to account for the average velocity.
10. The results of the analysis in this example indicate that the pressure-assisted-thermal-process improved the quality retention to 84.2%, as compared to 74.1% for a traditional thermal process.

Most likely, the analysis in Example 8.3 would focus on the product temperature at the beginning of the process and identify the temperature that would lead to maximum retention of product quality, while ensuring the desired reduction the target microbial population.

List of symbols

A = area, m^2
α = thermal diffusivity; m^2/s
C = concentration or intensity of product quality attribute
c_p = specific heat; kJ/kg K

D = decimal reduction time, min
d_c = characteristic dimension, m
E_A = activation energy constant, kJ/mole
F = thermal death time, min
F_R = thermal death time at reference temperature, min
f_h = heating rate constant, s or min
f_c = cooling rate constant, s or min
g = temperature difference between heating medium and product slowest heating location, and end of process, °C
h = convective heat transfer coefficient, W/m^2 K
J_o = Bessel function
j_c = heating lag constant at geometric center
j_m = mass average heating lag constant
j_{cc} = cooling lag constant at geometric center
k = thermal conductivity, W/m K
$\mathbf{k}$ = rate constant, 1/s
$\mathbf{k_o}$ = reference rate constant, 1/s
L = length, m
LR = lethal rate
m = mass flow rate, kg/s
N = microbial population
N_o = initial microbial population
N_{Bi} = Biot Number
N_{Re} = Reynolds Number
q = thermal energy transfer, W
R = radius, m
R_g = gas constant; kJ/kg K
R_1 = argument of Bessel function
r = radial direction
r_c = number of containers in a process
ρ = density, kg/m^3
T_{CM} = cooling medium temperature, °C
T_M = medium temperature, °C
T_o = initial temperature, °C
T_R = reference temperature, °C
$\mathbf{T}$ = absolute temperature, K
t = time, s
t_B = process time, min
t_p = operator process time, min
U = thermal death time at process temperature, min
u = mean velocity, m/s

u_{max} = maximum velocity, m/s
V = volume, m^3
z = Thermal Resistance Constant, C

Bibliography

Ball, C. O. (1923). Thermal process time for canned foods. Part 1. *Bulletin of the National Research Council*, *7*(37), 76.

Ball, C. O., & Olson, F. C. W. (1957). *Sterilization in food technology*. New York: McGraw-Hill.

Bigelow, W. D., Bohart, G. S., Richardson, A. C., & Ball, C. O. (1920). Heat penetration in processing canned foods. *National Canners Association Bulletin, No. 16L.*

Feliciotti, E., & Esselen, Jr. W. B., (1957). Thermal destruction rates of thiamine in pureed meats and vegetables. *Food Technology*, *11*(2), 77–84.

Heldman, D. R., & Hartel, R. W. (1997). *Principles of food processing*. New York: Chapman & Hall, Inc.

Lewis, M., & Heppel, N. (2000). *Continuous thermal processing of foods*. Gaithersburg, MD: Aspen Publishers, Inc.

Pflug, I. J., & Esselen, W. B. (1963). Food processing by heat sterilization . In J. L. Heid & M. A. Joslyn (Eds.), *Food processing operations*. (Vol. 2, pp. 410–479). Westport, CT: The AVI Pub Co., Inc.

Pflug, I. J., Blaisdell, J. L., & Kopelman, I. J. (1965). Developing temperature-time curves for objects that can be approximated by a sphere, infinite plate or infinite cylinder. *ASHRAE Transaction*, *71*(1), 238–248.

Singh, R. P., & Heldman, D. R. (2009). *Introduction to food engineering* (4th ed.). San Diego, CA: Elsevier Academic Press.

Somerville, J., & Balasubramaniam, V. M. (2009). Pressure-assisted thermal sterilization of low-acid shelf-stable foods. *Resource: Engineering & Technology for a Sustainable World*, *16*(7), 14–17.

Stumbo, C. R. (1965). *Thermobacteriology in food processing*. New York: Academic Press.

Teixeira, A. A. (2007). Thermal processing of canned foods. In D. R. Heldman & D. B. Lund (Eds.), *Handbook of food engineering* (2nd ed., pp. 745–797). Boca Raton, FL: CRC Press, Inc.

Teixeira, A. A., Dixon, J. R., Zahradnik, J. W., & Zinsmeinster, G. E. (1969). Computer optimization of nutrient retention in thermally processing of conduction-heating foods. *Food Technology*, *23*, 137.

9 Designing Processes in the Future

9.1 Assembly of kinetic parameters 268
9.2 Transport models 271
9.3 Process models 272
9.4 Opportunities for evolving process technologies 273
Bibliography 276

Preserving foods will continue to be a priority in the future. Reducing or eliminating food-borne illness is a high priority among consumers and regulatory agencies. As new pathogens continue to emerge, access to the appropriate kinetic information to characterize these pathogens is very important. As consumer expectations for product quality increase, more precise processes for extending the shelf life of all foods will be required. All processes must be applied in a way that enhances the quality attributes of the food products. New technologies for preserving food products will continue to evolve and be evaluated for food applications. New and innovative approaches to food preservation will be considered. The successful processes of the future are likely to include current processes with improvements, as well as alternative processes based on new technologies. Combinations of current and new technologies will be evaluated based on improvements in process efficiency and the effectiveness in delivery of desired product quality attributes.

Food Preservation Process Design
ISBN: 978-0-12-372486-1

9.1 Assembly of kinetic parameters

As has been demonstrated throughout this book, accurate and reliable kinetic parameters are key components of process design for any technology considered for food preservation processes. Chapter 3 presented the typical kinetic parameters for microbial pathogens and many of the limitations to currently available parameters. An emerging change is the use of alternative models for explaining microbial survivor curves—primarily describing deviations from the traditional first-order model. These models introduce the additional parameters needed to more accurately predict the number of survivors at any time during a preservation process. In addition, these models introduce different terminology for the kinetic parameters associated with survivor curves and the influence of the preservation technology intensity on the rate of reduction in microbial population. Hopefully, consistent models and kinetic parameters will be used during measurement and assembly of microbial survivor data in the future. These models and kinetic parameters should be consistent with those used to describe the retention of product quality attributes. A summary of the recommended parameters is presented in Table 9.1.

Consistency in the parameters used for process design will become more important in the future. The efficient comparison of different preservation technologies will depend on access to the appropriate parameters. The evaluation of the impact of the processes on product quality attributes requires that the kinetic parameters used as inputs to the process design be expressed in the same manner. Ultimately, the optimization of the processes is dependent on consistency in the kinetic parameters available for process design.

The kinetic data for predicting survivors of microbial populations during a preservation process clearly indicate that some components of the food have significant influence on the number

Table 9.1 Recommended kinetic parameters for preservation process design

Survivor curve	Rate constant	k or k′	1/min
Deviation		n	
Temperature intensity	Activation Energy Constant	E_A	kJ/mole
Pressure intensity	Activation Volume Constant	ΔV	m^3/mole

of survivors. In general, differences in the magnitude of kinetic parameters suggest that microbial populations are protected from the impact of the preservation agent (temperature, pressure, etc.), and the rates of reduction in microbial population are reduced when the microbial population is incorporated into a food product, as compared to a pure culture. The rate constants for microbial survivor curves indicate that rates are higher when the microbial population is suspended in a buffer solution. In addition, the results suggest that different food components (proteins, carbohydrates, fats, ash, etc.) may influence the survivor curve rate constants in a different manner. In the future, researchers should consider measuring kinetic parameters for survivors of microbial populations in substrates with different levels of typical food components. One objective of these investigations places emphasis on evaluating the influence of each product component on the kinetic parameters in a quantitative manner. This approach creates opportunities for predicting kinetic parameters based on product composition and eliminates the need for measuring the parameters for each new and different product.

Future investigations must include evaluating survivor curve kinetics for different preservation technologies. Although the current literature includes considerable amounts of kinetic data for ultra-high pressure (UHP) processes, more specific attention should be given to the combined influence of pressure and temperature, as well as the influence of product composition. In addition, there are modest amounts of kinetic data for the pulsed electric field (PEF) technology. Additional kinetic data are needed, and future measurements should include the influence of product composition. As other preservation technologies evolve and are considered for application to foods, the approach to measurement and assembly of kinetic data should be similar to the approach presented and described for thermal, UHP, and PEF technologies.

An additional potential impact on kinetic data for survivors of microbial populations is the influence of two or more microbial populations within the food product. Although the preservation process will always be based on the population of the most resistant pathogen or spoilage microorganism, some evidence suggests that multiple populations of microorganisms within a given product could influence the survival kinetics of the target microbial population. It may be difficult to predict the influence of multiple populations, but this factor should be considered when measuring kinetic data, and the potential impacts can be recognized during proccss design.

Kinetic data for food product quality attributes will become more important in the future. As indicated in Chapter 4, kinetic parameters for several key quality attributes have been measured and assembled. In general, considerably less kinetic data is available to describe the influence of the process on quality attributes than for microbial populations. Although the information is adequate for general comparisons of different attributes in the same product, the quantitative reliability of parameters could be improved. In many situations, the influence of product composition on the magnitude of kinetic parameters may not be as evident as with kinetic parameters for microbial survivors. The relationships between the magnitudes of kinetic parameters and product composition are obvious for many products. The most significant challenge is that most food products have multiple quality attributes, and the most sensitive attribute to the preservation process may be not evident. In addition, relationships among quality attributes in any given product may impact the retention of one or more of the attributes.

As suggested in Chapter 6, the development of process design models to account for the impact of multiple technologies on the survival of the microbial population deserves special attention. Combination processes, such as the simultaneous use of UHP and temperature, may receive more attention in the future, and the appropriate process design models need to be developed. More specifically, the appropriate kinetic parameters need to be measured and assembled. These models may require additional parameters, and investigations involving measurements should evaluate the use of multiple-parameter models. After these models and parameters have been identified, the assembly of the kinetic parameters ensures the most complete preservation process design possible.

Ultimately, the kinetic parameters for microbial survivors and for quality attributes need to be assembled into a database that is accessible to everyone involved in preservation process design. The organization of the database for these parameters needs to be user-friendly with a format that encourages process optimization. The easy access to kinetic parameters will continue to be a key component to food safety, as well as optimizing processes for enhanced product quality attributes.

9.2 Transport models

Although there have been significant developments in transport models for foods over the past 50 years, there is a continuing need for improvements in models and food property data. Future developments are needed in three areas: (1) predicting physical properties of foods, (2) improving current transport models for foods, and (3) developing appropriate transport models for foods when new and emerging preservation technologies are identified. The success of current process design models, as well as future process models, depends on these developments.

9.2.1 Physical properties of foods

Important developments about the physical properties of foods include the significant efforts in measuring properties, as well as the prediction of properties based on composition (Rahman, 1995; Rao, Rizvi & Datta, 2005; Sahin & Sumnu, 2006). The efforts of Choi and Okos (1986) illustrate the opportunities for predicting physical properties of foods and specifically the changes in these properties as a function of moisture content and temperature. The importance of these relationships must be emphasized due to the changes occurring during typical preservation processes. There is still room for improvement in the prediction of properties based on composition, with specific attention to contributions of various compositional components, and the impact product component phase changes during a process.

Physical properties prediction models need to be developed in areas where the influence of factors are unique to a new or alternative preservation technology. The most obvious current example is the application of UHP as a preservation process. Model coefficients are immediately needed to account for the influence of pressure on properties of basic food components. Similar needs will emerge for other preservation technologies. For example, the placement of a food in an electric field may influence the magnitude of the basic physical properties of the product. In many situations, new technologies will introduce new and unique physical properties to be incorporated into the process design. In most cases, these unique properties may change in magnitude as a function of temperature, as well as other parameters associated with the technology. Future efforts should be dedicated to the coefficients needed to encourage the prediction of physical properties as a function of product composition, as influenced by the preservation process parameters.

9.2.2 New transport models

The most evolved transport models for foods are those that predict temperature distribution histories within a food during the preservation process. These models are reliable for situations where heating (or cooling) of the product is by pure conduction, or during heating (or cooling) of a liquid food in a well-defined heat exchanger. More sophisticated convection models are needed for heating (and cooling) of foods within a container, as well as a range of variations when the liquid product contains solid pieces. Similar challenges exist for predicting temperature distribution histories within solid pieces in a liquid carrier during transportation through a thermal processing system.

There are significant needs for the development of appropriate transport models to predict the intensity of alternative technology agents, such as UHP or PEF. Although the distribution of elevated pressure within a food product structure is expected to be uniform and rapid, the ability to predict and confirm these distributions is critical to the process design. There are potential impacts of variability in the response of food product components to pressure changes on pressure distributions within the product structure. These impacts make predictions of pressure distributions and corresponding temperature distributions very challenging, but they have significant impact on process design. Similar challenges exist when evaluating electric field strength distribution history within the product structure during a PEF process. The models for predicting these distributions for a liquid flowing through a PEF system may not be overly complex, but models for a PEF process applied to a solid food structure require careful analysis. Consideration must be given to models required to predict distribution histories of any agent associated with a new or different preservation technology.

9.3 Process models

The preservation process models of the future will be more analytical and precise. The basis for these models will be the same as

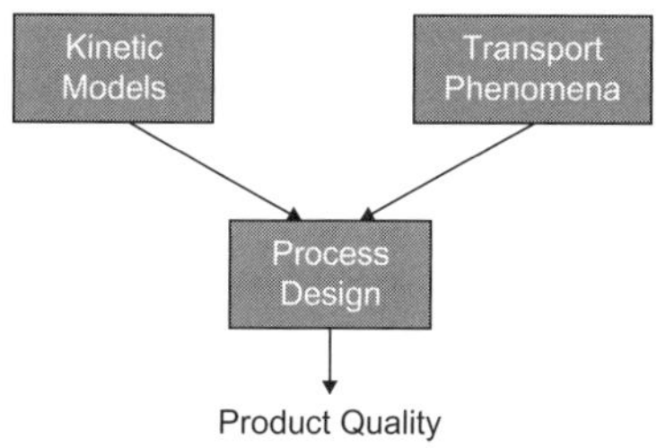

Figure 9.1 The preservation process model.

current models: the integration of kinetic models with transport models.

As illustrated in Figure 9.1, the integration of the appropriate kinetic parameters with the appropriate transport model provides the basis for the process design needed to ensure the desired product quality. Within the generalized model, product quality may be interpreted in different ways. Most often, the target is microbiological safety of the product being manufactured according to the process design. Future models will accommodate inputs for multiple kinetic parameters for different microbiological populations, as well as several key product quality attributes. The inputs will be integrated with the most appropriate transport models for the preservation technology. The process design model will provide multiple outcomes to correspond with key microbiological populations and quality attributes, and the capability to optimize the retention of quality attributes, while ensuring the safety or a target spoilage probability for the food product.

9.4 Opportunities for evolving process technologies

The IFT/FDA Task Force report (2001) evaluated the potential for each of the following preservation technologies for food:

- Microwave and Radio Frequency
- Ohmic and Inductive Heating
- High Pressure Processing
- Pulsed Electric Field

- High Voltage Arc Discharge
- Pulsed Light
- Oscillating Magnetic Fields
- Ultraviolet Light
- Ultrasound
- X-Rays

As indicated by the illustrations and examples presented in the previous chapters, there has been sufficient development for the first four to be given consideration for applications as food preservation processes. This conclusion is supported by the analysis provided by Sun (2005), where major attention was given to UHP, PEF, and alternative thermal processes. Although there have been modest developments for the other six technologies, these developments have not demonstrated sufficient potential to create incentive for the measurement of appropriate kinetic parameters and the transport models needed for process design.

The opportunities for new process technologies to succeed depend on numerous factors. In general, two pathways to commercial operations exist: (1) replacing an existing preservation technology or (2) applying a new technology to create a new or different food product. Both pathways contain many challenges, but the success of an evolving technology will require the following:

- The evolving preservation technology demonstrates that it meets all of the expectations provided by a currently used or available technology.
- The new or evolving preservation technology provides a significant improvement in efficiency, as expressed in increased capacity or reduced energy requirements.
- A significant improvement in key product quality attributes is accomplished through the use of the new or evolving preservation technology.

A successful, new process technology will meet all requirements of the current technology and provide, at minimum, an improvement in process efficiency or product quality. Most likely, the successful evolving technologies will meet or exceed all three of the preceding requirements.

The kinetic models, the transport models, and the process models presented and demonstrated throughout this book are key components of an effort to evaluate the potential of an evolving process technology. The process models provide mechanisms for assessing potential quality improvements to be achieved while meeting the expectations for product safety or spoilage probability. By varying the inputs to the models, the influence of capacity on effectiveness and quality optimization can be evaluated. These same inputs can be evaluated for the current technology, and the processes can be compared as a step in reaching a decision on a change in process. These process models are key tools in the evaluation of new technologies and should become a first step in any plan to consider a new technology.

The transport models are key inputs to the process models. Although significant basic information about foods and food components is available, unique characteristics of the process technology may require the measurement of new and different physical properties. These measurements should be accomplished in a basic and fundamental manner, and attempt to relate the magnitude of new physical properties to product composition, and the properties of key individual components of the product. These approaches provide the opportunity to accommodate changes in the composition of the product during the process and to evaluate the process for products with different compositions.

The kinetic models are the second key, but equally important, input to the process models. Although significant amount of kinetic data have been assembled to describe the impact of thermal process technologies on food components, including microbial populations, the amounts of data available for most other process technologies are limited. The evaluation of a new process technology may require measuring kinetic parameters and considering alternative kinetic models. These measurements should include the response of the target microbial populations and the key product quality attributes for the product being considered to the process technology. Whenever possible, measurements should consider product composition and evaluate the relationships of kinetic parameters to product composition. These kinetic parameters, and the appropriate models, will then provide potential opportunities for evaluating other products with different compositions. These kinetic parameters, for both survival of microbial populations and retention of key product quality attributes, are the primary inputs to the process models to be used to evaluate new process technologies.

The process models presented and described should be the first key step in the evaluation of a new process technology. Initially, the process model may be used with estimated values for kinetic parameters and transport properties to evaluate feasibility of a new technology. If the results are encouraging, experimental investigations to create improved inputs to the process model may be justified. When outcomes from the process model are sufficiently encouraging in terms of efficiency improvements and quality enhancement, modest scale-up of the new process technology will be warranted. The process model provides key information on pilot-scale investigations needed to confirm the previous predictions from the process model. Finally, the process model becomes a key tool in the scale-up of the new process technology to commercial operations, as well as an important tool in the continuous monitoring of a process technology and in the evaluation of unanticipated outcomes during commercial operation.

Bibliography

Ball, C. O., & Olson, F. C. W. (1957). *Sterilization in food technology*. New York: McGraw-Hill.

Choi, Y., & Okos, M. R. (1986). Effects of temperature and composition on thermal properties of food. In M. Le Maguer & P. Jelen, (Eds.), *Food engineering and process applications* (Vol. 1, pp. 93–101). London: Elsevier Applied Science Publishers.

Earle, R. L. (1983). *Unit operations in food processing* (2nd ed.). Oxford, UK: Pergamon Press.

Heldman, D. R., & Lund, D. B. (2007). *Handbook of food engineering* (2nd ed.). Boca Raton, FL: CRC Press, Taylor & Francis Group.

IFT/FDA. (2001). Kinetics of microbial inactivation for alternative food processing technologies. *Journal of Food Science* (Special Supplement), 1–108.

Manson, J. E., & Cullen, J. F. (1974). Thermal process simulation for aseptic processing of foods containing discrete particulate matter. *Journal of Food Science*, *39*, 1084.

Pflug, I. J., & Esselen, W. B. (1963). Food processing by heat sterilization. In J. L. Heid & M. A. Joslyn, (Eds.), *Food processing operations* (Vol. 2, pp. 410–479). Westport, CT: The AVI Pub Co., Inc.

Pflug, I. J., Blaisdell, J. L., & Kopelman, I. (1965). Developing temperature-time curves for objects that can be approximated by a sphere, infinite plate or infinite cylinder. *ASHRAE Transactions*, *71*(1), 238–248.

Potter, N. N., & Hotchkiss, J. H. (1995). *Food science* (5th ed.). New York: Chapman & Hall, p. 608.

Rahman, M. S. (1995). *Food properties handbook*. Boca Raton, FL: CRC Press.

Rao, M. A., Rizvi, S. S. H., & Datta, A. K. (2005). *Engineering properties of foods* (3rd ed.). Boca Raton, FL: CRC Press, Taylor & Francis Group.

Sahin, S., & Sumnu, S. G. (2006). *Physical properties of foods*. New York: Springer Science, LLC.

Singh, R. P., & Heldman, D. R. (2008). *Introduction to food engineering* (4th ed.). San Diego, CA: Elsevier Academic Press.

Stumbo, C. R. (1965). *Thermobacteriology in food processing*. New York: Academic Press.

Sun, D-W. (Ed.), (2005). *Emerging technologies for food processing*. San Diego, CA: Elsevier Academic Press.

Teixeira, A. A. (2007). Thermal processing on canned foods. In D. R. Heldman & D. B. Lund (Eds.), *Handbook of food engineering* (2nd ed., pp. 745–797). Boca Raton, FL: CRC Press, Taylor & Francis Group.

Teixeira, A. A., Dixon, J. R., Zahradnik, J. W., & Zinsmeinster, G. E. (1969). Computer optimization of nutrient retention in thermally processing of conduction-heating foods. *Food Technology*, *23*, 137.

Villota, R., & Hawkes, J. G. (2007). Reaction kinetics in food systems. In D. R. Heldman & D. B. Lund (Eds.), *Handbook of food engineering* (2nd ed., pp. 125–286). Boca Raton, FL: CRC Press, Taylor & Francis Group.

Appendix

Table A.3.1 Kinetic parameters for *salmonella spp*. survivor curves

Preservation technology	Microbial population	Time parameters D	k	n	Treatment intensity
Thermal	Salmonella	0.018–0.56 min	4.113–127.9/min		65.6°C
	Salmonella	0.645 min	3.569/min		60°C
	Salmonella	1.77 min	1.301/min		60°C
	Salmonella	0.87 min	2.647/min		60°C
Thermal	S. Senftenberg	276–480 min	0.005–0.008/min		70–71°C
	S. Senftenberg	0.56–1.11 min	2.075–4.113/min		65.5°C
	S. Senftenberg	4.42–5.63 min	0.409–0.521/min		60°C
	S. Senftenberg	3.828–8.5 min	0.271–0.602/min		60°C
	S. Senftenberg	15.51 min	0.149/min		60°C
	S. Senftenberg	5.07 min	0.454/min		60°C
Pressure	S. Senftenberg	6 min	0.384/min		345 MPa
Thermal	S. Typhimurium	4.7–18.3 min	0.126–0.49/min		55°C
	S. Typhimurium	396–1050 min	0.002–0.006/min		70–71°C
	S. Typhimurium	6.94 min	0.144/min	0.667	60°C
	S. Typhimurium	0.297 min	3.363/min	0.502	76°C
	S. Typhimurium	10 min	0.1/min	1.2	60°C
	S. Typhimurium	0.217 min	4.6/min	0.35	76°C
	S. Typhimurium	0.261 min	8.8/min		60°C
	S. Typhimurium	0.217 min	10.6/min		60°C
	S. Typhimurium	0.41–0.57 min	4.04–5.617/min		61°C
Pressure	S. Typhimurium	3 min	0.768/min		350 MPa
	S. Typhimurium	1.48 min	1.556/min		414 MPa
	S. Typhimurium	16.4 min	0.14/min		200 MPa
	S. Typhimurium	13.2 min	0.175/min		200 MPa
PEF	S. Typhimurium	4 microsec	0.5758/microsec		83 kV/cm
PEF	S. enteritidis		0.0303–0.0546/ microsec		30 kV/cm
Thermal	S. Eastbourne	270 min	0.0085/min		71°C
PEF	S. Dublin	4–42.4 microsec	0.054–0.52/microsec		15.40 kV/cm
	S. Dublin	360 microsec	0.0064/microsec		36.7 kV/cm
Thermal	Enter. Sakazakii	3.97–4.63 min	0.497–0.58/min		60°C
Pressure	Enter. Sakazakii	0.133 min	3.278/min	0.322	350 MPa
		0.190 min	2.281/min	0.446	350 MPa

Intensity coefficient		Related	Measurement	Other	
z (T, P,)	E_A or ΔV	information	substrate	information	References
4.4–5.6°C	392–499 kJ/mole		Milk	serovars	ICMSF (1996)
		aw = 0.996	Ground turkey	8 strains	Carlson et al. (2005)
		aw = 0.95	Ground turkey	8 strains	Carlson et al. (2005)
8.8°C	241.3 kJ/mole		White grape juice concentrate		Enache et al. (2006)
18.9°C	120 kJ/moles		Milk chocolate		ICMSF (1996)
4.4–5.6°C	392–499 kJ/mole		various foods		ICMSF (1996)
4.51–9.83°C	216–470.8 kJ/mole		Meat products	6 to 8 Strains	O'Bryan et al. (2006)
5.34–7.61°C	279–397.6 kJ/mole		Poultry products	6 to 8 Strains	O'Bryan et al. (2006)
6.0°C	353.9 kJ/mole	High fat content	Beef patties	6 to 8 Strains	Osaili et al. (2006)
5.89°C	360.5 kJ/mole		Ground pork	6 to 8 Strains	Murphy et al. (2004)
		T = 230°C	Buffer		Metrick et al. (1989)
4.5–4.6°C	448–45 kJ/mole		TBS + 10–42% MS		ICMSF (1996)
17.7°C	128 kJ/mole		Milk chocolate		ICMSF (1996)
			66% sugar		Mattick et al. (2001)
			66% sugar		Mattick et al. (2001)
			48% sugar		Mattick et al. (2001)
			48% sugar		Mattick et al. (2001)
7.357°C	288.6 kJ/mole		Ground beef	Multidrug Resistant	Stopforth et al. (2008)
6.993°C	303.6 kJ/mole		Ground beef	Non-multidrug-Resistant	Stopforth et al. (2008)
4.13–5.07°C	421.3–517.2 kJ/mole		Ground beef		O'Bryan et al. (2006)
			Milk		Paterson et al. (1995)
		T = 25°C	Meat		Ananth et al. (1998)
135 MPa	42.27 cm^3/mole	T = 25°C	Tryptic soy broth	Slow Decompression	Noma et al. (2002)
127 MPa	44.93 cm^3/mole	T = 25°C	Tryptic soy broth	Rapid Decompression	Noma et al. (2002)
			NaCl		Gupta & Murray (1989)
	30–46.85 kJ/mole	T = 20°C	Liquid egg whites		Amiali et al. (2008)
			Milk chocolate		ICMSF (1996)
			Skim milk	15–40°C	Sensoy et al. (1997)
			Milk	63°C	Dunn & Perlman (1987)
5.6°C	379.2 kJ/mole		Infant formula	Strain 607	Edelson-Mammel & Buchanan (2004)
		T < 30°C	Buffered peptone water		Pina Perez et al. (2007)
161 MPa	36.04 cm^3/mole	T < 30°C	Infant formula		Pina Perez et al. (2007)

Table A.3.2 Kinetic parameters for *E.coli spp.* survivor curves

Preservation technology	Microbial population	Time parameters D	k	n	Treatment intensity
Thermal	E. coli ATCC	1.3–5.1 min	0.45–1.77/min		57.2°C
	E. coli	0.58–0.92 min	2.5–3.97/min		60°C
	E. coli	0.45–0.67 min	3.44–5.12/min		60°C
Pressure	E. coli	1 min	2.303/min		400 MPa
	E. coli	2.5 min	0.92/min		400 MPa
	E. coli	3 min	0.768/min		450 MPa
Pressure	E. coli ATCC	9.22 min	0.25/min		250 MPa
	E. coli ATCC	5.65 min	0.41/min		250 MPa
PEF	E. coli	34.4–44.8 microsec	0.051–0.067/microsec		20–45 kV/cm
	E. coli	333 microsec	0.0069/microsec	0.6	22 kV/cm
	E. coli	64–96 microsec	0.024–0.036/microsec		40 kV/cm
	E. coli	27.4–49.6 microsec	0.046–0.084/microsec	0.6	50 kV/cm
	E. coli	270 microsec	0.00853/microsec		20 kV/cm
PEF	E. coli ATCC	166.5 microsec	0.0138/microsec		25 kV/cm
	E. coli ATCC	31.8 microsec	0.0724/microsec	0.6	25 kV/cm
	E. coli ATCC	41.9 microsec	0.055/microsec		40 kV/cm
	E. coli ATCC	4.0 microsec	0.576/microsec	0.6	40 kV/cm
Thermal	E. coli O111:B4	5.5–6.6 min	0.35–0.42/min		55°C
Thermal	E. coli O157:H7	4.1–6.4 min	0.36–0.56/min		57.2°C
	E. coli O157:H7	1.2–2.05 min	1.12–1.92/min		58°C
	E. coli O157:H7	0.45–0.47 min	4.9–5.21/min		60°C
	E. coli O157:H7	1.65–1.72 min	1.34–1.4/min		60°C
	E. coli O157:H7	0.9 min	2.56/min		60°C
	E. coli O157:H7	2.7 min	0.85/min		60°C
	E. coli O157:H7	1.63 min	1.41/min		60°C
	E. coli O157:H7	3.22 min	0.72/min		60°C
	E. coli O157:H7	3.02 min	0.76/min		60°C
	E. coli O157:H7	2.5 min	0.92/min		60°C
	E. coli O157:H7	2.41–2.7 min	0.85–0.96/min		60°C
	E. coli O157:H7	0.41–2.077 min	1.11–5.62/min		50°C
Pressure	E. coli O157:H7	3 min	0.768/min		400 MPa
	E. coli O157:H7	0.7 min	3.29/min		345 MPa
PEF	E. coli O157:H7		0.0236–0.245 microsec		30 kV/cm

Intensity coefficient		Related		Other	
z (T, P,)	E_A or ΔV	information	Substrates	information	References
			Dairy products		ICMSF (1996)
7.3–8.1°C	257.8–286 kJ/mole		Cow's milk		Sela et al. (2003)
7.5–10.5°C	202.1–283.1 kJ/mole		Camel milk		Sela et al. (2003)
		T = 50°C	Milk		Gervilla et al. (1997)
			Meat		Patterson & Kilpatrick (1998)
			Buffer		Patterson et al. (1995)
126 MPa	45.3 cm^3/mole	T = 25°C	Apple juice	BHIA; Pulse	Ramaswamy et al. (2003)
140 MPa	40.75 cm^3/mole	T = 25°C	Apple juice	VRBA; Pulse	Ramaswamy et al. (2003)
		T = 15°C	Skim milk		Martin-Belloso et al. (1997)
			Milk		Grahl et al. (1992)
		T = 15°C	Skim milk		Zhang et al. (1994)
		T = 30°C	Skim milk		Qin et al. (1995)
41 kV.cm		T = 20°C	Phosphate buffer		Hulsherger et al. (1980)
		T = 15–40°C	20% Carrot juice		Rodrigo et al. (2003)
		T = 15–40°C	20% Carrot juice		Rodrigo et al. (2003)
		T = 15–40°C	20% Carrot juice		Rodrigo et al. (2003)
		T = 15–40°C	20% Carrot juice		Rodrigo et al. (2003)
			Skim/whole milk		ICMSF (1996)
			Ground beef		Line et al. (1991)
3.6–3.79°C	553.5–582.7 kJ/mole	4.8–19.1% Fat	Ground beef		Smith et al. (2001)
4.35–4.78°C	444.2–488.1 kJ/mole	7–20% Fat	Beef		Ahmed et al. (1995)
7.6°C	279.4 kJ/mole	7% Fat; 0–4.5% Nal	Ground beef		Hunag & Jubeja (2003)
4.59–4.47°C	462.6–475 kJ/mole	3–11% Fat	Ground turkey		Kotrola & Conner (1997)
6.01°C	353.3 kJ/mole	11% Fat; 4% NaCl	Ground turkey		Kotrola & Conner (1997)
6.79°C	312.7 kJ/mole		Ground chicken		Juneja et al. (1997)
4.94°C	429.8 kJ/mole		Ground pork		Murphy et al. (2004)
5.4°C	393.2 kJ/mole		Breaded pork patties		Osaili et al. (2007)
5.2°C	408.3 kJ/mole		Chick-fried beef patties		Osaili et al. (2007)
9.2–9.9°C	214.5–230.8 kJ/mole	Stationary/ osmotic	White grape juice conc.		Enache et al. (2006)
6–26.5°C	801–353.9 kJ/mole	pH, MA, SO, BE	Apple cider		Steenstrup & Floros (2002)
		T = 50°C	Milk		Patterson & Kilpatrick (1998)
		T = 50°C			
	10.33–24.91 kJ/mole	T = 20°C	Liquid egg whites		Amiale et al. (2006)

Table A.3.3 Kinetic parameters for *Listeria spp.* survivor curves

Preservation technology	Microbial population	Time parameters D	k	n	Treatment intensity
Thermal	L. monocytogenes	0.22–0.58 min	3.97–10.47/min		63.3°C
	L. monocytogenes	1.6–16.7 min	0.14–1.44/min		60°C
	L. monocytogenes	3.8 min	0.61/min		60°C
	L. monocytogenes	6.27–8.32 min	0.28–0.37/min		60°C
	L. monocytogenes	4.18 min	0.55/min		60°C
	L. monocytogenes	4.67 min	0.49/min		60°C
	L. monocytogenes	14.34 min	0.16/min		60°C
	L. monocytogenes	3.12 min	0.74/min		60°C
	L. monocytogenes	0.15–0.31 min	7.43–15.34/min		60°C
	L. monocytogenes	5.61 min	0.41/min		60°C
	L. monocytogenes	20.61 min	0.11/min		60°C
	L. monocytogenes	22.98 min	0.1/min		60°C
	L. monocytogenes	1.46–1.91 min	1.21–1.58/min		60°C
	L. monocytogenes	2.04 min	1.13/min		60°C
	L. monocytogenes	16.7 min	0.14/min		60°C
	L. monocytogenes	8.49 min	0.27/min		60°C
Pressure	L. monocytogenes	3 min	0.768/min		375 MPa
	L. monocytogenes	2.17 min	1.061/min		414 MPa
	L. monocytogenes	1.89–4.17 min	0.552–1.219/min		414 MPa
	L. monocytogenes	0.37–0.63 min	3.656–6.224/min		414 MPa
PEF	L. monocytogenes	150–200 microsec	0.012–0.015/ microsec		30 kV/cm
	L. monocytogenes	540 microsec	0.0043/microsec		20 kV/cm
Thermal	L. innocus	7.36 min	0.31/min		60°C
	L. innocus	25.13 min	0.09/min		60°C
	L. innocus	20.84 min	0.11/min		60°C
	L. innocus	3.25 min	0.71/min		60°C
Pressure	L. innocus	3 min	0.768/min		450 MPa
PEF	L. innocus	76.9 microsec	0.03/microsec		50 kV/cm
	L. innocus	26.7 microsec	0.086/microsec		50 kV/cm
	L. innocus	18.8 microsec	0.1225/microsec		50 kV/cm
Thermal	S. aureus	2.5 min	0.921/min		60°C
	S. aureus	0.9 min	2.56 min		60°C
	S. aureus	3.0 min	0.768/min		60°C
	S. aureus	40.0 min	0.0576/min		60°C
Pressure	S. aureus	2.5 min	0.921/min		500 MPa
	S. aureus	3.0 min	0.768/min		700 MPa
	S. aureus	3.0 min	0.768/min		500 MPa
PEF	S. aureus	360 microsec	0.0064/microsec		20 kV/cm
	S. aureus	4000–6000 microsec	0.00038–0.000576/ microsec		16 kV/cm

Intensity coefficient				
z (T, P,)	**E_A or ΔV**	**Substrates**	**Other information**	**Reference**
5.5°C	386 kJ/mole	Milk		ICMSF (1996)
		Meat products		ICMSF (1996)
7.2°C	294.9 kJ/mole	Beef		Mackey et al. (1995)
5.98°C	355.1 kJ/mole	Beef	<7% Fat	Gaze et al. (1989)
6.1°C	348.1 kJ/mole	Ground beef	25% fat	Juneja (2003)
		Beef		Juneja (2003)
		Beef	4.8% Nal	Juneja (2003)
4.92°C	431.6 kJ/mole	Ground meat		Farber (1989)
4.2–4.9°C	433.3–505.6 kJ/mole	Minced beef		Bolton et al. (2000)
5.92°C	358.7 kJ/mole	Ground pork		Murphy et al. (2004)
5.9°C	359.9 kJ/mole	Breaded pork patties		Osaili et al. (2007)
6.1°C	348.1 kJ/mole	Chick-fried beef patties		Osaili et al. (2006)
		White grape juice conc.		Enache et al. (2006)
5.08°C	418 kJ/mole	Poultry meat		Murphy et al. (2004)
5.29°C	401.4 kJ/mole	Turkey breast		Murphy et al. (2003)
5.71°C	371.9 kJ/mole	Chicken breast		Murphy et al. (2003)
		Milk		Patterson et al. (1995)
		Meat	T = 25°C	Ananth et al. (1998)
		Ground pork	T = 25°C	Murano et al. (1999)
		Ground pork	T = 50°C	Murano et al. (1999)
		Milk	T = 10–50°C	Reina et al. (1998)
		Buffer		Hulsherger et al. (1983)
5.66°C	375.1 kJ/mole	Chicken breast		Murphy et al. (2003)
4.9°C	433.3 kJ/mole	Turkey breast		Murphy et al. (2003)
4.86°C	436.9 kJ/mole	Chicken pattie		Murphy et al. (2002)
8.67°C	244.9 kJ/mole	Beef pattie		Murphy et al. (2002)
		Eggs	T = 20°C	Ponce et al. (1998)
		Skim milk	T = 15–28°C	Fernandez et al. (1999)
		Skim milk	T = 22–34°C	Calderon-Miranda (1998b)
		Liquis whole egg	T = 26–36°C	Calderon-Miranda (1998a)
		Phosphate buffer	pH = 6.5	ICMSF (1996)
9.5°C	224 kJ/mole	Milk		ICMSF (1996)
		Pasta	aw = 0.92	ICMSF (1996)
		Pasta	aw = 0.8	ICMSF (1996)
		Milk	T = 50°C	Patterson & Kilpatrick (1998)
		Buffer		Patterson et al. (1995)
		Meat	T = 50°C	Patterson & Kilpatrick (1998)
	2.6 kV/cm	Phosphate buffer		Hulsherger et al. (1983)
		SMUF	T = 37°C	Hulsherger et al. (1983)

Table A.3.4 Kinetic parameters for *clostridium spp.* survivor curves

Preservation technology	Microbial population	Time parameters D	k	n	Treatment intensity
Thermal	Cl. botulinum 62A	0.61–2.48 min	0.929–3.775/min		110°C
	Cl. botulinum 62A	0.88–1.9 min	1.212–2.617/min		110°C
	Cl. botulinum 62A	1.79 min	1.287/min		110°C
	Cl. botulinum B	0.49–12.42 min	0.186–4.7 min		110°C
	Cl. botulinum B	1.19–2 min	1.152–1.935/min		110°C
	Cl. botulinum E	6.8–13 min	0.1770–0.339/min		75°C
	Cl. botulinum E	72–100 min	0.023–0.32/min		70°C
Pressure	Cl. botulinum 62A	10.59 min	0.218/min		689 MPa
	Cl. botulinum E	8.77 min	0.263/min		758 MPa
	Cl. botulinum E	3.88 in	0.681/min		758 MPa
	Cl. botulinum E	1.76 min	1.309/min		827 MPa
Thermal	Cl. perfringens	3.15 min	0.731/min		104.1°C
	Cl. perfringens	6.6 min	0.349/min		104.1°C
Thermal	Cl. sporogenes	12.73 min	0.08/min	0.99	105°C
	Cl. sporogenes	0.6 min	2.09/min	0.83	121°C
Pressure	Cl. sporogenes	6.756 min	0.341/min		700 MPa
	Cl. sporogenes	3.186 min	0.723/min		700 MPa
	Cl. sporogenes	0.901 min	2.556/min		700 MPa
	Cl. sporogenes	0.19 min	4.69/min	0.26	700 MPa
	Cl. sporogenes	0.1 min	8.16/min	0.48	700 MPa
Thermal	B. subtilis	32.8 min	0.0702/min		88°C
PEF	B. subtilis	17.5–26.3 microsec	8.76–13.16/microsec		50 kV/cm
	B. subtilis	11.3 microsec	20.38/microsec		33 kV/cm
Thermal	B. cereas	1.5–36.2 min	0.064–1.535/min		95°C
PEF	B. cereas	50–60 microsec	3.84–4.61/microsec		50 kV/cm

Intensity coefficient				
z (T, P,)	E_A or ΔV	Substrates	Other information	Reference
11.6°C	242.1 kJ/mole	Vegetable products		ICMSF (1996)
7.6–10°C	280.9–369.6 kJ/mole	Phosphate buffer		ICMSF (1996)
8.5°C	330.5 kJ/mole	Distilled water		ICMSF (1996)
7.4°C	380 kJ/mole	Vegetable products		ICMSF (1996)
7.7–11.3°C	248.6–364.8 kJ/mole	Phosphate buffer		ICMSF (1996)
9.78°C	237.1 kJ/mole	Seafood		ICMSF (1996)
6.8–7.5°C	300.4–331.3 kJ/mole	Oyster homogenate		ICMSF (1996)
1524 MPa	4.4 cm^3/mole	Buffer	T = 75°C	Reddy et al. (1999)
		Buffer	T = 35°C	Reddy et al. (1999)
		Crab meat	T = 35°C	Reddy et al. (1999)
		Crab meat	T = 35°C	Reddy et al. (1999)
		Phosphate buffer	pH = 7.0	ICMSF (1996)
		Beef gravy	pH = 7.0	ICMSF (1996)
		Water		Ahn et al. (2007)
		Water		Ahn et al. (2007)
725 MPa	9.7 cm^3/mole		T = 93°C	Rovere et al. (1996)
962 MPa	7.42 cm^3/mole		T = 100°C	Rovere et al. (1996)
752 MPa	9.7 cm^3/mole		T = 108°C	Rovere et al. (1996)
		Water	T = 105°C	Ahn et al. (2007)
		Water	T = 121°C	Ahn et al. (2007)
8.74°C	285.5 kJ/mole	0.1% NaCl		Cho et al. (1999)
		0.15% NaCl	T = 25°C	Marquez et al. (1997)
		Pea soup	T < 5.5°C	Vega-Mercado et al. (1996)
6.7–10.1°C	256.7–387 kJ/mole	Various		ICMSF (1996)
		0.15% NaCl	T = 25°C	Marquez et al. (1997)

Table A.4.2 Kinetic parameters for retention of heat-sensitive vitamins in foods

Preservation technology	Quality attribute	Time parameters k	Treatment intensity	Intensity coefficient E_A (kJ/mole)
Thermal	Ascorbic Acid	0.0967/min	150°C	117.57
Pressure	Ascorbic Acid	0.010289/min	850 Mpa	84.10
Thermal	Ascorbic Acid	0.002642/min	96°C	20.84
	Ascorbic Acid	0.01068/min	96°C	47.28
Thermal	Ascorbic Acid	0.00900/min	132.2°C	164.43
Thermal	Ascorbic Acid	0.0487/min	150°C	105.44
	Ascorbic Acid	0.0864/min	150°C	115.06
Pressure	Ascorbic Acid	0.005744/min	850 Mpa	74.56
Thermal	5-Methyl	0.06831/min	90°C	80.00
Pressure	5-Methyl	0.02508/min	100 Mpa	79.00
	5-Methyl	0.15037/min	800 Mpa	24.22
Thermal	Folic Acid	0.0137/min	140°C	80.29
	Folic Acid	0.01205/min	140°C	83.26
	Folic Acid	0.00345/min	140°C	74.48
	Folic Acid	0.0013/min	140°C	51.67
	5-Methyl	0.249/min	70°C	33.05
	5-Methyl	1.065/min	130°C	39.50
	5-Methyl	3.18/min	150°C	76.57
Thermal	Thiamine	0.1669/min	149°C	118.41
	Thiamine	0.1744/min	149°C	119.66
	Thiamine	0.0351/min	132°C	85.77
	Thiamine	0.1659/min	149°C	117.57
	Thiamine	0.0435/min	138°C	97.07
	Thiamine	0.03897/min	138°C	113.39
	Thiamine	0.228/min	149°C	117.99
	Thiamine	0.2171/min	149°C	116.73
	Thiamine	0.2326/min	149°C	119.24
	Thiamine	0.03673/min	138°C	115.06
	Thiamine	0.1935/min	149°C	115.90
	Thiamine	0.1693/min	149°C	114.64
	Thiamine	0.084/min	138°C	114.64
	Thiamine	0.002511/min	98°C	113.39
Thermal	Riboflavin	0.0266/min	133°C	86.60
	Riboflavin	0.01145/min	138°C	57.74
Thermal	Vitamin B6	0.4895/min	200°C	125.52
	Vitamin B6	0.0083/min	133°C	119.66
	Vitamin B6	0.0187/min	133°C	99.58
	Vitamin B6	0.0266/min	133°C	86.60
	Vitamin B6	0.01145/min	137.7°C	113.80
Thermal	Vitamin B12	0.01828/min	100°C	
Thermal	Vitamin A	0.09738/min	126.7°C	112.55
	Vitamin A	0.0011/min	80°C	49.79
	Vitamin A	0.00135/min	80°C	85.77
	Vitamin A	0.00116/min	80°C	43.93

Related information	Substrates	Other information	References
	Orange juice	120–150°C	Van den Broeck et al. (1998)
80°C	Orange juice	65–80°C	Van den Broeck et al. (1998)
11.2 Bx	Grapefruit juice	61–96°C	Saguy et al. (1978b)
62.5 Bx	Grapefruit juice	68–96°C	Saguy et al. (1978b)
	Canned peas	110–132.2°C	Lathrop & Leung (1980)
	Tomato juice	120–150°C	Van den Broeck et al. (1998)
	Tomato juice	120–150°C	Van den Broeck et al. (1998)
80°C	Tomato juice	65–80°C	Van den Broeck et al. (1998)
pH = 7.0	Phospate buffer	65–90°C	Nguyen et al. (2003)
65°C	Phospate buffer	50–65°C	Nguyen et al. (2003)
60°C	Phospate buffer	400–800 Mpa	Nguyen et al. (2003)
pH = 3.4	Apple juice	100–140°C	Mnkemi & Beveridge (1982)
pH = 4.3	Tomato juice	100–140°C	Mnkemi & Beveridge (1982)
pH = 5.0	Citrate buffer	100–140°C	Mnkemi & Beveridge (1982)
pH = 7.0	Phospate buffer	120–160°C	Nguyen et al. (2003)
pH = 3.4	Apple juice	50–70°C	Mnkemi & Beveridge (1982)
pH = 4.3	Tomato juice	100–130°C	Mnkemi & Beveridge (1982)
pH = 7.0	Phospate buffer	110–150°C	Viberg et al. (1997)
	Carrots	109–149°C	Feliciotti & Esselen (1957)
	Green beans	109–149°C	Feliciotti & Esselen (1957)
	Peas	104–132°C	Bendix et al. (1951)
	Peas	109–149°C	Feliciotti & Esselen (1957)
	Peas	115–138°C	Lenz & Lund (1977)
	Peas	121–138°C	Mulley et al. (1975)
	Spinach	109–149°C	Feliciotti & Esselen (1957)
	Beef heart	109–149°C	Feliciotti & Esselen (1957)
	Beef liver	109–149°C	Feliciotti & Esselen (1957)
	Beef	121–138°C	Mulley et al. (1975)
	Lamb	109–149°C	Feliciotti & Esselen (1957)
	Pork	109–149°C	Feliciotti & Esselen (1957)
	Pork	115–138°C	Lenz & Lund (1977)
	Meat loaf	70–98°C	Skjoldebrand et al. (1983)
	Casein liquid	105–133°C	Gregory & Hiner (1983)
	Cauliflower	106–138°C	Navankasattusus & Lund (1982)
Pyridoxine	Breakfast cereal	155–200°C	Evans et al. (1981)
Pyridoxine	Casein liquid	105–133°C	Gregory & Hines (1983)
Pyridoximine	Casein liquid	105–133°C	Gregory & Hines (1983)
Pyridoxal	Casein liquid	105–133°C	Gregory & Hines (1983)
Total B6	Cauliflower	105.9–137.7°C	Navankasattusas & Lind (1982)
	Milk		Watanabe et al. (1998)
	Beef liver	102.9–126.7°C	Wilkinson et al. (1982)
	Squash	60–80°C	Stefanovich & Karel (1982)
	Yellow corn	60–80°C	Stefanovich & Karel (1982)
	Sweet potato	60–80°C	Stefanovich & Karel (1982)

Table A.4.3 Kinetic parameters for retention of heat-sensitive vitamins in foods

Preservation technology	Quality attribute	Time parameters k	Treatment intensity	Intensity coefficient E_A (kJ/mole)
Thermal	Chlorophyll	0.01575/min	80°C	58.15
	Chlorophyll	0.0943/min	120°C	69.03
Pressure	Chlorophyll	0.02892/min	100 Mpa	76.27
	Chlorophyll	0.10239/min	850 Mpa	
Thermal	Chlorophyll	0.013/min	98°C	53.97
	Chlorophyll	0.12/min	115.6°C	51.46
	Chlorophyll	0.07599/min	100°C	38.49
	Chlorophyll	0.0032/min	90°C	43.93
	Chlorophyll	0.154/min	120°C	84.93
	Chlorophyll	0.0184/min	90°C	76.14
	Chlorophyll a	0.2666/min	126°C	114.22
	Chlorophyll a	0.1195/min	126°C	103.34
Thermal	Anthocyanins	0.00056661/min	80°C	68.49
	Anthocyanins	0.0009532/min	80°C	75.85
	Anthocyanins	0.0016192/min	80°C	80.08
	Anthocyanins	0.01925/min	108°C	92.88
	Anthocyanins	0.02666/min	108°C	105.01
	Anthocyanins	0.01879/min	121°C	54.81
	Anthocyanins	0.0018/min	92°C	104.6
	Anthocyanins	0.0204/min	108°C	88.70
Thermal	Betalains	0.01419/min	116°C	36.40
	Betalains	0.113/min	100°C	76.14
	Betalains	0.1177/min	100°C	83.26
Thermal	Carotenoids	3.47/min	100°C	94.14
	Carotenoids	0.024105/min	130°C	88.28
	Carotenoids	0.025783/min	150°C	29.28
Thermal	Browning	0.017/min	130°C	100.41
	Browning	0.01/min	130°C	82.84
	Browning	0.02647/min	190°C	89.11
		0.04516/min	190°C	97.90
		0.02414/min	190°C	93.30
		0.01476/min	190°C	93.72
	Browning	0.004451/min	95°C	34.26
		0.1643/min	96°C	127.19
	Browning	0.0245/min	95°C	63.17
		0.1481/min	96°C	99.57
	Browning	0.0029/min	122.5°C	107.52
	Browning	0.0301/min	98°C	102.08

Related information	Substrates	Other information	References
	Broccoli juice	70–90°C	Weemaes et al. (1999b)
	Broccoli juice	80–120°C	Weemaes et al. (1999b)
80°C	Broccoli juice	70–80°C	Weemaes et al. (1999a)
80°C	Broccoli juice	50–850 Mpa	Weemaes et al. (1999a)
	Asparagus	70–98°C	Lao et al. (2000)
Steam blanch	Brussel sprouts	87.8–115.6°C	Dietrich & Neumann (1965)
Steam blanch	Green beans	87.8–100°C	Dietrich et al. (1959)
Hunter-L	Olives	70–90°C	Sanchez et al. (1991)
Total Chlor	Peas	100–120°C	Steet & Tong (1996b)
Total Chlor	Peas	70–90°C	Steet & Tong (1996a)
	Peas	116–126°C	Schwartz & von Elbe (1983)
	Peas	116–126°C	Schwartz & von Elbe (1983)
15 Br	Cherry juice	50–80°C	Cemeroglu et al. (1994)
45 Br	Cherry juice	50–80°C	Cemeroglu et al. (1994)
71 Br	Cherry juice	50–80°C	Cemeroglu et al. (1994)
	Fruit juice	78–108°C	Tanchev (1983)
	Citrus buffer	78–108°C	Tanchev (1983)
pH = 3.4	Grape buffer	76.7–121°C	Sastry & Tischer (1952)
	Pomegranate juice	70–92°C	Mishkon & Saguy (1982)
Bulgarian Ruby	Raspberry juice	78–108°C	Tanchev (1972)
	Beet puree	102–116°C	von Elbe et al. (1974)
pH = 4.8	Beet juice	61.5–100°C	Saguy (1979)
pH = 6.2	Beet juice	61.5–100°C	Saguy (1979)
	Blue crab	76.6–100°C	Himelbloom et al. (1983)
Lycopene	Tomato juice	90–130°C	Miki & Akstsu (1070)
Lycopene	Tomato puree	90–150°C	Shi et al. (2003)
Maillard	Apple juice	40–130°C	Herrmann (1970)
Nonenzymatic	Apple juice	40–130°C	Herrmann (1970)
6% MC	Flour dough	150–190°C	Herrmann & Nour (1977)
14% MC	Flour dough	150–190°C	Herrmann & Nour (1977)
22% MC	Flour dough	150–190°C	Herrmann & Nour (1977)
30% MC	Flour dough	150–190°C	Herrmann & Nour (1977)
11.2 Br-Lag	Grapefruit juice	61–95°C	Saguy et al. (1978a)
62.5 Br-Lag	Grapefruit juice	68–96°C	Saguy et al. (1978a)
11.2 Br-Postlag	Grapefruit juice	61–95°C	Saguy et al. (1978a)
62.5 Br-Postlag	Grapefruit juice	68–96°C	Saguy et al. (1978a)
MinoltaL	Peaches	110–135°C	Avila & Silva (1981)
Macbeth-a	Pear	80–98°C	Ibarz et al. (1999)

TABLE A.6.1 f_h/U:g relationships when z = 12 F

f_h/U	Values of g when J of cooling curve is 0.40	0.60	0.80	1.00	1.20	1.40	1.60	1.80	2.00
0.50	0.024	0.026	0.028	0.030	0.032	0.034	0.036	0.038	0.040
0.60	0.048	0.054	0.060	0.066	0.072	0.078	0.084	0.090	0.096
0.70	0.88	0.99	0.110	0.121	0.132	0.143	0.154	0.165	0.174
0.80	0.153	0.163	0.172	0.182	0.192	0.201	0.211	0.220	0.230
0.90	0.202	0.218	0.234	0.250	0.266	0.282	0.298	0.314	0.330
1.00	0.260	0.280	0.300	0.320	0.340	0.360	0.380	0.400	0.420
1.50	0.498	0.574	0.649	0.724	0.799	0.874	0.950	1.02	1.10
2.00	0.770	0.890	1.01	1.13	1.25	1.37	1.49	1.61	1.73
3.00	1.24	1.45	1.66	1.87	2.08	2.29	2.50	2.71	2.92
4.00	1.59	1.88	2.17	2.46	2.75	3.04	3.33	3.62	3.91
5.00	1.96	2.30	2.64	2.98	3.32	3.66	4.00	4.34	4.68
6.00	2.36	2.73	3.10	3.47	3.84	4.21	4.58	4.95	5.32
7.00	2.67	3.08	3.49	3.90	4.31	4.72	5.13	5.54	5.95
8.00	2.98	3.42	3.86	4.30	4.74	5.18	5.62	6.06	6.50
9.00	3.32	3.77	4.22	4.67	5.12	5.57	6.02	6.47	6.92
10.0	3.59	4.06	4.53	5.00	5.47	5.94	6.41	6.88	7.35
12.5	4.35	4.84	5.33	5.82	6.31	6.80	7.29	7.78	8.27
15.0	4.90	5.43	5.96	6.49	7.02	7.55	8.08	8.61	9.14
20.0	5.93	6.49	7.05	7.61	8.17	8.73	9.29	9.85	10.4
25.0	6.70	7.30	7.90	8.50	9.10	9.70	10.3	10.9	11.5
30.0	7.39	7.99	8.59	9.19	9.80	10.4	11.0	11.6	12.2
40.0	8.50	9.10	9.70	10.3	10.9	11.5	12.1	12.7	13.3
50.0	9.50	10.1	10.7	11.3	11.9	12.5	13.1	13.7	14.3
60.0	9.90	10.6	11.3	12.0	12.7	13.4	14.1	14.8	15.5
70.0	10.4	11.1	11.8	12.5	13.2	13.9	14.6	15.3	16.0
80.0	11.0	11.7	12.4	13.1	13.8	14.5	15.2	15.9	16.6
100.0	11.5	12.3	13.1	13.9	14.7	15.5	16.3	17.1	17.9
150.0	13.2	14.0	14.8	15.6	16.4	17.2	18.0	18.8	19.6
200.0	14.1	15.0	15.9	16.8	17.7	18.6	19.5	20.4	21.3
250.0	15.1	16.0	16.9	17.8	18.7	19.6	20.5	21.4	22.3
300.0	15.8	16.7	17.6	18.5	19.4	20.3	21.2	22.1	23.0
350.0	16.6	17.5	18.4	19.3	20.2	21.1	22.0	22.9	23.8
400.0	17.2	18.1	19.0	19.9	20.8	21.7	22.6	23.5	24.4
500.0	18.4	19.3	20.2	21.1	22.0	22.9	23.8	24.7	25.6
750.0	20.3	21.2	22.1	23.0	23.9	24.8	25.7	26.6	27.5
1000.0	21.3	22.3	23.3	24.3	25.3	26.3	27.3	28.3	29.3

Source: From Stumbo (1965).

TABLE A.6.2 f_h/U:g relationships when z = 14 F

f_h/U	Values of g when J of cooling curve is								
	0.40	0.60	0.80	1.00	1.20	1.40	1.60	1.80	2.00
0.50	0.018	0.022	0.026	0.030	0.034	0.038	0.042	0.046	0.050
0.60	0.054	0.061	0.068	0.075	0.082	0.089	0.096	0.103	0.110
0.70	0.088	0.102	0.116	0.130	0.144	0.158	0.192	0.186	0.200
0.80	0.162	0.178	0.194	0.210	0.226	0.242	0.258	0.274	0.290
0.90	0.196	0.224	0.252	0.280	0.308	0.336	0.364	0.392	0.420
1.00	0.268	0.302	0.336	0.370	0.404	0.438	0.472	0.506	0.540
1.50	0.544	0.646	0.748	0.850	0.952	1.05	1.15	1.26	1.36
2.00	0.910	1.06	1.21	1.36	1.51	1.66	1.81	1.96	2.11
3.00	1.53	1.78	2.03	2.28	2.53	2.78	3.03	3.28	3.53
4.00	2.08	2.40	2.72	3.04	3.36	3.68	4.00	4.32	4.64
5.00	2.55	2.93	3.31	3.69	4.07	4.45	4.83	5.21	5.59
6.00	3.00	3.42	3.84	4.26	4.68	5.10	5.52	5.94	6.36
7.00	3.39	3.85	4.31	4.77	5.23	5.69	6.15	6.61	7.07
8.00	3.74	4.24	4.74	5.24	5.74	6.24	6.74	7.24	7.74
9.00	4.15	4.66	5.17	5.68	6.19	6.70	7.21	7.72	8.23
10.0	4.40	4.95	5.50	6.05	6.60	7.15	7.70	8.25	8.80
12.5	5.29	5.86	6.43	7.00	7.57	8.14	8.71	9.28	9.85
15.0	5.92	6.53	7.14	7.75	8.36	8.97	9.58	10.2	10.8
20.0	7.04	7.70	8.36	9.02	9.68	10.3	11.0	11.7	12.3
25.0	8.00	8.70	9.40	10.1	10.8	11.5	12.2	12.9	13.6
30.0	8.80	9.50	10.2	10.9	11.6	12.3	13.0	13.7	14.4
40.0	9.80	10.6	11.4	12.2	13.0	13.8	14.6	15.4	16.2
50.0	10.9	11.7	12.5	13.3	14.1	14.9	15.7	16.5	17.3
60.0	11.7	12.5	13.3	14.1	14.9	15.7	16.5	17.3	18.1
70.0	12.1	13.0	13.9	14.8	15.7	16.6	17.5	18.4	19.3
80.0	13.1	13.9	14.7	15.5	16.3	17.1	17.9	18.7	19.5
100.0	13.8	14.7	15.6	16.5	17.4	18.3	19.2	20.1	21.0
150.0	15.5	16.5	17.5	18.5	19.5	20.5	21.5	22.5	23.5
200.0	16.6	17.7	18.8	19.9	21.0	22.1	23.2	24.3	25.4
250.0	17.8	18.9	20.0	21.1	22.2	23.3	24.4	25.5	26.6
300.0	18.6	19.7	20.8	21.9	23.0	24.1	25.2	26.3	27.4
350.0	19.5	20.6	21.7	22.8	23.9	25.0	26.1	27.2	28.3
400.0	20.2	21.3	22.4	23.5	24.6	25.7	26.8	27.9	29.0
500.0	21.5	22.6	23.7	24.8	26.9	27.0	28.1	29.2	30.3
750.0	23.9	25.0	26.1	27.2	28.3	29.4	30.5	31.6	32.7
1000.0	25.0	26.2	27.4	28.6	29.8	31.0	32.2	33.4	34.6

Source: From Stumbo (1965).

TABLE A.6.3 f_h/U:g relationships when z = 16 F

f_h/U	Values of g when J of cooling curve is								
	0.40	0.60	0.80	1.00	1.20	1.40	1.60	1.80	2.00
0.50	0.028	0.032	0.036	0.040	0.044	0.048	0.052	0.056	0.060
0.60	0.053	0.062	0.071	0.080	0.089	0.098	0.107	0.116	0.126
0.70	0.102	0.118	0.134	0.150	0.166	0.182	0.198	0.214	0.230
0.80	0.144	0.171	0.198	0.225	0.252	0.279	0.306	0.333	0.360
0.90	0.220	0.255	0.290	0.325	0.360	0.395	0.430	0.465	0.500
1.00	0.270	0.320	0.370	0.420	0.470	0.520	0.570	0.620	0.670
1.50	0.611	0.737	0.863	0.989	1.12	1.24	1.37	1.49	1.62
2.00	1.06	1.24	1.42	1.60	1.78	1.96	2.14	2.32	2.50
3.00	1.89	2.16	2.43	2.70	2.97	3.24	3.51	3.78	4.05
4.00	2.58	2.93	3.28	3.63	3.98	4.33	4.68	5.03	5.38
5.00	3.14	3.56	3.98	4.40	4.82	5.24	5.66	6.08	6.50
6.00	3.65	4.12	4.59	5.06	5.53	6.00	6.47	6.94	7.41
7.00	4.09	4.61	5.13	5.65	6.17	6.69	7.21	7.73	8.25
8.00	4.50	5.06	5.62	6.18	6.74	7.30	7.86	8.42	8.98
9.00	4.94	5.52	6.10	6.68	7.26	7.84	8.42	9.00	9.58
10.0	5.29	5.90	6.51	7.12	7.73	8.34	8.95	9.56	10.2
12.5	6.22	6.87	7.52	8.17	8.82	9.47	10.1	10.8	11.4
15.0	6.90	7.60	8.30	9.00	9.70	10.4	11.1	11.8	12.5
20.0	8.40	9.10	9.80	10.5	11.2	11.9	12.6	13.3	14.0
25.0	9.20	10.0	10.8	11.6	12.4	13.2	14.0	14.8	15.6
30.0	10.2	11.0	11.8	12.6	13.4	14.2	15.0	15.8	16.6
40.0	11.4	12.3	13.2	14.1	15.0	15.9	16.8	17.7	18.6
50.0	12.6	13.5	14.4	15.3	16.2	17.1	18.0	18.9	19.8
60.0	13.6	14.5	15.4	16.3	17.2	18.1	19.0	19.9	20.8
70.0	14.5	15.4	16.3	17.2	18.1	19.0	19.9	20.8	21.7
80.0	15.3	16.2	17.1	18.0	18.9	19.8	20.7	21.6	22.5
100.0	16.1	17.1	18.1	19.1	20.1	21.1	22.1	23.1	24.1
150.0	17.7	18.9	20.1	21.3	22.5	23.7	24.9	26.1	27.3
200.0	19.4	20.6	21.8	23.0	24.2	25.4	26.6	27.8	29.0
250.0	20.7	21.9	23.1	24.3	25.5	26.7	27.9	29.1	30.3
300.0	21.4	22.7	24.0	25.3	26.6	27.9	29.2	30.5	31.8
350.0	22.4	23.7	25.0	26.3	27.6	28.9	30.2	31.5	32.8
400.0	23.2	24.5	25.8	27.1	28.4	29.7	31.0	32.3	33.6
500.0	24.7	26.0	27.3	28.6	29.9	31.2	32.5	33.8	35.1
750.0	27.1	28.5	29.9	31.3	32.7	34.1	35.5	36.9	38.3
1000.0	28.4	29.9	31.4	32.9	34.4	35.9	37.4	38.9	40.4

Source: From Stumbo (1965).

TABLE A.6.4 f_h/U:g relationships when z = 18 F

	Values of g when J of cooling curve is								
f_h/U	0.40	0.60	0.80	1.00	1.20	1.40	1.60	1.80	2.00
0.114	10^{-8}	10^{-8}	10^{-8}	10^{-8}	10^{-8}	10^{-8}	10^{-8}	10^{-8}	10^{-8}
0.130	10^{-7}	10^{-7}	10^{-7}	10^{-7}	10^{-7}	10^{-7}	10^{-7}	10^{-7}	10^{-7}
0.149	10^{-6}	10^{-6}	10^{-6}	10^{-6}	10^{-6}	10^{-6}	10^{-6}	10^{-6}	10^{-6}
0.176	10^{-5}	10^{-5}	10^{-5}	10^{-5}	10^{-5}	10^{-5}	10^{-5}	10^{-5}	10^{-5}
0.213	10^{-4}	10^{-4}	10^{-4}	10^{-4}	10^{-4}	10^{-4}	10^{-4}	10^{-4}	10^{-4}
0.271	10^{-3}	10^{-3}	10^{-3}	10^{-3}	10^{-3}	10^{-3}	10^{-3}	10^{-3}	10^{-3}
0.368	10^{-2}	10^{-2}	10^{-2}	10^{-2}	10^{-2}	10^{-2}	10^{-2}	10^{-2}	10^{-2}
0.50	0.035	0.040	0.045	0.050	0.055	0.060	0.065	0.070	0.075
0.60	0.063	0.077	0.091	0.105	0.119	0.133	0.147	0.161	0.175
0.70	0.111	0.135	0.159	0.183	0.207	0.231	0.255	0.279	0.303
0.80	0.159	0.197	0.235	0.273	0.311	0.349	0.387	0.425	0.463
0.90	0.234	0.281	0.328	0.375	0.422	0.469	0.516	0.563	0.610
1.00	0.288	0.352	0.416	0.480	0.544	0.608	0.672	0.736	0.800
1.50	0.740	0.880	1.02	1.16	1.30	1.44	1.58	1.72	1.86
2.00	1.20	1.41	1.62	1.83	2.04	2.25	2.46	2.67	2.88
3.00	2.23	2.53	2.83	3.13	3.43	3.73	4.03	4.33	4.63
4.00	3.04	3.43	3.82	4.21	4.60	4.99	5.38	5.77	6.16
5.00	3.74	4.20	4.66	5.12	5.58	6 04	6.50	6.96	7.42
6.00	4.30	4.82	5.34	5.86	6.38	6.90	7.42	7.94	8.46
7.00	4.81	5.38	5.95	6.52	7.09	7.66	8.23	8.80	9.37
8.00	5.30	5.91	6.52	7.13	7.74	8.35	8.96	9.57	10.2
9.00	5.76	6.40	7.04	7.68	8.32	8.96	9.60	10.2	10.9
10.0	6.14	6.82	7.50	8.18	8.86	9.54	10.2	10.9	11.6
12.5	7.17	7.90	8.63	9.36	10.1	10.8	11.5	12.2	12.9
15.0	7.94	8.72	9.50	10.3	11.1	11.9	12.7	13.5	14.3
20.0	9.50	10.3	11.1	11.9	12.7	13.5	14.3	15.1	15.9
25.0	10.6	11.4	12.3	13.2	14.1	15.0	15.9	16.8	17.7
30.0	11.5	12.4	13.3	14.2	15.1	16.0	16.9	17.8	18.7
40.0	12.9	13.9	14.9	15.9	16.9	17.9	18.9	19.9	20.9
50.0	14.4	15.4	16.4	17.4	18.4	19 4	20.4	21.4	22.4
60.0	15.1	16.2	17.3	18.4	19.5	20.6	21.7	22.8	23.9
70.0	16.0	17.1	18.2	19.3	20.4	21.5	22.6	23.7	24.8
80.0	16.6	17.8	19.0	20.2	21.4	22.6	23.8	25.0	26.2
100.0	18.4	19.6	20.8	22.0	23.2	24.4	25.6	26.8	28.0
150.0	20.3	21.6	22.9	24.2	25.5	26.8	28.1	29.4	30.7
200.0	21.9	23.3	24.7	26.1	27.5	28.9	30.3	31.7	33.1
250.0	23.4	24.8	26.2	27.6	29.0	30.4	31.8	33.2	34.6
300.0	24.6	26.0	27.4	28.8	30.2	31.6	33.0	34.4	35.8
350.0	25.3	26.8	28.3	29.8	31.3	32.8	34.3	35.8	37.3
400.0	26.3	27.8	29.3	30.8	32.3	33.8	35.3	36.8	38.3
500.0	27.8	29.3	30.8	32.3	33.8	35.3	36.8	38.3	39.8
750.0	30.6	32.2	33.8	35.4	37.0	38.6	40.2	41.8	43.4
1000.0	32.1	33.8	35.5	37.2	38.9	40.6	42.3	44.0	45.7

Source: From Stumbo (1965).

TABLE A.6.5 f_h/U:g relationships when z = 20 F

	Values of g when J of cooling curve is								
f_h/U	0.40	0.60	0.80	1.00	1.20	1.40	1.60	1.80	2.00
0.50	0.032	0.038	0.044	0.050	0.056	0.062	0.068	0.074	0.080
0.60	0.069	0.082	0.094	0.107	0.120	0.132	0.145	0.157	0.170
0.70	0.108	0.132	0.156	0.180	0.204	0.228	0.252	0.276	0.300
0.80	0.155	0.195	0.235	0.275	0.315	0.355	0.395	0.435	0.475
0.90	0.210	0.270	0.330	0.390	0.450	0.510	0.570	0.630	0.690
1.00	0.296	0.374	0.452	0.530	0.608	0.686	0.764	0.842	0.920
1.50	0.738	0.912	1.09	1.26	1.43	1.61	1.78	1.96	2.13
2.00	1.37	1.60	1.83	2.06	2.29	2.52	2.75	2.98	3.21
3.00	2.55	2.88	3.21	3.54	3.87	4.20	4.53	4.86	5.19
4.00	3.54	3.96	4.38	4.80	5.22	5.64	6.06	6.48	6.90
5.00	4.33	4.83	5.33	5.83	6.33	6.83	7.33	7.83	8.33
6.00	4.96	5.53	6.10	6.67	7.24	7.81	8.38	8.95	9.52
7.00	5.45	6.10	6.75	7.40	8.05	8.70	9.35	10.0	10.7
8.00	6.02	6.70	7.38	8.06	8.74	9.42	10.1	10.8	11.5
9.00	6.60	7.30	8.00	8.70	9.40	10.1	10.8	11.5	12.2
10.0	7.00	7.75	8.50	9.25	10.0	10.8	11.5	12.3	13.0
12.5	8.00	8.90	9.70	10.5	11.3	12.1	12.9	13.7	14.5
15.0	8.80	9.70	10.6	11.5	12.4	13.3	14.2	15.1	16.0
20.0	10.6	11.5	12.4	13.3	14.2	15.1	16.0	16.9	17.8
25.0	11.8	12.8	13.8	14.8	15.8	16.8	17.8	18.8	19.8
30.0	12.9	13.9	14.9	15.9	16.9	17.9	18.9	19.9	20.9
40.0	14.5	15.6	16.7	17.8	18.9	20.0	21.1	22.2	23.3
50.0	16.1	17.2	18.3	19.4	20.5	21.6	22.7	23.8	24.9
60.0	17.0	18.2	19.4	20.6	21.8	23.0	24.2	25.4	26.6
70.0	18.2	19.4	20.6	21.8	23.0	24.2	25.4	26.6	27.8
80.0	19.2	20.4	21.6	22.8	24.0	25.2	26.4	27.6	28.8
100.0	20.5	21.8	23.1	24.4	25.7	27.0	28.3	29.6	30.9
150.0	22.9	24.3	25.7	27.1	28.5	29.9	31.3	32.7	34.1
200.0	24.7	26.2	27.7	29.2	30.7	32.2	33.7	35.2	36.7
250.0	26.1	27.7	29.3	30.9	32.5	34.1	35.7	37.3	38.9
300.0	27.4	29.0	30.6	32.2	33.8	35.4	37.0	38.6	40.2
350.0	28.6	30.2	31.8	33.4	35.0	36.6	38.2	39.8	41.4
400.0	29.3	31.0	32.7	34.4	36.1	37.8	39.5	41.2	42.9
500.0	30.7	32.5	34.3	36.1	37.9	39.7	41.5	43.3	45.1
750.0	34.2	36.0	37.8	39.6	41.4	43.2	45.0	46.8	48.6
1000.0	35.5	37.5	39.5	41.5	43.5	45.5	47.5	49.5	51.5

Source: From Stumbo (1965).

TABLE A.6.6 f_h/U:g relationships when z = 22 F

f_h/U	Values of g when the J of cooling curve is 0.40	0.60	0.80	1.00	1.20	1.40	1.60	1.80	2.00
0.50	0.042	0.048	0.054	0.060	0.066	0.072	0.078	0.084	0.090
0.60	0.084	0.097	0.109	0.122	0.135	0.147	0.160	0.172	0.185
0.70	0.125	0.151	0.176	0.202	0.228	0.253	0.279	0.304	0.330
0.80	0.160	0.208	0.257	0.306	0.355	0.404	0.452	0.501	0.550
0.90	0.229	0.297	0.364	0.432	0.500	0.567	0.635	0.702	0.770
1.00	0.270	0.370	0.470	0.570	0.670	0.770	0.870	0.970	1.07
1.50	0.790	0.990	1.19	1.39	1.59	1.79	1.99	2.19	2.39
2.00	1.51	1.77	2.03	2.29	2.55	2.81	3.07	3.33	3.59
3.00	2.87	3.23	3.59	3.95	4.31	4.67	5.03	5.39	5.75
4.00	4.04	4.49	4.94	5.39	5.84	6.29	6.74	7.19	7.64
5.00	5.07	5.56	6.05	6.54	7.03	7.52	8.01	8.50	8.99
6.00	5.62	6.24	6.86	7.48	8.10	8.72	9.34	9.96	10.6
7.00	6.24	6.92	7.60	8.28	8.96	9.64	10.3	11.0	11.7
8.00	6.72	7.48	8.24	9.00	9.76	10.5	11.3	12.0	12.8
9.00	7.43	8.19	8.95	9.71	10.5	11.2	12.0	12.8	13.5
10.0	7.85	8.67	9.49	10.3	11.1	11.9	12.7	13.5	14.3
12.5	9.08	9.96	10.8	11.7	12.6	13.5	14.4	15.3	16.2
15.0	10.1	11.0	11.9	12.8	13.7	14.6	15.5	16.4	17.3
20.0	11.7	12.7	13.7	14.7	15.7	16.7	17.7	18.7	19.7
25.0	12.8	14.0	15.2	16.4	17.6	18.8	20.0	21.2	22.4
30.0	15.0	15.2	16.4	17.6	18.8	20.0	21.2	22.4	23.6
40.0	16.1	17.3	18.5	19.7	20.9	22.1	23.3	24.5	25.7
50.0	17.6	18.9	20.2	21.5	22.8	24.1	25.4	26.7	28.0
60.0	18.9	20.2	21.5	22.8	24.1	25.4	26.7	28.0	29.3
70.0	20.2	21.5	22.8	24.1	25.4	26.7	28.0	29.3	30.6
80.0	21.4	22.7	24.0	25.3	26.6	27.9	29.2	30.5	31.8
100.0	22.8	24.2	25.6	27.0	28.4	29.8	31.2	32.6	34.0
150.0	25.2	26.8	28.4	30.0	31.6	33.2	34.8	36.4	38.0
200.0	27.5	29.1	30.7	32.3	33.9	35.5	37.1	38.7	40.3
250.0	29.0	30.7	32.4	34.1	35.8	37.5	39.2	40.9	42.6
300.0	30.3	32.1	33.9	35.7	37.5	39.3	41.1	42.9	44.7
350.0	31.5	33.3	35.1	36.9	38.7	40.5	42.3	44.1	45.9
400.0	32.4	34.3	36.2	38.1	40.0	41.9	43.8	45.7	47.6
500.0	33.9	35.9	37.9	39.9	41.9	43.9	45.9	47.9	49.9
750.0	37.4	39.5	41.6	43.7	45.8	47.9	50.0	52.1	54.2
1000.0	38.9	41.2	43.5	45.8	48.1	50.4	52.7	55.0	57.3

Source: From Stumbo (1965).

Table A.6.7 Analysis of quality retention as influenced by location within a cylindrical container

						Region 1						Region 2				
		Time (min)	t/f	log(T_M-T)		Temp. (°C)		k (1/min)	Concen.	log(T_M-T)		Temp. (°C)		k (1/min)	Consen.	log(T_M-T)
Medium Temp	125°C															
Initial Temp	70°C								100						100	
j - value	1.5942															
Operator time	136.5 min	40	0.381898	1.942905	1.561007	88.60787	2.050957	0.001286117	99.74311	1.914713	1.532815	90.89522	1.894149	0.001504	99.40002	1.836229
Heating rate	104.74 min	42	0.400993	1.942905	1.541913	90.17328	1.943428	0.001432122	99.45783		1.51372	92.36224	1.794611	0.001662	98.74143	
k - ref	0.01/min	44	0.420088	1.942905	1.522818	91.67135	1.841388	0.001585971	99.14285		1.494625	93.76616	1.700101	0.001827	98.0226	
Act Energy	75 kJ/mole	46	0.439183	1.942905	1.503723	93.10499	1.74452	0.001747289	98.79699		1.475531	95.10968	1.610331	0.001998	97.24224	
R - value	8.314	48	0.458278	1.942905	1.484628	94.47696	1.652526	0.001915654	98.4192		1.456436	96.39542	1.525034	0.002176	96.39946	
Ref T	121°C	50	0.477373	1.942905	1.465533	95.78991	1.56513	0.002090608	98.00854		1.437341	97.62584	1.44396	0.00236	95.49377	
Cmed Temp	35°C	52	0.496467	1.942905	1.446438	97.04638	1.482074	0.00227166	97.56427		1.418246	98.80334	1.366876	0.002549	94.52505	
j2-value	1.494	54	0.515562	1.942905	1.427343	98.24881	1.403118	0.002458293	97.08576		1.399151	99.9302	1.293564	0.002743	93.49363	
j3-value	1.247	56	0.534657	1.942905	1.408248	99.39951	1.328035	0.002649975	96.57258		1.380056	101.0086	1.223819	0.002941	92.40019	
j4-value	0.721	58	0.553752	1.942905	1.389153	100.5007	1.256615	0.002846158	96.02442		1.360961	102.0406	1.15745	0.003143	91.24586	
j5-value	0.2331	60	0.572847	1.942905	1.370058	101.5546	1.18866	0.003046291	95.44116		1.341866	103.0282	1.094276	0.003348	90.0321	
		62	0.591942	1.942905	1.350964	102.5631	1.123986	0.003249818	94.82284		1.322771	103.9733	1.03413	0.003555	88.76078	
		64	0.611037	1.942905	1.331869	103.5282	1.062418	0.003456191	94.16965		1.303676	104.8778	0.976853	0.003765	87.43408	
		66	0.630132	1.942905	1.312774	104.4518	1.003793	0.003664867	93.48193		1.284582	105.7433	0.922296	0.003976	86.05451	
		68	0.649227	1.942905	1.293679	105.3357	0.947958	0.003875315	92.76019		1.265487	106.5716	0.870318	0.004188	84.62487	
		70	0.668322	1.942905	1.274584	106.1815	0.894769	0.004087022	92.00505		1.246392	107.3643	0.820789	0.004401	83.14822	
		72	0.687416	1.942905	1.255489	106.991	0.844089	0.00429949	91.21729		1.227297	108.1229	0.773583	0.004614	81.62785	
		74	0.706511	1.942905	1.236394	107.7657	0.795791	0.004512243	90.39781		1.208202	108.8489	0.728583	0.004826	80.06724	
		76	0.725606	1.942905	1.217299	108.507	0.749754	0.004724829	89.5476		1.189107	109.5436	0.685678	0.005037	78.47004	
		78	0.744701	1.942905	1.198204	109.2165	0.705864	0.004936816	88.66779		1.170012	110.2085	0.644765	0.005248	76.84001	
		80	0.763796	1.942905	1.179109	109.8954	0.664015	0.005147802	87.75959		1.150917	110.8448	0.605745	0.005457	75.18102	
		82	0.782891	1.942905	1.160015	110.5451	0.624105	0.005357407	86.82428		1.131822	111.4536	0.568524	0.005664	73.49698	
		84	0.801986	1.942905	1.14092	111.1669	0.586038	0.00556528	85.86324		1.112727	112.0363	0.533014	0.005868	71.79185	
		86	0.821081	1.942905	1.121825	111.7619	0.549723	0.005771094	84.87788		1.093633	112.594	0.499132	0.006071	70.06957	
		88	0.840176	1.942905	1.10273	112.3314	0.515076	0.005974551	83.8697		1.074538	113.1276	0.466799	0.00627	68.33407	
		90	0.859271	1.942905	1.083635	112.8763	0.482015	0.006175379	82.84022		1.055443	113.6383	0.435941	0.006467	66.58919	
		92	0.878365	1.942905	1.06454	113.3978	0.450463	0.00637333	81.79099		1.036348	114.127	0.406486	0.00666	64.8387	
		94	0.89746	1.942905	1.045445	113.8969	0.420348	0.006568181	80.72358		1.017253	114.5947	0.378368	0.00685	63.0863	
		96	0.916555	1.942905	1.02635	114.3745	0.391601	0.006759736	79.63958		0.998158	115.0423	0.351522	0.007036	61.33551	
		98	0.93565	1.942905	1.007255	114.8315	0.364157	0.00694782	78.54059		0.979063	115.4707	0.32589	0.007219	59.58975	
		100	0.954745	1.942905	0.98816	115.2689	0.337954	0.00713228	77.4282		0.959968	115.8806	0.301413	0.007398	57.85227	
		102	0.97384	1.942905	0.969066	115.6875	0.312933	0.007312986	76.30398		0.940873	116.2728	0.278037	0.007573	56.12615	
		104	0.992935	1.942905	0.949971	116.0881	0.289039	0.007489826	75.16949		0.921778	116.6482	0.25571	0.007744	54.41431	
		106	1.01203	1.942905	0.930876	116.4714	0.266219	0.007662711	74.02626		0.902683	117.0075	0.234384	0.007911	52.71947	
		108	1.031125	1.942905	0.911781	116.8383	0.244423	0.007831565	72.87582		0.883589	117.3513	0.214013	0.008073	51.04417	
		110	1.05022	1.942905	0.892686	117.1894	0.223602	0.007996332	71.71961		0.864494	117.6803	0.194551	0.008232	49.39075	
		112	1.069314	1.942905	0.873591	117.5253	0.203712	0.00815697	70.55907		0.845399	117.9952	0.175957	0.008387	47.76136	
		114	1.088409	1.942905	0.854496	117.8469	0.18471	0.008313454	69.39559		0.826304	118.2965	0.158191	0.008537	46.15797	
		116	1.107504	1.942905	0.835401	118.1546	0.166554	0.008465769	68.23051		0.807209	118.5848	0.141214	0.008683	44.58232	
		118	1.126599	1.942905	0.816306	118.449	0.149206	0.008613915	67.06511		0.788114	118.8608	0.124992	0.008825	43.03601	
		120	1.145694	1.942905	0.797211	118.7308	0.132629	0.008757901	65.90064		0.769019	119.1248	0.109488	0.008963	41.52043	
		122	1.164789	1.942905	0.778116	119.0005	0.116787	0.008897748	64.73828		0.749924	119.3776	0.094671	0.009097	40.03679	
		124	1.183884	1.942905	0.759022	119.2585	0.101647	0.009033486	63.57916		0.730829	119.6194	0.080509	0.009226	38.58613	
		126	1.202979	1.942905	0.739927	119.5055	0.087177	0.009165153	62.42435		0.711734	119.8509	0.066973	0.009352	37.16933	
		128	1.222074	1.942905	0.720832	119.7419	0.073346	0.009292794	61.27487		0.69264	120.0724	0.054034	0.009474	35.78712	
		130	1.241169	1.942905	0.701737	119.968	0.060126	0.009416463	60.13168		0.673545	120.2843	0.041665	0.009592	34.44006	
		132	1.260264	1.942905	0.682642	120.1845	0.047488	0.009536217	58.99569		0.65445	120.4872	0.02984	0.009706	33.12858	
		134	1.279358	1.942905	0.663547	120.3916	0.035408	0.00965212	57.86775		0.635355	120.6813	0.018536	0.009816	31.85298	
		136	1.298453	1.942905	0.644452	120.5899	0.023858	0.00976424	56.74864		0.61626	120.8671	0.007728	0.009923	30.61343	
		2	0.019095	2.134965		120.8	0.011628	0.009884392	55.63781	2.108177		121	0	0.01	29.41306	2.033135
		4	0.03819			121	0	0.01	54.5361			121.2	-0.01162	0.010117	28.24656	
		6	0.057285			121.2	-0.01162	0.01011684	53.44373			121.4	-0.02322	0.010235	27.1135	
		8	0.07638			121.4	-0.02322	0.010234926	52.36086			121.6	-0.03481	0.010354	26.01348	
		10	0.095475			121.6	-0.03481	0.010354267	51.28769			121.8	-0.04639	0.010475	24.94604	
		12	0.114569			121.8	-0.04639	0.010474877	50.22441			122	-0.05796	0.010597	23.91075	
		14	0.133664			122	-0.05796	0.010596768	49.17117			122.2	-0.06952	0.01072	22.90713	
		16	0.152759			122.2	-0.06952	0.010719951	48.12817			122	-0.05796	0.010597	21.95646	
		18	0.171854			122.4	-0.08107	0.01084444	47.09556		1.936323	121.3621	-0.02102	0.010212	21.07761	
		20	0.190949			122.2	-0.06952	0.010719951	46.09659		1.917228	117.6472	0.196507	0.008216	20.39618	
		22	0.210044		1.924921	119.1243	0.109523	0.008962614	45.27766		1.898133	114.0921	0.408588	0.006646	19.86112	
		24	0.229139		1.905826	115.5056	0.323798	0.00723396	44.6273		1.879038	110.69	0.615226	0.005405	19.43632	
		26	0.248234		1.886731	112.0427	0.532628	0.005870601	44.10638		1.859943	107.4342	0.816436	0.00442	19.0957	
		28	0.267329		1.867637	108.7287	0.736023	0.004790153	43.68585		1.840849	104.3184	1.012243	0.003634	18.82013	
		30	0.286424		1.848542	105.5572	0.934003	0.003929775	43.34384		1.821754	101.3367	1.202679	0.003004	18.59535	
		32	0.305518		1.829447	102.5222	1.126599	0.003241339	43.06377		1.802659	98.48319	1.387787	0.002496	18.4106	
		34	0.324613		1.810352	99.61775	1.313848	0.002687839	42.83289		1.783564	95.75245	1.567615	0.002085	18.25766	
		36	0.343708		1.791257	96.83821	1.495796	0.002240702	42.64137		1.764469	93.13918	1.742219	0.001751	18.13021	
		38	0.362803		1.772162	94.17823	1.672497	0.001877775	42.48153		1.745374	90.63832	1.911662	0.001478	18.02332	
		40	0.381898		1.753067	91.63268	1.844012	0.001581815	42.34735		1.726279	88.24503	2.076014	0.001254	17.93312	
		42	0.400993		1.733972	89.19662	2.010407	0.001339342	42.23406		1.707184	85.9547	2.235348	0.00107	17.85656	
		44	0.420088		1.714877	86.86535	2.171753	0.001139776	42.1379		1.688089	83.76288	2.389745	0.000917	17.79121	
		46	0.439183		1.695782	84.63436	2.328129	0.000974779	42.05583		1.668994	81.66534	2.539287	0.000789	17.73514	
		48	0.458278		1.676687	82.49933	2.479616	0.000837753	41.98542		1.6499	79.65802	2.684062	0.000683	17.68676	
		50	0.477373		1.657593	80.45614	2.626301	0.000723456	41.92472		1.630805	77.73706	2.824161	0.000594	17.64482	

	Region 3						Region 4						Region 5			
	Temp. (°C)		k (1/min)	Concen.	$\log(T_M\text{-}T)$		Temp. (°C)		k (1/min)	Concen.	$\log(T_M\text{-}T)$		Temp. (°C)		k (1/min)	Concen.
				100						100						100
1.454331	96.53369	1.515896	0.002196	99.1254	1.598298	1.2164	108.5411	0.747639	0.004735	98.12389	1.107905	0.726007	119.6788	0.077033	0.009259	96.3643
1.435236	97.75817	1.435273	0.002381	98.18601		1.197305	109.2491	0.703849	0.004947	96.20138		0.706912	119.9077	0.06365	0.009383	92.81446
1.416141	98.92998	1.358615	0.00257	97.18176		1.17821	109.9266	0.662093	0.005158	94.237		0.687817	120.1268	0.050857	0.009504	89.3522
1.397046	100.0514	1.285706	0.002765	96.11303		1.159115	110.575	0.622272	0.005367	92.2354		0.668722	120.3364	0.038628	0.009621	85.97887
1.377952	101.1246	1.216342	0.002963	94.98057		1.14002	111.1955	0.584289	0.005575	90.2013		0.649627	120.537	0.026937	0.009734	82.69546
1.358857	102.1516	1.150333	0.003165	93.78558		1.120925	111.7893	0.548055	0.005781	88.13952		0.630532	120.729	0.015761	0.009844	79.50263
1.339762	103.1344	1.087502	0.003371	92.52962		1.101831	112.3576	0.513484	0.005984	86.05483		0.611438	120.9127	0.005075	0.009949	76.40076
1.320667	104.0749	1.027679	0.003578	91.21464		1.082736	112.9014	0.480496	0.006185	83.95203		0.592343	121.0885	-0.00514	0.010052	73.38991
1.301572	104.975	0.970709	0.003788	89.84292		1.063641	113.4218	0.449013	0.006383	81.83584		0.573248	121.2568	-0.01491	0.01015	70.46989
1.282477	105.8364	0.916443	0.003999	88.41709		1.044546	113.9198	0.418964	0.006577	79.71089		0.554153	121.4178	-0.02425	0.010245	67.64027
1.263382	106.6607	0.864742	0.004212	86.94005		1.025451	114.3965	0.39028	0.006769	77.58169		0.535058	121.5719	-0.03318	0.010337	64.90041
1.244287	107.4496	0.815474	0.004424	85.41499		1.006356	114.8526	0.362896	0.006957	75.45264		0.515963	121.7193	-0.04172	0.010426	62.24945
1.225192	108.2045	0.768517	0.004637	83.84532		0.987261	115.2891	0.33675	0.007141	73.32793		0.496868	121.8604	-0.04989	0.010512	59.68635
1.206097	108.927	0.723753	0.004849	82.23463		0.968166	115.7068	0.311783	0.007321	71.21162		0.477773	121.9955	-0.0577	0.010594	57.20992
1.187002	109.6184	0.681073	0.005061	80.58669		0.949071	116.1065	0.287941	0.007498	69.10754		0.458678	122.1247	-0.06517	0.010673	54.81882
1.167908	110.28	0.640373	0.005271	78.9054		0.929976	116.4891	0.26517	0.007671	67.01931		0.439583	122.2484	-0.07232	0.01075	52.51158
1.148813	110.9132	0.601556	0.00548	77.19474		0.910881	116.8552	0.243421	0.007839	64.95035		0.420489	122.3668	-0.07915	0.010824	50.28662
1.129718	111.5191	0.564527	0.005686	75.45875		0.891787	117.2055	0.222645	0.008004	62.90384		0.401394	122.48	-0.08568	0.010895	48.14227
1.110623	112.099	0.529201	0.005891	73.7015		0.872692	117.5408	0.202798	0.008164	60.88272		0.382299	122.5884	-0.09194	0.010963	46.07676
1.091528	112.654	0.495494	0.006093	71.92705		0.853597	117.8617	0.183836	0.008321	58.88972		0.363204	122.6922	-0.09791	0.011029	44.08828
1.072433	113.185	0.463327	0.006292	70.13942		0.834502	118.1687	0.165719	0.008473	56.92731		0.344109	122.7914	-0.10363	0.011092	42.17495
1.053338	113.6932	0.432627	0.006488	68.34257		0.815407	118.4626	0.148408	0.008621	54.99773		0.325014	122.8864	-0.1091	0.011153	40.33483
1.034243	114.1796	0.403322	0.006681	66.54038		0.796312	118.7438	0.131867	0.008765	53.10301		0.305919	122.9774	-0.11433	0.011211	38.56598
1.015148	114.645	0.375347	0.006871	64.73661		0.777217	119.0129	0.116058	0.008904	51.24493		0.286824	123.0644	-0.11934	0.011267	36.8664
0.996053	115.0905	0.348639	0.007056	62.93491		0.758122	119.2704	0.10095	0.00904	49.42506		0.267729	123.1476	-0.12412	0.011322	35.23409
0.976959	115.5167	0.323136	0.007239	61.13876		0.739027	119.5169	0.086511	0.009171	47.64476		0.248634	123.2273	-0.1287	0.011374	33.66707
0.957864	115.9246	0.298783	0.007417	59.35149		0.719932	119.7527	0.07271	0.009299	45.90517		0.229539	123.3036	-0.13308	0.011423	32.16331
0.938769	116.315	0.275525	0.007592	57.57626		0.700838	119.9785	0.059518	0.009422	44.20726		0.210445	123.3765	-0.13727	0.011471	30.72082
0.919674	116.6886	0.253312	0.007762	55.81604		0.681743	120.1945	0.046907	0.009542	42.55179		0.19135	123.4464	-0.14128	0.011518	29.33762
0.900579	117.0461	0.232093	0.007929	54.07361		0.662648	120.4012	0.034852	0.009657	40.93936		0.172255	123.5132	-0.14512	0.011562	28.01173
0.881484	117.3883	0.211824	0.008091	52.35158		0.643553	120.599	0.023327	0.009769	39.3704		0.15316	123.5771	-0.14879	0.011604	26.74122
0.862389	117.7157	0.19246	0.008249	50.65232		0.624458	120.7883	0.012309	0.009878	37.84518		0.134065	123.6384	-0.1523	0.011645	25.52418
0.843294	118.029	0.173959	0.008403	48.97802		0.605363	120.9695	0.001775	0.009982	36.36383		0.11497	123.6969	-0.15566	0.011684	24.3587
0.824199	118.3289	0.156282	0.008553	47.33069		0.586268	121.1428	-0.0083	0.010083	34.92634		0.095875	123.753	-0.15887	0.011722	23.24295
0.805104	118.6158	0.13939	0.008699	45.71212		0.567173	121.3088	-0.18975	0.012089	33.27756		0.07678	123.8066	-1.69007	0.054199	18.71279
0.78601	118.8904	0.123248	0.00884	44.1239		0.548078	121.4675	-0.28696	0.013324	31.55048		0.057685	123.8579	-1.72027	0.05586	14.96577
0.766915	119.1532	0.107822	0.008978	42.56746		0.528983	121.6195	-0.37974	0.014619	29.75843		0.03859	123.9071	-1.74915	0.057497	11.89096
0.74782	119.4047	0.093079	0.009111	41.04403		0.509889	121.7649	-0.46832	0.015973	27.91656		0.019496	123.9541	-1.77676	0.059107	9.387244
0.728725	119.6454	0.078987	0.009241	39.55465		0.490794	121.9041	-0.55289	0.017383	26.04143		0.000401	123.9991	-1.80316	0.060688	7.36397
0.70963	119.8758	0.065518	0.009366	38.10022		0.471699	122.0372	-0.63365	0.018845	24.15062		-0.01869	124.0421	-1.82841	0.06224	5.74103
0.690535	120.0962	0.052643	0.009487	36.68145		0.452604	122.1647	-0.71076	0.020355	22.26217		-0.03779	124.0833	-1.85256	0.063761	4.448616
0.67144	120.3071	0.040335	0.009605	35.29892		0.433509	122.2866	-0.78441	0.021911	20.39408		-0.05688	124.1228	-1.87566	0.065251	3.426668
0.652345	120.509	0.028569	0.009718	33.95306		0.414414	122.4033	-0.85475	0.023508	18.5638		-0.07598	124.1605	-1.89775	0.066708	2.624144
0.63325	120.7022	0.017321	0.009828	32.64416		0.395319	122.515	-0.92194	0.025142	16.78771		-0.09507	124.1966	-1.91887	0.068133	1.998155
0.614155	120.887	0.006567	0.009935	31.37238		0.376224	122.6219	-0.98612	0.026808	15.08068		-0.11417	124.2312	-1.93908	0.069523	1.513056
0.595061	121.064	-0.00372	0.010037	30.13776		0.357129	122.7242	-1.04744	0.028503	13.45567		-0.13326	124.2642	-1.9584	0.07088	1.139525
0.575966	121.2333	-0.01355	0.010136	28.94025		0.338034	122.8221	-1.10603	0.030223	11.92346		-0.15236	124.2959	-1.97689	0.072202	0.853681
0.556871	121.3953	-0.02295	0.010232	27.77968		0.31894	122.9158	-1.16201	0.031963	10.49243		-0.17145	124.3262	-1.99457	0.07349	0.636254
0.537776	121.5503	-0.03194	0.010325	26.6558		0.299845	123.0055	-1.2155	0.03372	9.168517		-0.19055	124.3552	-2.01148	0.074744	0.471832
	121.7	-0.04061	0.010414	25.56819	1.802445	1.78335	95.72254	1.569599	0.002081	9.092504	1.318662	1.299567	54.93274	4.612714	9.92E-05	0.471645
	121.85	-0.04929	0.010505	24.51605		1.764255	93.11056	1.744145	0.001748	9.029153		1.280472	54.07533	4.684826	9.23E-05	0.471471
	122	-0.05796	0.010597	23.49861		1.74516	90.61093	1.913531	0.001476	8.976017		1.261377	53.25481	4.754191	8.62E-05	0.471308
	121.8	-0.04639	0.010475	22.53437		1.726065	88.21882	2.077826	0.001252	8.931177		1.242282	52.46957	4.820899	8.06E-05	0.471156
1.937661	121.6285	-0.03646	0.010371	21.61864		1.70697	85.92961	2.237105	0.001068	8.893116		1.223187	51.71812	4.885041	7.56E-05	0.471014
1.918566	117.9022	0.181445	0.008341	20.90929		1.687875	83.73887	2.391446	0.000915	8.860628		1.204093	50.99899	4.946701	7.11E-05	0.47088
1.899471	114.3361	0.393908	0.006744	20.35277		1.668781	81.64236	2.540934	0.000788	8.832745		1.184998	50.31079	5.005967	6.7E-05	0.470754
1.880376	110.9235	0.600927	0.005483	19.91125		1.649686	79.63604	2.685657	0.000682	8.808691		1.165903	49.6522	5.062919	6.33E-05	0.470635
1.861281	107.6576	0.802517	0.004482	19.55746		1.630591	77.71601	2.825704	0.000593	8.787833		1.146808	49.02193	5.11764	5.99E-05	0.470522
1.842186	104.5322	0.998701	0.003684	19.27141		1.611496	75.87858	2.961171	0.000518	8.769658		1.127713	48.41878	5.170209	5.68E-05	0.470415
1.823091	101.5413	1.189512	0.003044	19.0382		1.592401	74.12019	3.092153	0.000454	8.753745		1.108618	47.84157	5.220701	5.4E-05	0.470313
1.803996	98.67903	1.374992	0.002528	18.84663		1.573306	72.43743	3.21875	0.0004	8.739749		1.089523	47.28919	5.269191	5.15E-05	0.470216
1.784902	95.93987	1.555187	0.002111	18.68812		1.554211	70.82706	3.341062	0.000354	8.727382		1.070428	46.76057	5.315753	4.91E-05	0.470124
1.765807	93.31854	1.730156	0.001773	18.55608		1.535116	69.28595	3.45919	0.000315	8.716408		1.051333	46.25468	5.360456	4.7E-05	0.470036
1.746712	90.80996	1.899958	0.001496	18.4454		1.516021	67.81114	3.573238	0.000281	8.706629		1.032238	45.77056	5.403369	4.5E-05	0.469951
1.727617	88.40929	2.064665	0.001269	18.35203		1.496926	66.39977	3.683308	0.000251	8.697878		1.013143	45.30727	5.444558	4.32E-05	0.46987
1.708522	86.11189	2.224348	0.001081	18.27282		1.477832	65.0491	3.789503	0.000226	8.690016		0.994049	44.8639	5.484088	4.15E-05	0.469792
1.689427	83.91331	2.379087	0.000926	18.20524		1.458737	63.75654	3.891929	0.000204	8.682926		0.974954	44.4396	5.522022	4E-05	0.469717
1.670332	81.8093	2.528967	0.000797	18.14726		1.439642	62.51957	3.990687	0.000185	8.676508		0.955859	44.03356	5.558418	3.85E-05	0.469644
1.651237	79.79579	2.674073	0.00069	18.09727		1.420547	61.33582	4.085882	0.000168	8.670676		0.936764	43.64498	5.593336	3.72E-05	0.469574
1.632142	77.8689	2.814496	0.000599	18.05393		1.401452	60.20298	4.177615	0.000153	8.665359		0.917669	43.27311	5.626833	3.6E-05	0.469507
1.613047	76.02489	2.950332	0.000523	18.01619		1.382357	59.11887	4.265989	0.00014	8.660495		0.898574	42.91724	5.658963	3.49E-05	0.469441
1.593953	74.2602	3.081675	0.000459	17.98315		1.363262	58.0814	4.351102	0.000129	8.65603		0.879479	42.57668	5.689778	3.38E-05	0.469378
1.574858	72.57142	3.208625	0.000404	17.95411		1.344167	57.08855	4.433056	0.000119	8.651918		0.860384	42.25077	5.71933	3.28E-05	0.469316
1.555763	70.95528	3.331281	0.000357	17.92845		1.325072	56.13841	4.511948	0.00011	8.64812		0.841289	41.93888	5.747668	3.19E-05	0.469256

Table A.6.8.a Reduction of microbial population and quality attribute during a microwave process

			Heat rate (C/s)	Time (s)	Temp (°C)		k (1/min)
Microwave Constant	55.61		0.001575	0	20		
Electric Intensitiy	12 V/cm			2	21.57493	18.18662	6.95E-09
MW frequency	2450 Hz			4	23.14986	17.34845	1.61E-08
Dielectric constant	64			6	24.72479	16.51915	3.68E-08
Loss tangent	0.219			8	26.29972	15.69858	8.37E-08
Density	900 kg/m^3			10	27.87465	14.88659	1.88E-07
Specific Heat	3880 J/kg K			12	29.44958	14.08307	4.21E-07
Microbial				14	31.02451	13.28787	9.32E-07
Initial population	1000			16	32.59944	12.50086	2.05E-06
Rate constant	0.55/min			18	34.17437	11.72193	4.46E-06
E-constant	386 kJ/mole			20	35.7493	10.95094	9.65E-06
Vitamin				22	37.32423	10.18778	2.07E-05
Rate constant	0.068/min			24	38.89916	9.43232	4.41E-05
E-constant	80 kJ/mole			26	40.47409	8.684456	9.31E-05
Temp	90°C			28	42.04902	7.944069	0.000195
j - value	2.048			30	43.62395	7.211047	0.000406
Heating rate - f	38.2 min			32	45.19888	6.485282	0.000839
Med Temp	65.5°C			34	46.77381	5.766666	0.001722
Cool Med	20°C			36	48.34874	5.055093	0.003507
cooling rate - f	62.7 min			38	49.92367	4.350462	0.007095
cooling j	1.815			40	51.4986	3.65267	0.014257
jm-value	0.551			42	53.07353	2.961619	0.028454
jm cool	0.813			44	54.64847	2.277211	0.056413
				46	56.2234	1.599351	0.111115
				48	57.79833	0.927946	0.217451
				50	59.37326	0.262904	0.422849
				50.8	60.00323	-0.00135	0.550744
				2.5	62	-0.83237	1.264323
	0.079744817	1.913425	1.83368	5	62	-0.83237	1.264323
	0.159489633		1.753935	10	62	-0.83237	1.264323
	0.23923445		1.674191	15	62	-0.83237	1.264323
	0.318979266		1.594446	20	59.30481	0.291675	0.410856
	0.398724083		1.514701	25	52.71153	3.119867	0.02429
	0.4784689		1.434956	30	47.22426	5.562434	0.002112
	0.558213716		1.355211	35	42.65746	7.660015	0.000259
	0.637958533		1.275466	40	38.85673	9.452572	4.32E-05

N		k (1/min)	Conc.			
1000			100			
1000	6.157335	0.000144	99.99952			
1000	5.983622	0.000171	99.99895			
1000	5.811746	0.000203	99.99827			
1000	5.641679	0.000241	99.99747			
1000	5.473392	0.000285	99.99652			
1000	5.306858	0.000337	99.99539			
999.9999	5.14205	0.000398	99.99407			
999.9999	4.97894	0.000468	99.99251			
999.9997	4.817503	0.00055	99.99067			
999.9994	4.657713	0.000645	99.98852			
999.9987	4.499544	0.000756	99.986			
999.9972	4.342973	0.000884	99.98306			
999.9941	4.187975	0.001032	99.97962			
999.9876	4.034527	0.001203	99.97561			
999.9741	3.882606	0.001401	99.97094			
999.9461	3.732188	0.001628	99.96552			
999.8888	3.583252	0.001889	99.95922			
999.7719	3.435776	0.00219	99.95193			
999.5354	3.289738	0.002534	99.94348			
999.0605	3.145118	0.002928	99.93373			
998.1134	3.001895	0.003379	99.92247			
996.2383	2.860049	0.003894	99.9095			
992.5552	2.71956	0.004481	99.89458			
985.3868	2.580409	0.005151	99.87743			
974.338	2.442576	0.005912	99.86169			
970.7672	2.387808	0.006244	99.85753			
41.32611	2.215576	0.007418	98.02512			
0.074265	3.555685	0.001942	98.89251	1.512186	1.432441	47.06704
0.000133	3.988297	0.00126	98.27138		1.352696	42.52662
2.4E-07	4.357949	0.000871	97.84448		1.272951	38.74784
3.07E-08	4.672494	0.000636	97.53396		1.193206	35.60294
2.72E-08	4.939205	0.000487	97.2968		1.113462	32.98558
2.69E-08	5.16468	0.000389	97.10793		1.033717	30.80729
2.69E-08	5.354812	0.000321	96.95204		0.953972	28.99439
2.69E-08	5.514799	0.000274	96.8194		0.874227	27.48561

Table A.6.8.b Reduction of microbial population and quality attribute during a traditional thermal process

		Time (min)	t/f	$\log(T_M\text{-}T)$		Temp (°C)	
		0				20	
		5	0.130208	1.969341	1.839133	20	19.0338
		10	0.260417		1.708925	20	19.0338
		15	0.390625		1.578716	27.59327	15.03104
Microbial		20	0.520833		1.448508	37.41283	10.14507
Rate constant	0.55 min	25	0.651042		1.3183	44.68868	6.719609
E-constant	386 kJ/mole	30	0.78125		1.188091	50.07975	4.281006
Vitamin		35	0.911458		1.057883	54.07429	2.525961
Rate constant	0.068/min	40	1.041667		0.927675	57.03407	1.252954
E-constant	80 kj/mole	45	1.171875		0.797466	59.22713	0.324343
Temp	90°C	50	1.302083		0.667258	60.85209	−0.35585
j - value	2.048	53	1.380208		0.589133	61.61731	−0.67387
heating rate - f	38.4 min	5				61.4	−0.58371
Med Temp	65.5°C	10				61.4	−0.58371
Cool Med	20°C	15				61.4	−0.58371
cooling rate - f	62.7°C	20	0.318979	1.878151	1.559171	59.3	0.293698
cooling j	1.815	25	0.398724		1.479427	50.15967	4.24547
jm-value	0.551	30	0.478469		1.399682	45.10046	6.530427
jm cool	0.813	35	0.558214		1.319937	40.88993	8.488248
		40	0.637959		1.240192	37.3857	10.15815
		45	0.717703		1.160447	34.46929	11.57695

k (1/min)	N	Time (min)	t/f	$\log(T_M\text{-}T)$		Temp. (°C)		k (1/min)	Concen.
	1000	0				20			100
2.98E-09	1000	5	0.130208	1.399163	1.268955	46.92389	3.569136	0.001916	99.04645
2.98E-09	1000	10	0.260417		1.138746	51.73595	3.123446	0.002992	97.57557
1.63E-07	999.9992	15	0.390625		1.008538	55.30146	2.801636	0.004128	95.58209
2.16E-05	999.8912	20	0.520833		0.87833	57.94334	2.567662	0.005217	93.12126
0.000664	996.5778	25	0.651042		0.748121	59.90086	2.396694	0.006189	90.28364
0.007606	959.3904	30	0.78125		0.617913	61.35129	2.271305	0.007016	87.17137
0.04399	769.9683	35	0.911458		0.487705	62.42599	2.179097	0.007694	83.88167
0.157113	350.9997	40	1.041667		0.357496	63.2223	2.111155	0.008235	80.49811
0.397651	48.06382	45	1.171875		0.227288	63.81233	2.06102	0.008658	77.08769
0.785064	0.94858	50	1.302083		0.09708	64.24951	2.023986	0.008985	73.70128
1.079	0.037262	53	1.380208		0.018955	64.45539	2.006579	0.009142	71.70732
0.985969	0.000269	5	0.079745	1.558015	1.47827	50.07947	3.27537	0.002571	70.79156
0.985969	1.95E-06	10	0.15949		1.398525	45.03372	3.747893	0.001603	70.22658
0.985969	1.41E-08	15	0.239234		1.318781	40.83438	4.152737	0.001069	69.8522
0.410026	1.81E-09	20	0.318979		1.239036	37.33946	4.498022	0.000757	69.58834
0.007881	1.74E-09	25	0.398724		1.159291	34.43082	4.791373	0.000564	69.39222
0.000802	1.73E-09	30	0.78125		0.776765	25.98088	5.675964	0.000233	69.3114
0.000113	1.73E-09	35	0.911458		0.646557	24.43156	5.843609	0.000197	69.24313
2.13E-05	1.73E-09	40	1.041667		0.516348	23.28359	5.968957	0.000174	69.18296
5.16E-06	1.73E-09	45	1.171875		0.38614	22.43299	6.062463	0.000158	69.12821

Table A.6.9.a Changes in temperature and microbial population during an ohmic heating process

		Heat rate (C/s)	Time (s)	Temp (°C)		k (1/s)	N
Temperature	100°C	20.94758	0	80			1000
Voltage	5000 V		0.5	90.47379	7.691567	0.00063	999.685
Elec Cond	0.32 S/m		1	100.9476	4.911014	0.010164	994.6176
T coefficient	0.035		1.5	111.4214	2.281976	0.140874	926.9703
Length	1.175 m		2	121.8952	−0.2076	1.698397	396.5186
Density	1000 kg/m^3		2.5	121.8952	−0.20761	1.698416	169.6123
Specific Heat	3880 j/kg K		3	121.8952	−0.20761	1.698416	72.55228
Microbial			3.5	121.8952	−0.20761	1.698416	31.0345
Initial population	1000		4	121.8952	−0.20761	1.698416	13.27512
Rate constant	1.38/min		4.5	121.8952	−0.20761	1.698416	5.678481
E-constant	300 kJ/mole		5	121.8952	−0.20761	1.698416	2.428991
			5.5	121.8952	−0.20761	1.698416	1.03901
			6	121.8952	−0.20761	1.698416	0.44444
			6.5	121.8952	−0.20761	1.698416	0.190111
			7	121.8952	−0.20761	1.698416	0.081321
			7.5	121.8952	−0.20761	1.698416	0.034785
			8	121.8952	−0.20761	1.698416	0.014879
			8.5	121.8952	−0.20761	1.698416	0.006365
			9	121.8952	−0.20761	1.698416	0.002723

Table A.6.9.b (Table A.6.9.a Continued)

		Time (s)	Temp (°C)		k (1/s)	N
Temperature	100°C	9.5	121.8952	−0.20761	1.698416	0.001165
Voltage	5000V	10	121.8952	−0.20761	1.698416	0.000498
Elec Cond	0.32 S/m	10.5	121.8952	−0.20761	1.698416	0.000213
T coefficient	0.035	11	121.8952	−0.20761	1.698416	9.12E-05
Length	1.175 m	11.5	121.8952	−0.20761	1.698416	3.9E-05
Density	1000 kg/m^3	12	121.8952	−0.20761	1.698416	1.67E-05
Specific Heat	3880 j/kg K	12.5	121.8952	−0.20761	1.698416	7.14E-06
Microbial		13	121.8952	−0.20761	1.698416	3.05E-06
Initial population	1000	13.5	121.8952	−0.20761	1.698416	1.31E-06
Rate constant	1.38/min	14	121.8952	−0.20761	1.698416	5.59E-07
E-constant	300 kJ/mole	14.5	121.8952	−0.20761	1.698416	2.39E-07
		15	121.8952	−0.20761	1.698416	1.02E-07
		15.5	121.8952	−0.20761	1.698416	4.37E-08
		16	121.8952	−0.20761	1.698416	1.87E-08
		16.5	120.5902	0.095355	1.254489	9.99E-09
		17	119.2852	0.400338	0.924729	6.29E-09
		17.5	117.9802	0.707357	0.680265	4.48E-09
		18	116.6752	1.016432	0.4994	3.49E-09
		18.5	115.3702	1.327584	0.365861	2.9E-09
		19	114.0652	1.640835	0.267469	2.54E-09
		19.5	112.7602	1.956205	0.195124	2.3E-09
		20	111.4552	2.273715	0.142042	2.15E-09

Table A.6.9.c (Table A.6.9.a and A.6.9.b Continued)

		Time (s)	Temp (°C)		k (1/s)	N
						2.15E-09
Temperature	100°C	20.5	110.1502	2.593389	0.103177	2.04E-09
Voltage	5000V	21	108.8452	2.915248	0.074783	1.97E-09
Elec Cond	0.32S/m	21.5	107.5402	3.239314	0.054083	1.91E-09
T coefficient	0.035	22	106.2352	3.565611	0.039026	1.88E-09
Length	1.175m	22.5	104.9302	3.894161	0.028097	1.85E-09
Density	1000kg/m^3	23	103.6252	4.224988	0.020183	1.83E-09
Specific Heat	3880j/kg K	23.5	102.3202	4.558115	0.014465	1.82E-09
Microbial		24	101.0152	4.893567	0.010343	1.81E-09
Initial population	1000	24.5	99.7102	5.231369	0.007378	1.8E-09
Rate constant	1.38/min	25	98.4052	5.571544	0.00525	1.8E-09
E-constant	300kJ/mole	25.5	97.1002	5.914118	0.003727	1.8E-09
		26	95.7952	6.259116	0.00264	1.79E-09
		26.5	94.4902	6.606565	0.001865	1.79E-09
		27	93.1852	6.95649	0.001314	1.79E-09
		27.5	91.8802	7.308918	0.000924	1.79E-09
		28	91.1952	7.49492	0.000767	1.79E-09
		29	88.5852	8.210086	0.000375	1.79E-09
		30	85.9752	8.935652	0.000182	1.79E-09
		32	80.9752	10.35551	4.39E-05	1.79E-09
		34	75.5352	11.94658	8.94E-06	1.79E-09
		36	70.3152	13.52072	1.85E-06	1.79E-09
		38	65.0952	15.14346	3.66E-07	1.79E-09
		40	59.8752	16.8171	6.86E-08	1.79E-09

Table A.6.10.a Quality retention during a PAPT

		Time (min)	Temp (°C)	mid-T (°C)		k (1/min)	N		k (1/min)	Retention (%)
j - value	2.048	0	0	80			1000			100
heating rate - f	38.4 min	0.5	150	87.5	4.704134	0.009058	995.4813631	2.127622	0.001191	99.94046
Med Temp	130°C	1	300	95	3.072175	0.04632	972.6907912	1.617635	0.001984	99.84138
Cool Med	30°C	1.5	450	102.5	1.505406	0.221927	870.5304207	1.12802	0.003237	99.67993
cooling rate - f	62.5 min	2	600	110	0	1	528.0033904	0.657581	0.005181	99.42204
cooling rate	1.815	2.5	750	117.5	−1.44758	4.252812	62.97242176	0.205212	0.008145	99.01798
jm-value	0.551	3	750	117.5	−1.44758	4.252812	7.510417499	0.205212	0.008145	98.61557
jm cool	0.813	3.5	750	117.5	−1.44758	4.252812	0.895731328	0.205212	0.008145	98.21478
Density	900 kg/m^3	4	750	117.5	−1.44758	4.252812	0.106829562	0.205212	0.008145	97.81563
Specific Heat	3500 kJ/kg K	4.5	750	117.5	−1.44758	4.252812	0.012741047	0.205212	0.008145	97.4181
Microbial		5	750	117.5	−1.44758	4.252812	0.001519563	0.205212	0.008145	97.02218
Initial population	1000	5.5	750	117.5	−1.44758	4.252812	0.000181231	0.205212	0.008145	96.62787
Rate constant	1/min	6	750	117.5	−1.44758	4.252812	2.16145E-05	0.205212	0.008145	96.23517
E-constant	240 kJ/mole	6.5	750	117.5	−1.44758	4.252812	2.57786E-06	0.205212	0.008145	95.84406
Temp	110°C	7	750	117.5	−1.44758	4.252812	3.07449E-07	0.205212	0.008145	95.45454
Vitamin		7.5	750	117.5	−1.44758	4.252812	3.6668E-08	0.205212	0.008145	95.0666
Rate constant	0.01/min	8	750	117.5	−1.44758	4.252812	4.37322E-09	0.205212	0.008145	94.68024
E-constant	75 kJ/mole	8.5	500	105	0.996967	0.368997	3.63642E-09	0.969133	0.003794	94.5008
Temp	121°C	9	250	92.5	3.608719	0.027087	3.5875E-09	1.785305	0.001677	94.42157
		9.5	0	80	6.40544	0.001653	3.58454E-09	2.659281	0.0007	94.38853

Table A.6.10.b Comparison of thermal ptocess to PATP

		Time (min)	t/f	log	t/f - log	T		k (1/min)	N	log	t/f - log	T		k (1/min)	Retention (%)
j – value	2.048	0				80			1000			80			100
heating rate	38.4 min	1	0.026042	2.0103	1.984258	80	6.40544	0.001653	998.3488	1.440122	1.41408	104.0534	1.029044	0.003573	98.22913
Med Temp	130°C	2	0.052083	2.0103	1.958217	80	6.40544	0.001653	996.7004		1.388038	105.5635	0.933607	0.003931	96.31712
Cool Med	30°C	3	0.078125	2.0103	1.932175	80	6.40544	0.001653	995.0546		1.361997	106.9858	0.844417	0.004298	94.26931
cooling rate	62.5 min	4	0.104167	2.0103	1.906133	80	6.40544	0.001653	993.4116		1.335955	108.3252	0.761027	0.004672	92.09277
cooling j	1.815	5	0.130208	2.0103	1.880092	80	6.40544	0.001653	991.7713		1.309913	109.5867	0.683025	0.005051	89.79615
jm-value	0.551	6	0.15625	2.0103	1.85405	80	6.40544	0.001653	990.1337		1.283872	110.7748	0.610031	0.005433	87.38952
jm cool	0.813	7	0.182292	2.0103	1.828008	80	6.40544	0.001653	988.4989		1.25783	111.8937	0.541697	0.005818	84.88415
Density	900 kg/m^3	8	0.208333	2.0103	1.801967	80	6.40544	0.001653	986.8667		1.231788	112.9475	0.477703	0.006202	82.80423
Specific Heat	3500 kJ/kgK	9	0.234375	2.0103	1.775925	80	6.40544	0.001653	985.2372		1.205747	113.94	0.417752	0.006585	81.1844
Microbial		10	0.260417	2.0103	1.749883	80	6.40544	0.001653	983.6104		1.179705	114.8747	0.361571	0.006966	80.06123
Initial population	1000	11	0.286458	2.0103	1.723842	80	6.40544	0.001653	981.9863		1.153663	115.755	0.308906	0.007342	79.47553
Rate constant	1/min	12	0.3125	2.0103	1.6978	80.13453	6.374288	0.001705	980.3136		1.127622	116.584	0.259525	0.007714	78.8648
E-constant	240 kJ/mole	13	0.338542	2.0103	1.671758	83.03673	5.70795	0.003319	977.0648		1.10158	117.3649	0.213209	0.00808	78.23015
Temp	110°C	14	0.364583	2.0103	1.645717	85.77003	5.09025	0.006156	971.068		1.075538	118.1002	0.169758	0.008439	77.57277
Vitamin		15	0.390625	2.0103	1.619675	88.34425	4.517046	0.010921	960.5205		1.049497	118.7928	0.128985	0.00879	76.8939
Rate constant	0.01/min	16	0.416667	2.0103	1.593633	90.76865	3.984621	0.018599	942.8204		1.023455	119.4451	0.090716	0.009133	76.19485
E-constant	75 kJ/mole	17	0.442708	2.0103	1.567592	93.05194	3.489632	0.030512	914.4874		0.997413	120.0594	0.054791	0.009467	75.47693
Temp	121°C	18	0.46875	2.0103	1.54155	95.20235	2.032099	0.13106	802.1564		0.971372	120.6379	0.021059	0.009792	74.74149
		19	0.494792	2.0103	1.515508	97.2276	1.60323	0.201245	655.9328		0.94533	121.1828	-0.01062	0.010107	73.9899
		20	0.520833	2.0103	1.489467	99.13498	1.20359	0.300115	485.8712		0.919288	121.696	-0.04037	0.010412	73.22352
		21	0.546875	2.0103	1.463425	100.9313	0.830938	0.435641	314.2859		0.893247	122.1793	-0.06832	0.010707	72.44369
		22	0.572917	2.0103	1.437383	102.6232	1.4802	0.227592	250.3127		0.867205	122.6345	-0.09459	0.010992	71.65174

23	0.598958	2.0103	1.411342	104.2165	1.155584	0.314874	182.6985		0.841163	123.0631	-0.11927	0.011267	70.84899
24	0.625	2.0103	1.3853	105.7171	0.852358	0.426408	119.2747		0.815122	123.4669	-0.14246	0.011531	70.03672
25	0.651042	2.0103	1.359258	107.1304	0.568969	0.566109	67.71585		0.78908	123.8471	-0.16426	0.011785	69.21617
26	0.677083	2.0103	1.333217	108.4614	0.303994	0.737865	32.37723		0.763038	124.2052	-0.18475	0.012029	68.38854
27	0.703125	2.0103	1.307175	109.715	0.056126	0.94542	12.57909		0.736997	124.5425	-0.20402	0.012263	67.55499
28	0.729167	2.0103	1.281133	110.8956	−0.17584	1.192242	3.818258		0.710955	124.8601	-0.22214	0.012487	65.88871
29	0.755208	2.0103	1.255092	112.0075	−0.393	1.481412	0.867953		0.684913	125.1592	-0.23917	0.012702	64.23596
1	0.026042	2.17273	2.146689	113	−0.58578	1.796396	0.14399	1.888542	1.8625	102.8618	1.104895	0.003312	63.81181
2	0.052083		2.120647	113	−0.58578	1.796396	0.023887		1.836458	98.62118	1.37877	0.002519	63.49115
3	0.078125		2.094605	113	−0.58578	1.796396	0.003963		1.810417	94.62738	1.642482	0.001935	62.87983
4	0.104167		2.068564	113	−0.58578	1.796396	0.000657		1.784375	90.86601	1.896139	0.001501	62.69129
5	0.130208		2.042522	113	−0.58578	1.796396	0.000109		1.758333	87.32356	2.139875	0.001177	62.32353
6	0.15625		2.01648	113	−0.58578	1.796396	1.81E-05		1.732292	83.98729	2.373849	0.000931	62.20757
7	0.182292		1.990439	113	−0.58578	1.796396	3E-06		1.70625	80.84519	2.59824	0.000744	62.11507
8	0.208333		1.964397	113	−0.58578	1.796396	4.98E-07		1.680208	77.88596	2.813246	0.0006	62.04056
9	0.234375		1.938355	113	−0.58578	1.796396	8.26E-08		1.654167	75.09896	3.019081	0.000488	61.88922
10	0.260417		1.912314	111.7172	−0.33642	1.399928	2.04E-08		1.628125	72.47417	3.215973	0.000401	61.76521
11	0.286458		1.886272	106.9612	0.602789	0.547283	1.18E-08		1.602083	70.00214	3.404162	0.000332	61.66266
12	0.3125		1.86023	102.482	1.509091	0.221111	9.45E-09		1.576042	67.67398	3.583894	0.000278	61.57711
13	0.338542		1.834189	98.26349	2.382643	0.092306	8.61E-09		1.55	65.48133	3.755427	0.000234	61.50513
14	0.364583		1.808147	94.29051	3.223702	0.039807	8.28E-09		1.523958	63.41629	3.919022	0.000199	61.44409
15	0.390625		1.782105	90.54875	4.03262	0.017728	8.13E-09		1.497917	61.47143	4.074942	0.00017	61.3919
16	0.416667		1.756064	87.02477	4.809833	0.008149	8.07E-09		1.471875	59.63977	4.223454	0.000146	61.34695
17	0.442708		1.730022	83.70588	5.555854	0.003865	8.04E-09		1.445833	57.91471	4.364827	0.000127	61.30796
17	0.442708		1.730022	83.70588	5.555854	0.003865	8E-09		1.445833	57.91471	4.364827	0.000127	61.26899

Table A.8.1.a Temperature and retention distribution histories for heating medium at 145°C

						Center						Center + 1				
		Time (min)	f_h/t	log term		Temp (°C)		k (1/min)	Retention (%)	log term		Temp (°C)		k (1/min)	Retention (%)	log term
Input Conditions																
Medium Temp	145°C															
Initial Temp	70°C								100						100	
j - value	1.959	2	0.020202	2.167096	2.146894	70	5.627749	0.000156	99.96871	2.154295	2.134093	70	5.627749	0.000156	99.96871	2.075884
Heating rate	99 min	4	0.040404	2.167096	2.126692	70	5.627749	0.000156	99.93744		2.113891	70	5.627749	0.000156	99.93744	
k - ref	0.0435/min	6	0.060606	2.167096	2.10649	70	5.627749	0.000156	99.90617		2.093689	70	5.627749	0.000156	99.90617	
Act Energy	97 kJ/mole	8	0.080808	2.167096	2.086288	70	5.627749	0.000156	99.87491		2.073487	70	5.627749	0.000156	99.87491	
R - value	8.314	10	0.10101	2.167096	2.066086	70	5.627749	0.000156	99.84367		2.053285	70	5.627749	0.000156	99.84367	
Ref T	138°C	12	0.121212	2.167096	2.045884	70	5.627749	0.000156	99.81243		2.033082	70	5.627749	0.000156	99.81243	
Time Increment	2 min	14	0.141414	2.167096	2.025682	70	5.627749	0.000156	99.7812		2.01288	70	5.627749	0.000156	99.7812	
Cmed Temp	35°C	16	0.161616	2.167096	2.00548	70	5.627749	0.000156	99.74999		1.992678	70	5.627749	0.000156	99.74999	
j2-value	1.9021	18	0.181818	2.167096	1.985278	70	5.627749	0.000156	99.71878		1.972476	70	5.627749	0.000156	99.71878	
j3-value	1.5879	20	0.20202	2.167096	1.965075	70	5.627749	0.000156	99.68758		1.952274	70	5.627749	0.000156	99.68758	
j4-value	0.918	22	0.222222	2.167096	1.944873	70	5.627749	0.000156	99.65639		1.932072	70	5.627749	0.000156	99.65639	
j5-value	0.2968	24	0.242424	2.167096	1.924671	70	5.627749	0.000156	99.62521		1.91187	70	5.627749	0.000156	99.62521	
Cooling rate	100 min	26	0.262626	2.167096	1.904469	70	5.627749	0.000156	99.59404		1.891668	70	5.627749	0.000156	99.59404	
Cooling - j	1.958	28	0.282828	2.167096	1.884267	70	5.627749	0.000156	99.56288		1.871466	70.61826	5.566547	0.000166	99.56092	
		30	0.30303	2.167096	1.864065	70	5.627749	0.000156	99.53174		1.851264	73.99903	5.235742	0.000232	99.51482	
		32	0.323232	2.167096	1.843863	75.19872	5.119897	0.00026	99.48		1.831062	77.22613	4.92593	0.000316	99.45202	
		34	0.343434	2.167096	1.823661	78.3713	4.817359	0.000352	99.41002		1.81086	80.30656	4.635479	0.000422	99.36812	
		36	0.363636	2.167096	1.803459	81.39967	4.533624	0.000467	99.31716		1.790658	83.24697	4.362916	0.000554	99.25803	
		38	0.383838	2.167096	1.783257	84.29041	4.267273	0.00061	99.19609		1.770456	86.05374	4.106905	0.000716	99.116	
		40	0.40404	2.167096	1.763055	87.04975	4.017017	0.000783	99.04082		1.750254	88.73294	3.866237	0.000911	98.93563	
		42	0.424242	2.167096	1.742853	89.68368	3.781688	0.000991	98.84469		1.730052	91.29037	3.63981	0.001142	98.70988	
		44	0.444444	2.167096	1.722651	92.19789	3.560222	0.001237	98.60049		1.70985	93.73155	3.42662	0.001414	98.4312	
		46	0.464646	2.167096	1.702449	94.59783	3.351648	0.001524	98.30048		1.689648	96.06178	3.225751	0.001728	98.09159	
		48	0.484848	2.167096	1.682247	96.88869	3.155079	0.001855	97.93653		1.669446	98.2861	3.036363	0.002088	97.68274	
		50	0.505051	2.167096	1.662045	99.07542	2.969702	0.002232	97.50025		1.649244	100.4093	2.857689	0.002497	97.19614	
		52	0.525253	2.167096	1.641843	101.1628	2.794772	0.002659	96.9831		1.629042	102.436	2.689021	0.002956	96.62326	
		54	0.545455	2.167096	1.621641	103.1552	2.629604	0.003137	96.3766		1.60884	104.3706	2.529709	0.003466	95.95575	
		56	0.565657	2.167096	1.601439	105.0571	2.473567	0.003666	95.67248		1.588638	106.2173	2.379154	0.004029	95.18558	
		58	0.585859	2.167096	1.581237	106.8726	2.32608	0.004249	94.86291		1.568436	107.98	2.236804	0.004646	94.30525	
		60	0.606061	2.167096	1.561035	108.6056	2.186606	0.004885	93.94062		1.548234	109.6626	2.102147	0.005315	93.30802	
		62	0.626263	2.167096	1.540833	110.2597	2.054647	0.005574	92.89919		1.528032	111.2688	1.974711	0.006038	92.18803	
		64	0.646465	2.167096	1.520631	111.8387	1.929744	0.006316	91.73315		1.50783	112.8019	1.854056	0.006812	90.94056	
		66	0.666667	2.167096	1.500429	113.346	1.811471	0.007108	90.4382		1.487628	114.2654	1.739777	0.007637	89.56211	
		68	0.686869	2.167096	1.480227	114.7847	1.699431	0.007951	89.01139		1.467426	115.6623	1.631495	0.00851	88.05063	
		70	0.707071	2.167096	1.460025	116.158	1.593257	0.008842	87.45116		1.447224	116.9958	1.528858	0.00943	86.40554	
		2	0.02	2.201144	2.181144	117.6	1.482579	0.009877	85.74064	2.193025	2.173025	118.4	1.421527	0.010499	84.6102	2.13845
		4	0.04	2.201144	2.161144	118.9	1.383496	0.010905	83.89081		2.153025	119.8	1.315285	0.011675	82.65739	
		6	0.06	2.201144	2.141144	120.2	1.285069	0.012033	81.89591		2.133025	121.1	1.217307	0.012877	80.55578	
		8	0.08	2.201144	2.121144	121.4	1.194789	0.01317	79.76687		2.113025	122.4	1.119974	0.014194	78.30119	
		10	0.1	2.201144	2.101144	122.5	1.112513	0.0143	77.51788		2.093025	123.6	1.030694	0.015519	75.9082	
		12	0.12	2.201144	2.081144	123.5	1.038113	0.015404	75.16607		2.073025	124.8	0.941953	0.016959	73.3767	
		14	0.14	2.201144	2.061144	124.4	0.971473	0.016466	72.73104		2.053025	125.9	0.861075	0.018388	70.72725	
		16	0.16	2.201144	2.041144	125.2	0.912491	0.017466	70.23423		2.033025	127	0.780643	0.019928	67.9638	
		18	0.18	2.201144	2.021144	125.9	0.861075	0.018388	67.69825		2.013025	128	0.707906	0.021431	65.11224	
		20	0.2	2.201144	2.001144	126.5	0.817148	0.019213	65.14616		1.993025	129	0.63553	0.02304	62.17995	
		22	0.22	2.201144	1.981144	127	0.780643	0.019928	62.60077		1.973025	128.9777	0.637142	0.023003	59.38412	
		24	0.24	2.201144	1.961144	126.4417	0.821413	0.019132	60.2507		1.953025	124.748	0.945788	0.016894	57.41114	
		26	0.26	2.201144	1.941144	122.3261	1.125488	0.014115	58.57355		1.933025	120.7087	1.246733	0.012504	55.99324	
		28	0.28	2.201144	1.921144	118.3958	1.421847	0.010495	57.35689		1.913025	116.8511	1.539955	0.009326	54.95853	
		30	0.3	2.201144	1.901144	114.6424	1.710478	0.007864	56.46185		1.893025	113.1672	1.825449	0.00701	54.19341	
		32	0.32	2.201144	1.881144	111.0579	1.991385	0.005938	55.79527		1.873025	109.6491	2.103225	0.00531	53.62095	
		34	0.34	2.201144	1.861144	107.6347	2.264587	0.004518	55.29332		1.853025	106.2894	2.373309	0.004053	53.18806	
		36	0.36	2.201144	1.837508	103.7872	2.577579	0.003304	54.92913		1.829388	102.5131	2.682639	0.002975	52.87257	
		38	0.38	2.201144	1.821144	101.2436	2.788034	0.002677	54.63582		1.813025	100.0167	2.890578	0.002416	52.61769	
		40	0.4	2.201144	1.801144	98.26218	3.038388	0.002084	54.40855		1.793025	97.09044	3.137884	0.001887	52.4195	
		42	0.42	2.201144	1.781144	95.41491	3.281257	0.001635	54.23095		1.773025	94.29591	3.377738	0.001484	52.26411	
		44	0.44	2.201144	1.761144	92.69579	3.516725	0.001292	54.09102		1.753025	91.62715	3.610228	0.001176	52.14128	
		46	0.46	2.201144	1.741144	90.09906	3.744888	0.001028	53.97989		1.733025	89.07851	3.835455	0.000939	52.04342	
		48	0.48	2.201144	1.721144	87.61919	3.965849	0.000824	53.89096		1.713025	86.64457	4.053524	0.000755	51.96488	
		54	0.54	2.201144	1.661144	80.8294	4.586683	0.000443	53.71341		1.653025	79.98054	4.665979	0.000409	51.8072	
		56	0.56	2.201144	1.641144	78.76674	4.780032	0.000365	53.67419		1.633025	77.95609	4.856642	0.000338	51.77216	
		58	0.58	2.201144	1.621144	76.79691	4.966807	0.000303	53.64168		1.613025	76.02274	5.040789	0.000281	51.74303	
		60	0.6	2.201144	1.601144	74.91574	5.14715	0.000253	53.61454		1.593025	74.17642	5.218562	0.000236	51.71866	
		62	0.62	2.201144	1.581144	73.11923	5.321207	0.000213	53.59175		1.573025	72.41319	5.390109	0.000198	51.69814	
		64	0.64	2.201144	1.561144	71.40359	5.489124	0.00018	53.57249		1.553025	70.72932	5.555577	0.000168	51.68075	
		66	0.66	2.201144	1.541144	69.76515	5.651054	0.000153	53.55612		1.533025	69.12123	5.715118	0.000143	51.66594	
		68	0.68	2.201144	1.521144	68.20046	5.807147	0.000131	53.54211		1.513025	67.58552	5.868886	0.000123	51.65323	
		70	0.7	2.201144	1.501144	66.7062	5.957557	0.000113	53.53007		1.493025	66.11893	6.017033	0.000106	51.64229	
		72	0.72	2.201144	1.481144	65.27918	6.102438	9.73E-05	53.51965		1.473025	64.71835	6.159713	9.19E-05	51.63279	

Center + 2						Center + 3						Center + 4				
	Temp (°C)		k (1/min)	Retention (%)	log term		Temp (°C)		k (1/min)	Retention	log term		Temp (°C)		k (1/min)	Retention
				100						100						100
2.055682	70	5.627749	0.000156	99.96871	1.837904	1.817702	79.27934	4.73177	0.000383	99.92337	1.347525	1.327323	123.7518	1.019442	0.015695	96.90983
2.03548	70	5.627749	0.000156	99.93744		1.7975	82.26644	4.453305	0.000506	99.82224		1.307121	124.7175	0.948035	0.016856	93.6972
2.015278	70	5.627749	0.000156	99.90617		1.777298	85.11778	4.19183	0.000658	99.69102		1.286919	125.6394	0.880197	0.01804	90.37695
1.995076	70	5.627749	0.000156	99.87491		1.757096	87.83952	3.946094	0.000841	99.52351		1.266717	126.5194	0.815733	0.019241	86.96519
1.974874	70	5.627749	0.000156	99.84367		1.736894	90.43755	3.714961	0.00106	99.31285		1.246515	127.3593	0.754465	0.020456	83.479
1.954672	70	5.627749	0.000156	99.81243		1.716692	92.9175	3.497395	0.001317	99.0516		1.226313	128.1611	0.69622	0.021683	79.93617
1.93447	70	5.627749	0.000156	99.7812		1.69649	95.28473	3.292451	0.001617	98.73187		1.206111	128.9265	0.640839	0.022918	76.35493
1.914268	70	5.627749	0.000156	99.74999		1.676288	97.54437	3.099265	0.001961	98.34538		1.185909	129.657	0.588172	0.024157	72.75356
1.894066	70	5.627749	0.000156	99.71878		1.656086	99.7013	2.917044	0.002353	97.88364		1.165707	130.3544	0.538077	0.025398	69.15021
1.873864	70.20644	5.607289	0.00016	99.68693		1.635884	101.7602	2.745063	0.002795	97.33807		1.145505	131.0201	0.49042	0.026638	65.56257
1.853662	73.60592	5.273875	0.000223	99.64251		1.615682	103.7255	2.582652	0.003287	96.70019		1.125303	131.6555	0.445075	0.027874	62.00763
1.83346	76.85089	4.96166	0.000305	99.58183		1.59548	105.6015	2.429196	0.003833	95.96178		1.105101	132.262	0.401924	0.029103	58.50146
1.813258	79.94838	4.668992	0.000408	99.50059		1.575278	107.3922	2.284126	0.004431	95.11511		1.084899	132.841	0.360854	0.030323	55.05902
1.793056	82.90507	4.394377	0.000537	99.39376		1.555076	109.1016	2.146919	0.005083	94.15313		1.064697	133.3936	0.321761	0.031532	51.69401
1.772854	85.72738	4.136467	0.000695	99.25568		1.534874	110.7332	2.017088	0.005787	93.06962		1.044495	133.9211	0.284543	0.032728	48.41873
1.752652	88.42141	3.894038	0.000886	99.08		1.514672	112.2907	1.894184	0.006544	91.85943		1.024293	134.4247	0.249107	0.033908	45.24402
1.73245	90.993	3.665974	0.001113	98.85975		1.49447	113.7774	1.777789	0.007352	90.51861		1.004091	134.9054	0.215363	0.035072	42.17919
1.712248	93.4477	3.451263	0.001379	98.58744		1.474268	115.1965	1.667517	0.008209	89.04459		0.983889	135.3642	0.183227	0.036217	39.23199
1.692046	95.79083	3.248977	0.001688	98.25509		1.454066	116.5511	1.563006	0.009113	87.43628		0.963687	135.8021	0.152618	0.037343	36.40866
1.671844	98.02746	3.058268	0.002043	97.85441		1.433864	117.8441	1.463921	0.010063	85.69417		0.943485	136.2202	0.123462	0.038448	33.71394
1.651642	100.1624	2.87836	0.002446	97.3769		1.413662	119.0784	1.369949	0.011054	83.82039		0.923283	136.6193	0.095687	0.039531	31.15112
1.63144	102.2004	2.708539	0.002899	96.81402		1.39346	120.2566	1.280799	0.012085	81.81875		0.903081	137.0002	0.069225	0.040591	28.72216
1.611238	104.1457	2.548149	0.003403	96.15737		1.373257	121.3812	1.196197	0.013152	79.69467		0.882879	137.3638	0.044011	0.041627	26.42776
1.591036	106.0026	2.396585	0.00396	95.39886		1.353055	122.4547	1.11589	0.014252	77.45518		0.862677	137.7109	0.019985	0.042639	24.26746
1.570834	107.7751	2.253288	0.00457	94.53093		1.332853	123.4794	1.039639	0.015381	75.1088		0.842475	138.0422	-0.00291	0.043627	22.23978
1.550632	109.467	2.117744	0.005233	93.54669		1.312651	124.4576	0.96722	0.016536	72.66543		0.822273	138.3584	-0.02473	0.044589	20.34233
1.53043	111.082	1.989474	0.005949	92.4402		1.292449	125.3913	0.898424	0.017714	70.13615		0.802071	138.6603	-0.04553	0.045526	18.57193
1.510228	112.6237	1.868036	0.006718	91.20656		1.272247	126.2825	0.833055	0.01891	67.5331		0.781869	138.9484	-0.06536	0.046438	16.92473
1.490026	114.0952	1.753021	0.007536	89.84214		1.252045	127.1333	0.770929	0.020122	64.86921		0.761667	139.2235	-0.08425	0.047324	15.39631
1.469824	115.4999	1.644046	0.008404	88.34468		1.231843	127.9453	0.711873	0.021347	62.15803		0.741465	139.486	-0.10227	0.048184	13.98185
1.449622	116.8407	1.540756	0.009319	86.71344		1.211641	128.7205	0.655724	0.022579	59.41349		0.721263	139.7366	-0.11944	0.049019	12.67615
1.42942	118.1206	1.442822	0.010277	84.94927		1.191439	129.4604	0.602328	0.023818	56.64965		0.701061	139.9759	-0.13582	0.049828	11.4738
1.409218	119.3423	1.349935	0.011278	83.05466		1.171237	130.1667	0.551543	0.025059	53.88049		0.680858	140.2042	-0.15143	0.050612	10.36922
1.389016	120.5085	1.261808	0.012317	81.03374		1.151035	130.8409	0.503231	0.026299	51.11973		0.660656	140.4222	-0.16632	0.051371	9.356765
1.368814	121.6217	1.178173	0.013391	78.89228		1.130833	131.4845	0.457265	0.027536	48.38058		0.640454	140.6303	-0.18051	0.052106	8.430771
2.11845	122.7	1.097603	0.014515	76.63502	1.909685	1.889685	112.5684	1.872369	0.006689	47.7377	1.496252	1.476252	64.94003	6.137051	9.4E-05	8.427601
2.09845	123.8	1.015866	0.015751	74.25851		1.869685	109.0773	2.148858	0.005073	47.25581		1.456252	63.59251	6.275265	8.19E-05	8.424842
2.07845	124.8	0.941953	0.016959	71.78203		1.849685	105.7433	2.417661	0.003877	46.8908		1.436252	62.30564	6.408296	7.17E-05	8.422426
2.05845	125.8	0.86841	0.018253	69.20875		1.829685	102.5593	2.678822	0.002986	46.6116		1.416252	61.07668	6.536297	6.31E-05	8.420302
2.03845	126.7	0.802535	0.019496	66.56206		1.809685	99.51862	2.932395	0.002317	46.39608		1.396252	59.90304	6.659418	5.58E-05	8.418424
2.01845	127.6	0.736957	0.020818	63.84762		1.789685	96.6148	3.178451	0.001812	46.22826		1.376252	58.78221	6.777812	4.95E-05	8.416756
1.99845	128.4	0.678912	0.022062	61.09168		1.769685	93.84168	3.417069	0.001427	46.0965		1.356252	57.71184	6.891625	4.42E-05	8.415268
1.97845	129.2	0.621099	0.023375	58.30139		1.749685	91.19337	3.648339	0.001132	45.99221		1.336252	56.68964	7.001007	3.96E-05	8.413934
1.95845	125.8761	0.862829	0.018356	56.19989		1.729685	88.66425	3.872363	0.000905	45.90902		1.316252	55.71344	7.1061	3.57E-05	8.412733
1.93845	121.786	1.165866	0.013557	54.69657		1.709685	86.24896	4.089248	0.000729	45.84216		1.296252	54.78118	7.207048	3.22E-05	8.411648
1.91845	117.88	1.461184	0.01009	53.60382		1.689685	83.94238	4.299111	0.000591	45.78803		1.276252	53.89088	7.30399	2.93E-05	8.410664
1.89845	114.1498	1.748773	0.007568	52.79853		1.669685	81.73961	4.502077	0.000482	45.74389		1.256252	53.04065	7.397063	2.67E-05	8.409766
1.87845	110.5874	2.028641	0.005721	52.19787		1.649685	79.63598	4.698275	0.000396	45.70765		1.236252	52.22869	7.486401	2.44E-05	8.408946
1.85845	107.1854	2.300808	0.004358	51.74492		1.629685	77.62703	4.887841	0.000328	45.67768		1.216252	51.45327	7.572136	2.24E-05	8.408193
1.83845	103.9366	2.565311	0.003345	51.3999		1.609685	75.7085	5.070913	0.000273	45.65275		1.196252	50.71275	7.654395	2.06E-05	8.4075
1.81845	100.8339	2.822202	0.002587	51.13463		1.589685	73.87631	5.247636	0.000229	45.63186		1.176252	50.00556	7.733304	1.91E-05	8.406859
1.79845	97.87089	3.071543	0.002016	50.92885		1.569685	72.12659	5.418157	0.000193	45.61425						
1.774813	94.54061	3.356589	0.001516	50.77465		1.546049	70.15999	5.61189	0.000159	45.59976						
1.75845	92.33894	3.547888	0.001252	50.64766		1.529685	68.85985	5.741192	0.00014	45.58702						
1.73845	89.75826	3.775075	0.000998	50.5467		1.509685	67.33591	5.89401	0.00012	45.57609						
1.71845	87.29373	3.995074	0.000801	50.46582		1.489685	65.88055	6.041234	0.000103	45.56666						
1.69845	84.94013	4.208	0.000647	50.40055		1.469685	64.4907	6.183016	8.98E-05	45.55848						
1.67845	82.69245	4.413973	0.000527	50.34749		1.449685	63.1634	6.319512	7.83E-05	45.55134						
1.65845	80.54594	4.61312	0.000432	50.30405		1.429685	61.89584	6.450874	6.87E-05	45.54508						
1.63845	78.49603	4.805575	0.000356	50.26825		1.409685	60.68533	6.577256	6.05E-05	45.53957						
1.61845	76.53839	4.991476	0.000296	50.23854		1.389685	59.5293	6.698808	5.36E-05	45.53469						
1.59845	74.66885	5.170963	0.000247	50.21372												
1.57845	72.88346	5.344184	0.000208	50.19286												
1.55845	71.17842	5.511286	0.000176	50.17522												
1.53845	69.55013	5.67242	0.00015	50.16021												
1.51845	67.99511	5.827739	0.000128	50.14736												
1.49845	66.51009	5.977395	0.00011	50.1363												
1.47845	65.0919	6.121543	9.55E-05	50.12672												
1.45845	63.73754	6.260336	8.31E-05	50.11839												
1.43845	62.44414	6.393929	7.27E-05	50.1111												
1.41845	61.20895	6.522475	6.39E-05	50.1047												

Table A.8.1.b Temperature and retention distribution histories for heating medium at 135°C

						Center						Center + 1				
		Time (t) (min)	f_h/t	log term		Temp (°C)		k (1/min)	Retention (%)	log term		Temp (°C)		k (1/min)	Retention (%)	log term
Input values																
Medium Temp	135°C								100						100	
Initial Temp	70°C	2	0.020202	2.104948	2.084746	70	5.627749	0.000156	99.96871	2.092147	2.071945	70	5.627749	0.000156	99.96871	2.013737
j - value	1.959	4	0.040404		2.064544	70	5.627749	0.000156	99.93744		2.051743	70	5.627749	0.000156	99.93744	
Heating rate	99 min	6	0.060606		2.044342	70	5.627749	0.000156	99.90617		2.031541	70	5.627749	0.000156	99.90617	
k - ref	0.0435/min	8	0.080808		2.02414	70	5.627749	0.000156	99.87491		2.011339	70	5.627749	0.000156	99.87491	
Act Energy	97 kJ/mole	10	0.10101		2.003938	70	5.627749	0.000156	99.84367		1.991137	70	5.627749	0.000156	99.84367	
R - value	8.314	12	0.121212		1.983736	70	5.627749	0.000156	99.81243		1.970935	70	5.627749	0.000156	99.81243	
Ref T	138°C	14	0.141414		1.963534	70	5.627749	0.000156	99.7812		1.950733	70	5.627749	0.000156	99.7812	
Time Increment	2 min	16	0.161616		1.943332	70	5.627749	0.000156	99.74999		1.930531	70	5.627749	0.000156	99.74999	
Cmed Temp	35°C	18	0.181818		1.92313	70	5.627749	0.000156	99.71878		1.910329	70	5.627749	0.000156	99.71878	
j2-value	1.9021	20	0.20202		1.902928	70	5.627749	0.000156	99.68758		1.890127	70	5.627749	0.000156	99.68758	
j3-value	1.5879	22	0.222222		1.882726	70	5.627749	0.000156	99.65639		1.869924	70	5.627749	0.000156	99.65639	
j4-value	0.918	24	0.242424		1.862524	70	5.627749	0.000156	99.62521		1.849722	70	5.627749	0.000156	99.62521	
j5-value	0.2968	26	0.262626		1.842322	70	5.627749	0.000156	99.59404		1.82952	70	5.627749	0.000156	99.59404	
Cooling rate	100 min	28	0.282828		1.82212	70	5.627749	0.000156	99.56288		1.809318	70.53583	5.574694	0.000165	99.56119	
Cooling - j	1.958	30	0.30303		1.801917	70	5.627749	0.000156	99.53174		1.789116	73.46582	5.287486	0.00022	99.51742	
		32	0.323232		1.781715	74.50556	5.186732	0.000243	99.48334		1.768914	76.26265	5.017828	0.000288	99.46013	
		34	0.343434		1.761513	77.25512	4.923172	0.000317	99.42039		1.748712	78.93235	4.764424	0.000371	99.38636	
		36	0.363636		1.741311	79.87972	4.675423	0.000405	99.33979		1.72851	81.48071	4.526098	0.000471	99.29283	
		38	0.383838		1.721109	82.38502	4.442348	0.000512	99.23813		1.708308	83.91324	4.30178	0.000589	99.17589	
		40	0.40404		1.700907	84.77645	4.222911	0.000638	99.11168		1.688106	86.23522	4.09049	0.000728	99.03164	
		42	0.424242		1.680705	87.05919	4.016168	0.000784	98.9564		1.667904	88.45165	3.891337	0.000888	98.85588	
		44	0.444444		1.660503	89.23817	3.821252	0.000953	98.76804		1.647702	90.56735	3.703501	0.001072	98.64421	
		46	0.464646		1.640301	91.31812	3.63737	0.001145	98.54212		1.6275	92.58688	3.52623	0.00128	98.39209	
		48	0.484848		1.620099	93.30353	3.463794	0.001362	98.27405		1.607298	94.51462	3.358834	0.001513	98.09485	
		50	0.505051		1.599897	95.1987	3.299853	0.001605	97.95917		1.587096	96.35474	3.200676	0.001772	97.74783	
		52	0.525253		1.579695	97.00773	3.144931	0.001874	97.59279		1.566894	98.11123	3.05117	0.002058	97.34638	
		54	0.545455		1.559493	98.73454	2.998456	0.002169	97.17034		1.546692	99.78788	2.909774	0.00237	96.88601	
		56	0.565657		1.539291	100.3829	2.859903	0.002491	96.68736		1.52649	101.3883	2.775985	0.00271	96.3624	
		58	0.585859		1.519089	101.9563	2.728784	0.00284	96.13963		1.506288	102.916	2.649341	0.003075	95.77152	
		60	0.606061		1.498887	103.4581	2.604647	0.003216	95.52326		1.486086	104.3743	2.529409	0.003467	95.1097	
		62	0.626263		1.478685	104.8918	2.487072	0.003617	94.83471		1.465884	105.7663	2.415789	0.003884	94.37367	
		64	0.646465		1.458483	106.2602	2.375671	0.004043	94.07089		1.445682	107.095	2.30811	0.004326	93.56066	
		66	0.666667		1.438281	107.5665	2.27008	0.004494	93.22922		1.42548	108.3633	2.206025	0.004791	92.66845	
		68	0.686869		1.418079	108.8134	2.169962	0.004967	92.30768		1.405278	109.574	2.109212	0.005278	91.69538	
		70	0.707071		1.397877	110.0036	2.075004	0.005462	91.30486		1.385076	110.7297	2.017368	0.005786	90.64045	
		72	0.727273		1.377675	111.1397	1.98491	0.005977	90.21997		1.364874	111.8328	1.930214	0.006313	89.50329	
		74	0.747475		1.357473	112.2242	1.899407	0.00651	89.0529		1.344672	112.8858	1.847486	0.006857	88.28422	
		76	0.767677		1.337271	113.2594	1.818237	0.007061	87.80422		1.32447	113.8909	1.768938	0.007417	86.98421	
		78	0.787879		1.317069	114.2476	1.741162	0.007626	86.47514		1.304268	114.8503	1.69434	0.007992	85.60494	
		80	0.808081		1.296867	115.1908	1.667957	0.008205	85.06758		1.284066	115.7662	1.623476	0.008579	84.1487	
		82	0.828283		1.276665	116.0912	1.598409	0.008796	83.58408		1.263864	116.6404	1.556143	0.009176	82.61845	
		2	0.02	2.200786	2.180786	117	1.528532	0.009433	82.02194	2.191138	2.171138	117.5	1.490228	0.009801	81.01465	2.128569
		4	0.04		2.160786	117.9	1.459655	0.010106	80.38079		2.151138	118.4	1.421527	0.010499	79.33132	
		6	0.06		2.140786	118.7	1.398697	0.010741	78.67247		2.131138	119.2	1.360724	0.011157	77.58078	
		8	0.08		2.120786	119.5	1.337987	0.011413	76.897		2.111138	120	1.300169	0.011853	75.76326	
		10	0.1		2.100786	120.2	1.285069	0.012033	75.06842		2.091138	120.7	1.247385	0.012496	73.89331	
		12	0.12		2.080786	120.9	1.232339	0.012685	73.18788		2.071138	121.4	1.194789	0.01317	71.97231	
		14	0.14		2.060786	121.5	1.18729	0.01327	71.27109		2.051138	122	1.149854	0.013776	70.01644	
		16	0.16		2.040786	122	1.149854	0.013776	69.33427		2.031138	122.6	1.105056	0.014407	68.02779	
		18	0.18		2.020786	122.4	1.119974	0.014194	67.39375		2.011138	123.1	1.067828	0.014953	66.02343	
		20	0.2		2.000786	122.7	1.097603	0.014515	65.46548		1.991138	123.6	1.030694	0.015519	64.00567	
		22	0.22		1.980786	122.9	1.082708	0.014732	63.56469		1.971138	124	1.001054	0.015986	61.99166	
		24	0.24		1.960786	123	1.075266	0.014843	61.7055		1.951138	124.359	0.974502	0.016416	59.98939	
		26	0.26		1.940786	122.2542	1.130861	0.01404	59.99693		1.931138	120.3372	1.27472	0.012159	58.54821	
		28	0.28		1.920786	118.3271	1.427081	0.01044	58.75714		1.911138	116.4964	1.567213	0.009075	57.49512	
		30	0.3		1.900786	114.5767	1.715574	0.007824	57.84487		1.891138	112.8284	1.851978	0.006826	56.71549	
		32	0.32		1.880786	110.9952	1.996342	0.005909	57.16532		1.871138	109.3256	2.129027	0.005174	56.13158	
		34	0.34		1.860786	107.5749	2.269407	0.004497	56.65351		1.851138	105.9804	2.398387	0.003953	55.68959	
		36	0.36		1.840786	104.3084	2.534804	0.003449	56.26411		1.831138	102.7857	2.660101	0.003042	55.35176	
		38	0.38		1.820786	101.1891	2.792581	0.002665	55.96503		1.811138	99.73488	2.914224	0.00236	55.09115	
		40	0.4		1.800786	98.21006	3.0428	0.002075	55.73325		1.791138	96.82133	3.160824	0.001844	54.88834	
		42	0.42		1.780786	95.36514	3.285536	0.001628	55.5521		1.771138	94.03891	3.399979	0.001452	54.7292	
		44	0.44		1.760786	92.64826	3.520873	0.001286	55.40936		1.751138	91.38172	3.63178	0.001151	54.60332	
		46	0.46		1.740786	90.05366	3.748906	0.001024	55.29598		1.731138	88.84412	3.856327	0.00092	54.50296	
		48	0.48		1.720786	87.57584	3.969739	0.000821	55.20523		1.711138	86.42074	4.073726	0.00074	54.42234	
		50	0.5		1.700786	85.20953	4.183485	0.000663	55.13206		1.691138	84.10643	4.284096	0.0006	54.35711	
		52	0.52		1.680786	82.94973	4.390264	0.000539	55.07263		1.671138	81.89627	4.487559	0.000489	54.30394	
		54	0.54		1.660786	80.79164	4.590202	0.000442	55.02402		1.651138	79.78559	4.684244	0.000402	54.26031	
		56	0.56		1.640786	78.73068	4.783432	0.000364	54.98398		1.631138	77.76991	4.874287	0.000332	54.22425	
		58	0.58		1.620786	76.76247	4.970091	0.000302	54.95078		1.611138	75.84495	5.057826	0.000277	54.19426	
		60	0.6		1.600786	74.88285	5.15032	0.000252	54.92307		1.591138	74.00662	5.235006	0.000232	54.16915	
		58	0.58		1.620786	76.76247	4.970091	0.000302	54.8899		1.611138	75.84495	5.057826	0.000277	54.13919	
		60	0.6		1.600786	74.88285	5.15032	0.000252	54.86222		1.591138	74.00662	5.235006	0.000232	54.1141	
		62	0.62		1.580786	73.08783	5.324266	0.000212	54.83897		1.571138	72.25103	5.405973	0.000195	54.09297	
		64	0.64		1.560786	71.37359	5.492075	0.000179	54.81933		1.551138	70.57446	5.570875	0.000166	54.07506	
		66	0.66		1.540786	69.73651	5.653899	0.000152	54.80262		1.531138	68.97335	5.729866	0.000141	54.05978	
		68	0.68		1.520786	68.17311	5.809889	0.00013	54.78833		1.511138	67.44429	5.883097	0.000121	54.04668	
		70	0.7		1.500786	66.68007	5.960198	0.000112	54.77603		1.491138	65.98406	6.030721	0.000105	54.03538	
		72	0.72		1.480786	65.25424	6.104982	9.71E-05	54.7654		1.471138	64.58955	6.172894	9.07E-05	54.02558	
		74	0.74		1.460786	63.89257	6.244393	8.44E-05	54.75615		1.451138	63.2578	6.309768	7.91E-05	54.01703	
		76	0.76		1.440786	62.59219	6.378585	7.38E-05	54.74806		1.431138	61.98599	6.441498	6.93E-05	54.00954	
		78	0.78		1.420786	61.35034	6.507713	6.49E-05	54.74096		1.411138	60.77142	6.568237	6.11E-05	54.00294	
		80	0.8		1.400786	60.16438	6.631927	5.73E-05	54.73468		1.391138	59.61151	6.690135	5.41E-05	53.9971	

	Center + 2						Center + 3						Center + 4			
	Temp (°C)		k (1/min)	Retention (%)	log term		Temp (°C)		k (1/min)	Retention (%)	log term		Temp (°C)		k (1/min)	Retention (%)
				100						100						100
1.993534	70	5.627749	0.000156	99.96871	1.775756	1.755554	78.04209	4.848497	0.000341	99.93181	1.285377	1.265175	116.5849	1.560411	0.009137	98.18916
1.973332	70	5.627749	0.000156	99.93744		1.735352	80.63092	4.60519	0.000435	99.84491		1.244973	117.4218	1.496209	0.009743	96.29436
1.95313	70	5.627749	0.000156	99.90617		1.71515	83.10208	4.376242	0.000547	99.73576		1.224771	118.2208	1.435181	0.010356	94.3204
1.932928	70	5.627749	0.000156	99.87491		1.694948	85.46092	4.160644	0.000678	99.60051		1.204569	118.9834	1.377159	0.010975	92.27266
1.912726	70	5.627749	0.000156	99.84367		1.674746	87.71255	3.957476	0.000831	99.43505		1.184367	119.7114	1.321984	0.011597	90.15705
1.892524	70	5.627749	0.000156	99.81243		1.654544	89.86183	3.765894	0.001007	99.23501		1.164165	120.4063	1.269508	0.012222	87.97994
1.872322	70	5.627749	0.000156	99.7812		1.634342	91.91343	3.585126	0.001206	98.99586		1.143963	121.0696	1.21959	0.012848	85.74804
1.85212	70	5.627749	0.000156	99.74999		1.61414	93.87178	3.414459	0.001431	98.71296		1.123761	121.7028	1.172097	0.013473	83.46838
1.831918	70	5.627749	0.000156	99.71878		1.593938	95.74112	3.253241	0.001681	98.3816		1.103559	122.3072	1.126904	0.014096	81.14818
1.811716	70.17891	5.610015	0.000159	99.68702		1.573736	97.5255	3.100868	0.001958	97.99711		1.083357	122.8841	1.083894	0.014715	78.79479
1.791514	73.12513	5.320632	0.000213	99.64462		1.553534	99.22877	2.956783	0.002261	97.55489		1.063155	123.4347	1.042956	0.01533	76.41562
1.771312	75.93744	5.048961	0.000279	99.58902		1.533332	100.8546	2.820472	0.002592	97.05054		1.042953	123.9604	1.003985	0.015939	74.01804
1.75111	78.62193	4.793691	0.00036	99.51729		1.51313	102.4066	2.691458	0.002949	96.47992		1.022751	124.4622	0.966881	0.016542	71.60935
1.730908	81.1844	4.553634	0.000458	99.42617		1.492928	103.888	2.569299	0.003332	95.83918		1.002549	124.9411	0.931551	0.017136	69.19666
1.710706	83.6304	4.327705	0.000574	99.31207		1.472726	105.3021	2.453584	0.00374	95.12492		0.982347	125.3983	0.897905	0.017723	66.7869
1.690504	85.96523	4.114918	0.00071	99.1711		1.452524	106.6519	2.343934	0.004174	94.33416		0.962145	125.8347	0.865861	0.0183	64.38669
1.670302	88.19393	3.914368	0.000868	98.9991		1.432322	107.9404	2.239991	0.004631	93.46447		0.941943	126.2513	0.835339	0.018867	62.00238
1.6501	90.32134	3.725229	0.001049	98.79168		1.41212	109.1703	2.141428	0.005111	92.514		0.921741	126.649	0.806264	0.019424	59.63992
1.629898	92.35205	3.546742	0.001254	98.5443		1.391918	110.3443	2.047933	0.005612	91.48151		0.901539	127.0285	0.778564	0.019969	57.30492
1.609696	94.29047	3.378208	0.001484	98.25231		1.371716	111.4649	1.959221	0.006132	90.36641		0.881337	127.3908	0.752171	0.020503	55.00256
1.589494	96.14078	3.218985	0.00174	97.91102		1.351514	112.5346	1.875023	0.006671	89.16879		0.861135	127.7367	0.727023	0.021026	52.7376
1.569292	97.90699	3.068482	0.002022	97.5158		1.331312	113.5557	1.795086	0.007226	87.8894		0.840933	128.0668	0.703059	0.021535	50.51436
1.54909	99.59293	2.92615	0.002332	97.06209		1.31111	114.5304	1.719175	0.007796	86.52969		0.820731	128.3819	0.68022	0.022033	48.33673
1.528888	101.2022	2.791483	0.002668	96.54557		1.290908	115.4608	1.647069	0.008379	85.09176		0.800529	128.6827	0.658453	0.022518	46.20815
1.508686	102.7384	2.664014	0.003031	95.96217		1.270706	116.3489	1.578563	0.008973	83.57836		0.780327	128.9699	0.637706	0.02299	44.13161
1.488484	104.2047	2.543306	0.003419	95.30815		1.250504	117.1966	1.513461	0.009576	81.99283		0.760125	129.2439	0.617929	0.023449	42.1097
1.468282	105.6044	2.428957	0.003834	94.5802		1.230301	118.0058	1.451581	0.010188	80.33911		0.739923	129.5056	0.599076	0.023895	40.14458
1.44808	106.9405	2.320592	0.004272	93.77548		1.210099	118.7782	1.392753	0.010805	78.6216		0.719721	129.7553	0.581104	0.024329	38.23801
1.427878	108.2158	2.21786	0.004735	92.89169		1.189897	119.5155	1.336814	0.011427	76.84522		0.699519	129.9937	0.563968	0.024749	36.39137
1.407676	109.4332	2.120437	0.005219	91.92711		1.169695	120.2193	1.283614	0.012051	75.01524		0.679317	130.2212	0.547631	0.025157	34.60568
1.387474	110.5953	2.028019	0.005724	90.88065		1.149493	120.8911	1.233008	0.012677	73.13728		0.659115	130.4384	0.532053	0.025552	32.88163
1.367272	111.7045	1.940322	0.006249	89.75187		1.129291	121.5324	1.184864	0.013302	71.21722		0.638913	130.6458	0.517199	0.025934	31.21959
1.34707	112.7633	1.857082	0.006792	88.54101		1.109089	122.1445	1.139054	0.013925	69.26114		0.618711	130.8437	0.503034	0.026304	29.61964
1.326868	113.774	1.77805	0.00735	87.24896		1.088887	122.7288	1.095458	0.014546	67.27525		0.598509	131.0326	0.489526	0.026662	28.08158
1.306666	114.7388	1.702995	0.007923	85.87732		1.068685	123.2865	1.053963	0.015162	65.26581		0.578307	131.2129	0.476644	0.027008	26.60499
1.286464	115.6597	1.631698	0.008508	84.42831		1.048483	123.8189	1.014464	0.015773	63.23907		0.558105	131.385	0.464358	0.027341	25.18921
1.266262	116.5387	1.563957	0.009105	82.90481		1.028281	124.3271	0.976858	0.016377	61.20124		0.537903	131.5493	0.45264	0.027664	23.83341
1.24606	117.3778	1.499579	0.00971	81.3103		1.008079	124.8122	0.941051	0.016974	59.1584		0.5177	131.7062	0.441463	0.027975	22.53657
1.225858	118.1788	1.438385	0.010323	79.64877		0.987877	125.2753	0.906953	0.017563	57.11645		0.497498	131.8559	0.430803	0.028274	21.29751
1.205656	118.9433	1.380206	0.010941	77.92476		0.967675	125.7173	0.874479	0.018143	55.08108		0.477296	131.9988	0.420635	0.028563	20.11495
1.185454	119.6731	1.324882	0.011564	76.14323		0.947473	126.1392	0.843548	0.018713	53.05772		0.457094	132.1352	0.410935	0.028842	18.98748
2.108569	120.5	1.262447	0.012309	74.29165	1.922548	1.902548	114.9002	1.690472	0.008023	52.21317	1.459841	1.439841	62.53218	6.384804	7.34E-05	18.98469
2.088569	121.3	1.202291	0.013072	72.37455		1.882548	111.3041	1.971921	0.006055	51.58471		1.419841	61.29302	6.513695	6.45E-05	18.98225
2.068569	122	1.149854	0.013776	70.40774		1.862548	107.8698	2.245663	0.004605	51.11182		1.399841	60.10964	6.637681	5.7E-05	18.98008
2.048569	122.7	1.097603	0.014515	68.39324		1.842548	104.5902	2.511733	0.003529	50.75234		1.379841	58.97952	6.756912	5.06E-05	18.97816
2.028569	123.3	1.052963	0.015177	66.34839		1.822548	101.4581	2.770179	0.002725	50.47646		1.359841	57.90027	6.871536	4.51E-05	18.97645
2.008569	123.9	1.008458	0.015868	64.27582		1.802548	98.46698	3.021062	0.002121	50.26283		1.339841	56.86958	6.981702	4.04E-05	18.97492
1.988569	124.4	0.971473	0.016466	62.19359		1.782548	95.6105	3.264453	0.001662	50.09599		1.319841	55.88529	7.087554	3.63E-05	18.97354
1.968569	124.9	0.934582	0.017085	60.10437		1.762548	92.88258	3.500438	0.001313	49.96461		1.299841	54.9453	7.189235	3.28E-05	18.97229
1.948569	123.8319	1.013506	0.015788	58.23616		1.742548	90.27743	3.729111	0.001045	49.86033		1.279841	54.04761	7.286886	2.98E-05	18.97116
1.928569	119.8338	1.312732	0.011705	56.88867		1.722548	87.78954	3.950574	0.000837	49.77692		1.259841	53.19033	7.380643	2.71E-05	18.97013
1.908569	116.0156	1.604232	0.008745	55.90229		1.702548	85.41362	4.164939	0.000676	49.70971		1.239841	52.37163	7.470642	2.48E-05	18.96919
1.888569	112.3693	1.888005	0.006585	55.17091		1.682548	83.14463	4.372327	0.000549	49.65515		1.219841	51.58978	7.557013	2.27E-05	18.96833
1.868569	108.8871	2.164064	0.004996	54.62236		1.662548	80.97777	4.572862	0.000449	49.61056		1.199841	50.84311	7.639887	2.09E-05	18.96754
1.848569	105.5616	2.432439	0.00382	54.2066		1.642548	78.90843	4.766678	0.00037	49.57385		1.179841	50.13006	7.719388	1.93E-05	18.96681
1.828569	102.3859	2.693175	0.002943	53.88843		1.622548	76.93222	4.95391	0.000307	49.54342		1.159841	49.44909	7.79564	1.79E-05	18.96613
1.808569	99.35299	2.946327	0.002285	53.6427		1.602548	75.04496	5.1347	0.000256	49.51805		1.139841	48.79877	7.86876	1.66E-05	18.96549
1.788569	96.45663	3.191965	0.001787	53.45128		1.582548	73.24264	5.309193	0.000215	49.49674		1.119841	48.17773	7.938867	1.55E-05	18.96491
1.768569	93.69062	3.430171	0.001409	53.30091		1.562548	71.52144	5.477536	0.000182	49.47875		1.099841	47.58463	8.006071	1.45E-05	18.96436
1.748569	91.04911	3.661034	0.001118	53.18184		1.542548	69.8777	5.639881	0.000155	49.46346		1.079841	47.01823	8.070483	1.36E-05	18.96384
1.728569	88.52649	3.884656	0.000894	53.08682		1.522548	68.30795	5.796379	0.000132	49.45038		1.059841	46.47732	8.13221	1.28E-05	18.96336
1.708569	86.1174	4.101146	0.00072	53.01042		1.502548	66.80884	5.947183	0.000114	49.43914		1.039841	45.96076	8.191354	1.21E-05	18.9629
1.688569	83.81673	4.310621	0.000584	52.94854		1.482548	65.37721	6.092447	9.83E-05	49.42942		1.019841	45.46744	8.248015	1.14E-05	18.96247
1.668569	81.61962	4.513206	0.000477	52.89806		1.462548	64.01001	6.232325	8.55E-05	49.42097		0.999841	44.99633	8.30229	1.08E-05	18.96206
1.648569	79.52139	4.70903	0.000392	52.8566		1.442548	62.70434	6.366971	7.47E-05	49.41359		0.979841	44.54642	8.354272	1.02E-05	18.96167
1.628569	77.5176	4.898229	0.000325	52.8223		1.422548	61.45744	6.496538	6.56E-05	49.4071		0.959841	44.11676	8.404053	9.74E-06	18.9613
1.608569	75.60399	5.080943	0.00027	52.79376		1.402548	60.26666	6.621179	5.79E-05	49.40138		0.939841	43.70644	8.451719	9.29E-06	18.96095
1.588569	73.77651	5.257316	0.000227	52.76983		1.382548	59.12947	6.741045	5.14E-05	49.3963		0.919841	43.31458	8.497355	8.87E-06	18.96061
1.568569	72.03128	5.427496	0.000191	52.74966												
1.548569	70.3646	5.59163	0.000162	52.73255												
1.528569	68.77293	5.749872	0.000138	52.71795												
1.548569	70.3646	5.59163	0.000162	52.70085												
1.528569	68.77293	5.749872	0.000138	52.68626												
1.508569	67.2529	5.902374	0.000119	52.67373												
1.488569	65.80128	6.049289	0.000103	52.66292												
1.468569	64.41499	6.190773	8.91E-05	52.65354												
1.448569	63.0911	6.326978	7.78E-05	52.64535												
1.428569	61.82679	6.458058	6.82E-05	52.63817												
1.408569	60.61939	6.584166	6.01E-05	52.63184												
1.388569	59.46632	6.705454	5.33E-05	52.62624												

Table A.8.1.c Temperature and retention distribution histories for heating medium at 125°C

							Center						Center + 1			
		Time (t)	f_h/t	log term		Temp. (°C)		Rate Const	Retention (%)	log term		Temp. (°C)		Rate Const	Retention (%)	log term
Input Values									100						100	
Medium Temp	125°C	2	0.020202	2.032397	2.012195	70	5.627749	0.000156	99.96871	2.019596	1.999394	70	5.627749	0.000156	99.96871	1.941186
Initial Temp	70°C	4	0.040404	2.388944	1.991993	70	5.627749	0.000156	99.93744		1.979192	70	5.627749	0.000156	99.93744	
j - value	1.959	6	0.060606	2.388944	1.971791	70	5.627749	0.000156	99.90617		1.95899	70	5.627749	0.000156	99.90617	
Heating rate	99 min	8	0.080808	2.388944	1.951589	70	5.627749	0.000156	99.87491		1.938788	70	5.627749	0.000156	99.87491	
k - ref	0.0435/min	10	0.10101	2.388944	1.931387	70	5.627749	0.000156	99.84367		1.918586	70	5.627749	0.000156	99.84367	
Act Energy	97 kJ/mole	12	0.121212	2.388944	1.911185	70	5.627749	0.000156	99.81243		1.898384	70	5.627749	0.000156	99.81243	
R - value	8.314	14	0.141414	2.388944	1.890983	70	5.627749	0.000156	99.7812		1.878182	70	5.627749	0.000156	99.7812	
Ref T	138°C	16	0.161616	2.388944	1.870781	70	5.627749	0.000156	99.74999		1.85798	70	5.627749	0.000156	99.74999	
Time Increment	2 min	18	0.181818	2.388944	1.850579	70	5.627749	0.000156	99.71878		1.837778	70	5.627749	0.000156	99.71878	
Cmed Temp	35°C	20	0.20202	2.388944	1.830377	70	5.627749	0.000156	99.68758		1.817576	70	5.627749	0.000156	99.68758	
j2-value	1.9021	22	0.222222	2.388944	1.810175	70	5.627749	0.000156	99.65639		1.797374	70	5.627749	0.000156	99.65639	
j3-value	1.5879	24	0.242424	2.388944	1.789973	70	5.627749	0.000156	99.62521		1.777172	70	5.627749	0.000156	99.62521	
j4-value	0.918	26	0.262626	2.388944	1.769771	70	5.627749	0.000156	99.59404		1.75697	70	5.627749	0.000156	99.59404	
j5-value	0.2968	28	0.282828	2.388944	1.749569	70	5.627749	0.000156	99.56288		1.736768	70.45339	5.582846	0.000164	99.56145	
Cooling rate - f	100	30	0.30303	2.388944	1.729367	70	5.627749	0.000156	99.53174		1.716566	72.93262	5.339391	0.000209	99.5199	
Cooling - j	1.958	32	0.323232	2.388944	1.709165	73.8124	5.253835	0.000227	99.48648		1.696364	75.29916	5.110234	0.000263	99.46766	
		34	0.343434	2.388944	1.688963	76.13895	5.029663	0.000285	99.42988		1.676162	77.55814	4.89438	0.000326	99.40288	
		36	0.363636	2.388944	1.668761	78.35976	4.818449	0.000351	99.36002		1.65596	79.71445	4.690915	0.000399	99.32354	
		38	0.383838	2.388944	1.648559	80.47963	4.61931	0.000429	99.27482		1.635758	81.77275	4.499005	0.000484	99.22749	
		40	0.40404	2.388944	1.628357	82.50315	4.431439	0.000518	99.17212		1.615556	83.73749	4.317884	0.00058	99.1125	
		42	0.424242	2.388944	1.608155	84.4347	4.254091	0.000618	99.04963		1.595354	85.61294	4.146847	0.000688	98.97623	
		44	0.444444	2.388944	1.587953	86.27845	4.086582	0.000731	98.90499		1.575152	87.40314	3.985244	0.000809	98.8163	
		46	0.464646	2.388944	1.567751	88.03841	3.928283	0.000856	98.73582		1.55495	89.11197	3.832477	0.000942	98.6303	
		48	0.484848	2.388944	1.547549	89.71837	3.778612	0.000994	98.53969		1.534748	90.74314	3.687992	0.001088	98.41582	
		50	0.505051	2.388944	1.527347	91.32197	3.637031	0.001145	98.31422		1.514546	92.30017	3.551278	0.001248	98.1705	
		52	0.525253	2.388944	1.507145	92.85269	3.503043	0.00131	98.05706		1.494344	93.78643	3.42186	0.00142	97.89202	
		54	0.545455	2.388944	1.486943	94.31384	3.376187	0.001487	97.76592		1.474141	95.20513	3.2993	0.001606	97.57819	
		56	0.565657	2.388944	1.466741	95.70857	3.256035	0.001677	97.43865		1.453939	96.55935	3.183187	0.001803	97.22691	
		58	0.585859	2.388944	1.446539	97.03991	3.142188	0.001879	97.07323		1.433737	97.85202	3.073144	0.002013	96.83626	
		60	0.606061	2.388944	1.426337	98.31074	3.034278	0.002093	96.66778		1.413535	99.08594	2.968815	0.002234	96.4045	
		62	0.626263	2.388944	1.406134	99.52381	2.931959	0.002318	96.22062		1.393333	100.2638	2.869872	0.002467	95.93007	
		64	0.646465	2.388944	1.385932	100.6817	2.834911	0.002554	95.73028		1.373131	101.3881	2.776006	0.002709	95.41163	
		66	0.666667	2.388944	1.36573	101.787	2.742833	0.002801	95.19553		1.352929	102.4613	2.686932	0.002962	94.8481	
		68	0.686869	2.388944	1.345528	102.8421	2.655445	0.003057	94.61535		1.332727	103.4857	2.602379	0.003223	94.23864	
		70	0.707071	2.388944	1.325326	103.8492	2.572485	0.003321	93.98899		1.312525	104.4636	2.522098	0.003493	93.58264	
		72	0.727273	2.388944	1.305124	104.8106	2.493709	0.003593	93.31596		1.292323	105.397	2.445853	0.003769	92.8798	
		74	0.747475	2.388944	1.284922	105.7282	2.418887	0.003872	92.59604		1.272121	106.288	2.373423	0.004053	92.13005	
		76	0.767677	2.388944	1.26472	106.6041	2.347802	0.004158	91.82926		1.251919	107.1384	2.304602	0.004341	91.3336	
		78	0.787879	2.388944	1.244518	107.4402	2.280254	0.004448	91.01593		1.231717	107.9503	2.239196	0.004635	90.4909	
		80	0.808081	2.388944	1.224316	108.2384	2.216052	0.004743	90.1566		1.211515	108.7252	2.177022	0.004932	89.60269	
		82	0.828283	2.388944	1.204114	109.0002	2.155019	0.005042	89.25208		1.191313	109.4649	2.117909	0.005232	88.66993	
		84	0.848485	2.388944	1.183912	109.7274	2.096987	0.005343	88.30343		1.171111	110.171	2.061695	0.005535	87.69379	
		86	0.868687	2.388944	1.16371	110.4216	2.041797	0.005646	87.3119		1.150909	110.845	2.00823	0.005839	86.67569	
		88	0.888889	2.388944	1.143508	111.0842	1.989302	0.00595	86.27897		1.130707	111.4884	1.957369	0.006143	85.61722	
		90	0.909091	2.388944	1.123306	111.7167	1.939362	0.006255	85.20633		1.110505	112.1025	1.908979	0.006448	84.52018	
		92	0.929293	2.388944	1.103104	112.3204	1.891844	0.00656	84.0958		1.090303	112.6887	1.862932	0.006752	83.3865	
		94	0.949495	2.388944	1.082902	112.8967	1.846625	0.006863	82.9494		1.070101	113.2483	1.819108	0.007054	82.21827	
		96	0.969697	2.388944	1.0627	113.4469	1.803587	0.007165	81.76925		1.049899	113.7824	1.777395	0.007355	81.01771	
		98	0.989899	2.388944	1.042498	113.972	1.76262	0.007464	80.55761		1.029697	114.2923	1.737684	0.007653	79.78712	
		100	1.010101	2.388944	1.022296	114.4732	1.723618	0.007761	79.31681		1.009495	114.779	1.699876	0.007948	78.5289	
		102	1.030303	2.388944	1.002094	114.9517	1.686482	0.008055	78.04927		0.989293	115.2435	1.663875	0.008239	77.24549	
		2	0.02	2.19464	2.17464	115.5	1.644036	0.008404	76.74836	2.183643	2.163643	115.7	1.628584	0.008535	75.9381	2.113869
		4	0.04		2.15464	116	1.605436	0.008735	75.41923		2.143643	116.2	1.590024	0.008871	74.60276	
		6	0.06		2.13464	116.5	1.566935	0.009078	74.06231		2.123643	116.7	1.551562	0.009218	73.23992	
		8	0.08		2.11464	116.9	1.536205	0.009361	72.68861		2.103643	117.1	1.520863	0.009506	71.86068	
		10	0.1		2.09464	117.3	1.505538	0.009653	71.2988		2.083643	117.5	1.490228	0.009801	70.46571	
		12	0.12		2.07464	117.6	1.482579	0.009877	69.90422		2.063643	117.8	1.467293	0.010029	69.06641	
		14	0.14		2.05464	117.9	1.459655	0.010106	68.50554		2.043643	118.1	1.444392	0.010261	67.66345	
		16	0.16		2.03464	118.1	1.444392	0.010261	67.11397		2.023643	118.3	1.429145	0.010419	66.26809	
		18	0.18		2.01464	118.3	1.429145	0.010419	65.72994		2.003643	118.5	1.413913	0.010579	64.88075	
		20	0.2		1.99464	118.4	1.421527	0.010499	64.36419		1.983643	118.6	1.406303	0.01066	63.51218	
		22	0.22		1.97464	118.5	1.413913	0.010579	63.01671		1.963643	118.7	1.398697	0.010741	62.16237	

	Center + 2						Center + 3						Center + 4			
	Temp. (°C)		Rate Const	Retention (%)	log term		Temp. (°C)		Rate Const	Retention (%)	log term		Temp. (°C)		Rate Const	Retention (%)
				100						100						100
1.920984	70	5.627749	0.000156	99.96871	1.703205	1.683003	76.80485	4.96605	0.000303	99.93937	1.212827	1.192625	109.418	2.121656	0.005213	98.96287
1.900782	70	5.627749	0.000156	99.93744		1.662801	78.99539	4.758486	0.000373	99.86481		1.172423	110.1262	2.065259	0.005515	97.87727
1.88058	70	5.627749	0.000156	99.90617		1.642599	81.08637	4.562753	0.000454	99.77421		1.152221	110.8022	2.01162	0.005819	96.74477
1.860378	70	5.627749	0.000156	99.87491		1.622397	83.08231	4.37806	0.000546	99.66533		1.132019	111.4475	1.960594	0.006124	95.56712
1.840176	70	5.627749	0.000156	99.84367		1.602195	84.98754	4.203683	0.00065	99.53587		1.111816	112.0635	1.912048	0.006428	94.34631
1.819974	70	5.627749	0.000156	99.81243		1.581993	86.80617	4.038954	0.000766	99.38344		1.091614	112.6515	1.865852	0.006732	93.0845
1.799772	70	5.627749	0.000156	99.7812		1.561791	88.54214	3.883259	0.000895	99.20562		1.071412	113.2128	1.821888	0.007035	91.78401
1.77957	70	5.627749	0.000156	99.74999		1.541589	90.1992	3.736028	0.001037	99		1.05121	113.7485	1.78004	0.007335	90.44728
1.759368	70	5.627749	0.000156	99.71878		1.521387	91.78095	3.596737	0.001192	98.76418		1.031008	114.2599	1.740203	0.007634	89.07689
1.739166	70.15139	5.612742	0.000159	99.68711		1.501185	93.29081	3.4649	0.001361	98.4958		1.010806	114.7481	1.702275	0.007929	87.67551
1.718964	72.64434	5.367519	0.000203	99.64665		1.480983	94.73204	3.340064	0.001541	98.19262		0.990604	115.214	1.666159	0.00822	86.24586
1.698762	75.02399	5.13672	0.000256	99.59572		1.460781	96.10777	3.221812	0.001735	97.85251		0.970402	115.6588	1.631765	0.008508	84.79073
1.67856	77.29548	4.919335	0.000318	99.53245		1.440579	97.42096	3.109754	0.001941	97.47345		0.9502	116.0834	1.599009	0.008791	83.31294
1.658358	79.46372	4.714445	0.00039	99.45485		1.420377	98.67447	3.003528	0.002158	97.05364		0.929998	116.4887	1.567807	0.00907	81.81529
1.638156	81.53341	4.521206	0.000473	99.36079		1.400175	99.87101	2.902797	0.002387	96.59145		0.909796	116.8755	1.538085	0.009343	80.30061
1.617954	83.50904	4.338842	0.000568	99.24803		1.379973	101.0132	2.807245	0.002626	96.08545		0.889594	117.2448	1.509768	0.009612	78.77169
1.597751	85.39487	4.166643	0.000674	99.11425		1.359771	102.1034	2.716578	0.002875	95.53447		0.869392	117.5973	1.482788	0.009875	77.23126
1.577549	87.19498	4.003952	0.000794	98.95706		1.339569	103.1441	2.630523	0.003134	94.93758		0.84919	117.9337	1.45708	0.010132	75.68202
1.557347	88.91328	3.850166	0.000926	98.77406		1.319367	104.1375	2.548823	0.003401	94.29409		0.828988	118.2549	1.432582	0.010383	74.1266
1.537145	90.55347	3.704725	0.00107	98.56283		1.299165	105.0857	2.471236	0.003675	93.60359		0.808786	118.5615	1.409234	0.010628	72.56754
1.516943	92.11912	3.567114	0.001228	98.32099		1.278963	105.9908	2.397538	0.003956	92.86593		0.788584	118.8541	1.386982	0.010868	71.0073
1.496741	93.61361	3.436855	0.001399	98.04623		1.258761	106.8548	2.327517	0.004243	92.08122		0.768382	119.1335	1.365772	0.0111	69.44824
1.476539	95.04017	3.313502	0.001583	97.73633		1.238559	107.6796	2.260975	0.004535	91.24985		0.74818	119.4001	1.345555	0.011327	67.89262
1.456337	96.40189	3.196645	0.001779	97.38918		1.218357	108.4668	2.197726	0.004831	90.37246		0.727978	119.6546	1.326282	0.011548	66.34259
1.436135	97.70172	3.0859	0.001987	97.00283		1.198155	109.2183	2.137595	0.00513	89.44992		0.707776	119.8976	1.307908	0.011762	64.80019
1.415933	98.94247	2.980911	0.002207	96.57551		1.177953	109.9356	2.080417	0.005432	88.48336		0.687574	120.1295	1.29039	0.01197	63.26735
1.395731	100.1268	2.881344	0.002439	96.10564		1.157751	110.6203	2.026038	0.005736	87.47412		0.667372	120.3509	1.273688	0.012171	61.74586
1.375529	101.2573	2.786892	0.00268	95.59187		1.137549	111.2738	1.974311	0.00604	86.42374		0.64717	120.5622	1.257763	0.012367	60.23742
1.355327	102.3365	2.697263	0.002931	95.03306		1.117347	111.8977	1.925098	0.006345	85.33396		0.626968	120.7639	1.242577	0.012556	58.7436
1.335125	103.3666	2.612187	0.003192	94.42835		1.097145	112.4932	1.878272	0.006649	84.20668		0.606766	120.9564	1.228097	0.012739	57.26584
1.314923	104.3498	2.531411	0.00346	93.77711		1.076943	113.0617	1.833708	0.006952	83.04394		0.586564	121.1402	1.214287	0.012916	55.80548
1.294721	105.2884	2.454699	0.003736	93.07898		1.056741	113.6043	1.791292	0.007253	81.84794		0.566362	121.3156	1.201117	0.013087	54.36374
1.274519	106.1844	2.381827	0.004019	92.33389		1.036539	114.1223	1.750915	0.007552	80.62095		0.54616	121.4831	1.188557	0.013253	52.94174
1.254317	107.0396	2.312588	0.004307	91.54199		1.016337	114.6167	1.712473	0.007848	79.36537		0.525958	121.6429	1.176578	0.013412	51.54046
1.234115	107.8559	2.246786	0.0046	90.70374		0.996135	115.0886	1.675871	0.008141	78.08364		0.505756	121.7955	1.165152	0.013567	50.1608
1.213913	108.6351	2.184238	0.004897	89.81981		0.975933	115.5391	1.641014	0.00843	76.77825		0.485554	121.9412	1.154253	0.013715	48.80357
1.193711	109.3789	2.12477	0.005197	88.89114		0.955731	115.9691	1.607818	0.008714	75.45173		0.465352	122.0802	1.143858	0.013859	47.46945
1.173509	110.0889	2.068221	0.005499	87.9189		0.935529	116.3796	1.576199	0.008994	74.10663		0.44515	122.2129	1.133941	0.013997	46.15905
1.153307	110.7667	2.014437	0.005803	86.90446		0.915327	116.7714	1.546078	0.009269	72.74549		0.424948	122.3396	1.124482	0.01413	44.87288
1.133105	111.4136	1.963275	0.006107	85.84941		0.895125	117.1454	1.517384	0.009539	71.37082		0.404746	122.4605	1.115458	0.014258	43.61138
1.112903	112.0311	1.914598	0.006412	84.75552		0.874923	117.5024	1.490045	0.009803	69.98512		0.384544	122.5759	1.10685	0.014381	42.37489
1.092701	112.6206	1.868279	0.006716	83.62471		0.854721	117.8432	1.463995	0.010062	68.59081		0.364342	122.6861	1.098638	0.0145	41.1637
1.072499	113.1832	1.824197	0.007019	82.45905		0.834519	118.1685	1.439171	0.010315	67.19029		0.34414	122.7913	1.090803	0.014614	39.978
1.052297	113.7203	1.782239	0.007319	81.26075		0.814316	118.479	1.415514	0.010562	65.78587		0.323938	122.8917	1.083328	0.014723	38.81795
1.032095	114.233	1.742296	0.007618	80.0321		0.794114	118.7754	1.392968	0.010803	64.37979		0.303736	122.9875	1.076196	0.014829	37.68361
1.011893	114.7224	1.704267	0.007913	78.7755		0.773912	119.0583	1.371478	0.011037	62.9742		0.283534	123.079	1.069392	0.01493	36.57501
0.991691	115.1895	1.668057	0.008205	77.4934		0.75371	119.3283	1.350993	0.011266	61.57116		0.263332	123.1663	1.0629	0.015027	35.49213
0.971489	115.6354	1.633573	0.008493	76.18828		0.733508	119.5861	1.331466	0.011488	60.17264		0.24313	123.2496	1.056705	0.015121	34.43487
0.951287	116.061	1.60073	0.008776	74.86268		0.713306	119.8322	1.312851	0.011704	58.78051		0.222928	123.3292	1.050795	0.01521	33.40312
0.931085	116.4673	1.569447	0.009055	73.51912		0.693104	120.0671	1.295103	0.011913	57.39652		0.202726	123.4051	1.045155	0.015296	32.39671
0.910883	116.8552	1.539647	0.009329	72.16014		0.672902	120.2913	1.278182	0.012117	56.02233		0.182524	123.4776	1.039774	0.015379	31.41544
2.093869	117.3	1.505538	0.009653	70.78044	1.893747	1.873747	109.7734	2.093322	0.005363	55.4247	1.419297	1.399297	60.07826	6.640982	5.68E-05	31.41187
2.073869	117.7	1.474934	0.009953	69.38548		1.853747	106.4081	2.363684	0.004092	54.97293		1.379297	58.94955	6.760085	5.04E-05	31.4087
2.053869	118.1	1.444392	0.010261	67.97603		1.833747	103.1942	2.626392	0.003147	54.62805		1.359297	57.87164	6.874587	4.5E-05	31.40588
2.033869	118.4	1.421527	0.010499	66.56362		1.813747	100.1249	2.881502	0.002438	54.36231		1.339297	56.84225	6.984634	4.03E-05	31.40335
2.013869	118.7	1.398697	0.010741	65.14896		1.793747	97.19384	3.129079	0.001903	54.15575		1.319297	55.85918	7.09037	3.62E-05	31.40107
1.993869	118.9	1.383496	0.010905	63.74338		1.773747	94.39465	3.3692	0.001497	53.99383		1.299297	54.92037	7.19194	3.27E-05	31.39902
1.973869	119.1	1.368311	0.011072	62.34732		1.753747	91.72145	3.601955	0.001186	53.86588		1.279297	54.0238	7.289483	2.97E-05	31.39715
1.953869	119.2	1.360724	0.011157	60.97155		1.733747	89.16857	3.827442	0.000947	53.76398		1.259297	53.16759	7.383136	2.7E-05	31.39546
1.933869	119.2	1.360724	0.011157	59.62614		1.713747	86.73058	4.045768	0.000761	53.6822		1.239297	52.34991	7.473035	2.47E-05	31.3939
1.913869	117.0104	1.527731	0.009441	58.51087		1.693747	84.40232	4.257048	0.000616	53.61609		1.219297	51.56904	7.55931	2.27E-05	31.39248
1.893869	113.3194	1.81355	0.007094	57.68661		1.673747	82.17885	4.461404	0.000502	53.56226		1.199297	50.82331	7.64209	2.09E-05	31.39117

(Continued)

Table A.8.1.c Temperature and retention distribution histories for heating medium at 125°C (Continued)

		Center					Center + 1					
Time (t)	f_h/t	log term	Temp. (°C)		Rate Const	Retention (%)	log term	Temp. (°C)		Rate Const	Retention (%)	log term
24	0.24	1.95464	118.5	1.413913	0.010579	61.69744	1.943643	118.7	1.398697	0.010741	60.84125	
26	0.26	1.93464	118.5	1.413913	0.010579	60.40579	1.923643	118.7	1.398697	0.010741	59.5482	
28	0.28	1.91464	117.1562	1.516557	0.009547	59.26337	1.903643	118.7	1.398697	0.010741	58.28263	
30	0.3	1.89464	113.4585	1.802675	0.007171	58.41944	1.883643	111.4968	1.956705	0.006148	57.57043	
32	0.32	1.87464	109.9273	2.081073	0.005429	57.7886	1.863643	108.0539	2.230868	0.004673	57.03483	
34	0.34	1.85464	106.555	2.351778	0.004141	57.31195	1.843643	104.7659	2.497358	0.00358	56.6279	
36	0.36	1.83464	103.3345	2.614827	0.003183	56.94822	1.823643	101.6259	2.75622	0.002764	56.31577	
38	0.38	1.81464	100.259	2.870274	0.002466	56.66807	1.803643	98.62728	3.007515	0.00215	56.07419	
40	0.4	1.79464	97.32183	3.118186	0.001924	56.4504	1.783643	95.76358	3.251315	0.001684	55.88559	
42	0.42	1.77464	94.51689	3.358638	0.001513	56.27983	1.763643	93.02877	3.487703	0.00133	55.73715	
44	0.44	1.75464	91.83818	3.59172	0.001198	56.14509	1.743643	90.41704	3.716773	0.001058	55.61939	
46	0.46	1.73464	89.28005	3.817529	0.000956	56.03782	1.723643	87.92286	3.938628	0.000847	55.52523	
48	0.48	1.71464	86.83704	4.036172	0.000768	55.95177	1.703643	85.54094	4.15338	0.000683	55.44938	
50	0.5	1.69464	84.50399	4.247764	0.000622	55.88222	1.683643	83.26623	4.361146	0.000555	55.38784	
52	0.52	1.67464	82.27594	4.452427	0.000507	55.82561	1.663643	81.09389	4.562054	0.000454	55.33756	
54	0.54	1.65464	80.14818	4.650289	0.000416	55.7792	1.643643	79.01932	4.756233	0.000374	55.29618	
56	0.56	1.63464	78.11617	4.841485	0.000343	55.7409	1.623643	77.03812	4.943822	0.00031	55.2619	
58	0.58	1.61464	76.17563	5.026153	0.000286	55.70908	1.603643	75.1461	5.124962	0.000259	55.23332	
60	0.6	1.59464	74.32242	5.204436	0.000239	55.68247	1.583643	73.33923	5.299796	0.000217	55.20933	
62	0.62	1.57464	72.55262	5.37648	0.000201	55.66007	1.563643	71.61368	5.468472	0.000183	55.18908	
64	0.64	1.55464	70.86247	5.542433	0.00017	55.6411	1.543643	69.96579	5.631142	0.000156	55.17187	
66	0.66	1.53464	69.24839	5.702448	0.000145	55.62495	1.523643	68.39207	5.787956	0.000133	55.15716	
68	0.68	1.51464	67.70696	5.856676	0.000124	55.61111	1.503643	66.88918	5.939067	0.000115	55.14452	
70	0.7	1.49464	66.23491	6.005271	0.000107	55.59918	1.483643	65.45393	6.084631	9.91E-05	55.1336	

	Center + 2						Center + 3						Center + 4			
	Temp. (°C)		Rate Const	Retention (%)	log term		Temp. (°C)		Rate Const	Retention (%)	log term		Temp. (°C)		Rate Const	Retention (%)
1.873869	109.7944	2.091652	0.005372	57.0702		1.653747	80.05545	4.658966	0.000412	53.51812		1.179297	50.11114	7.721502	1.93E-05	31.38996
1.853869	106.4281	2.36206	0.004099	56.60427		1.633747	78.02762	4.849867	0.000341	53.48168		1.159297	49.43103	7.797667	1.79E-05	31.38884
1.833869	103.2133	2.624815	0.003152	56.24859		1.613747	76.09106	5.034247	0.000283	53.45139		1.139297	48.78153	7.870704	1.66E-05	31.38779
1.813869	100.1432	2.879971	0.002442	55.97455		1.593747	74.24166	5.212248	0.000237	53.42606		1.119297	48.16126	7.94073	1.55E-05	31.38682
1.793869	97.21129	3.127593	0.001906	55.76155		1.573747	72.47549	5.384017	0.0002	53.40473		1.099297	47.5689	8.007857	1.45E-05	31.38591
1.773869	94.41132	3.36776	0.001499	55.59459		1.553747	70.78882	5.549702	0.000169	53.38666		1.079297	47.00321	8.072195	1.36E-05	31.38506
1.753869	91.73737	3.600559	0.001188	55.46267		1.533747	69.17805	5.709455	0.000144	53.37127		1.059297	46.46297	8.13385	1.28E-05	31.38426
1.733869	89.18377	3.82609	0.000948	55.3576		1.513747	67.63979	5.863429	0.000124	53.35808		1.039297	45.94706	8.192925	1.2E-05	31.38351
1.713869	86.74509	4.044459	0.000762	55.27329												
1.693869	84.41618	4.255782	0.000617	55.20514												
1.673869	82.19209	4.46018	0.000503	55.14964												
1.653869	80.06809	4.657783	0.000413	55.10414												
1.633869	78.0397	4.848724	0.000341	55.06657												
1.613869	76.10259	5.033143	0.000284	55.03536												
1.593869	74.25267	5.211183	0.000237	55.00924												
1.573869	72.48601	5.382989	0.0002	54.98726												
1.553869	70.79886	5.548711	0.000169	54.96864												
1.533869	69.18764	5.7085	0.000144	54.95278												
1.513869	67.64895	5.862508	0.000124	54.93918												

Table A.8.1.d Temperature and retention distribution histories for heating medium at 120°C

							Center						Center + 1			
		Time (t)	f_h/t	log term		Temp (°C)	Rate Const		Retention (%)	log term		Temp (°C)		Rate Const	Retention (%)	log term
Input Values									100						100	
Medium Temp	120°C	2	0.020202	1.991004	1.970802	70	5.627749	0.000156	99.96871	1.978203	1.958001	70	5.627749	0.000156	99.96871	1.89979
Initial Temp	70°C	4	0.040404	1.991004	1.9506	70	5.627749	0.000156	99.93744		1.937799	70	5.627749	0.000156	99.93744	
j - value	1.959	6	0.060606	1.991004	1.930398	70	5.627749	0.000156	99.90617		1.917597	70	5.627749	0.000156	99.90617	
Heating rate	99 min	8	0.080808	1.991004	1.910196	70	5.627749	0.000156	99.87491		1.897395	70	5.627749	0.000156	99.87491	
k - ref	0.0435/min	10	0.10101	1.991004	1.889994	70	5.627749	0.000156	99.84367		1.877193	70	5.627749	0.000156	99.84367	
Act Energy	97 kJ/mole	12	0.121212	1.991004	1.869792	70	5.627749	0.000156	99.81243		1.856991	70	5.627749	0.000156	99.81243	
R - value	8.314	14	0.141414	1.991004	1.84959	70	5.627749	0.000156	99.7812		1.836789	70	5.627749	0.000156	99.7812	
Ref T	138°C	16	0.161616	1.991004	1.829388	70	5.627749	0.000156	99.74999		1.816587	70	5.627749	0.000156	99.74999	
Time Increment	2 min	18	0.181818	1.991004	1.809186	70	5.627749	0.000156	99.71878		1.796385	70	5.627749	0.000156	99.71878	
Cmed Temp	35°C	20	0.20202	1.991004	1.788984	70	5.627749	0.000156	99.68758		1.776183	70	5.627749	0.000156	99.68758	
j2-value	1.9021	22	0.222222	1.991004	1.768782	70	5.627749	0.000156	99.65639		1.755981	70	5.627749	0.000156	99.65639	
j3-value	1.5879	24	0.242424	1.991004	1.74858	70	5.627749	0.000156	99.62521		1.735779	70	5.627749	0.000156	99.62521	
j4-value	0.918	26	0.262626	1.991004	1.728378	70	5.627749	0.000156	99.59404		1.715577	70	5.627749	0.000156	99.59404	
j5-value	0.2968	28	0.282828	1.991004	1.708176	70	5.627749	0.000156	99.56288		1.695375	70.41217	5.586923	0.000163	99.56159	
Cooling rate - f	100	30	0.30303	1.991004	1.687974	70	5.627749	0.000156	99.53174		1.675173	72.66602	5.365403	0.000203	99.5211	
Cooling - j	1.958	32	0.323232	1.991004	1.667772	70	5.627749	0.000156	99.5006		1.654971	74.81742	5.156629	0.000251	99.47123	
		34	0.343434	1.991004	1.64757	70	5.627749	0.000156	99.46947		1.634769	76.87104	4.95974	0.000305	99.41054	
		36	0.363636	1.991004	1.627368	77.59978	4.890427	0.000327	99.40443		1.614567	78.83132	4.773944	0.000367	99.33751	
		38	0.383838	1.991004	1.607166	79.52694	4.708509	0.000392	99.32646		1.594365	80.7025	4.598514	0.000438	99.25055	
		40	0.40404	1.991004	1.586964	81.3665	4.536706	0.000466	99.23397		1.574163	82.48863	4.43278	0.000517	99.14801	
		42	0.424242	1.991004	1.566762	83.12245	4.374367	0.000548	99.12528		1.553961	84.19358	4.276125	0.000604	99.02821	
		44	0.444444	1.991004	1.54656	84.7986	4.220893	0.000639	98.99872		1.533759	85.82104	4.127979	0.000701	98.88947	
		46	0.464646	1.991004	1.526358	86.39855	4.07573	0.000739	98.85258		1.513557	87.37452	3.987815	0.000806	98.73009	
		48	0.484848	1.991004	1.506156	87.92579	3.938366	0.000847	98.68519		1.493355	88.8574	3.855144	0.000921	98.54841	
		50	0.505051	1.991004	1.485954	89.38361	3.808325	0.000965	98.4949		1.473153	90.27288	3.729513	0.001044	98.34282	
		52	0.525253	1.991004	1.465752	90.77518	3.685167	0.001092	98.28011		1.452951	91.62402	3.610503	0.001176	98.11175	
		54	0.545455	1.991004	1.44555	92.10349	3.568483	0.001227	98.0393		1.432749	92.91375	3.497721	0.001317	97.85375	
		56	0.565657	1.991004	1.425348	93.37143	3.457891	0.00137	97.77102		1.412547	94.14487	3.390805	0.001465	97.56743	
		58	0.585859	1.991004	1.405146	94.58174	3.353037	0.001522	97.47395		1.392345	95.32002	3.289416	0.001621	97.25153	
		60	0.606061	1.991004	1.384944	95.73704	3.253592	0.001681	97.14686		1.372143	96.44177	3.193236	0.001785	96.90493	
		62	0.626263	1.991004	1.364742	96.83983	3.159246	0.001847	96.78868		1.351941	97.51252	3.101971	0.001956	96.52662	
		64	0.646465	1.991004	1.34454	97.89249	3.069711	0.00202	96.39846		1.331739	98.53461	3.015345	0.002133	96.11576	
		66	0.666667	1.991004	1.324338	98.89731	2.984719	0.002199	95.97542		1.311537	99.51025	2.933099	0.002316	95.67166	
		68	0.686869	1.991004	1.304136	99.85646	2.904017	0.002384	95.51891		1.291335	100.4415	2.854993	0.002504	95.19379	
		70	0.707071	1.991004	1.283934	100.772	2.82737	0.002574	95.02848		1.271133	101.3305	2.780799	0.002697	94.68179	
		72	0.727273	1.991004	1.263732	101.646	2.754556	0.002768	94.50381		1.250931	102.1791	2.710306	0.002893	94.13545	
		74	0.747475	1.991004	1.24353	102.4802	2.685367	0.002967	93.94478		1.230729	102.989	2.643313	0.003094	93.55475	
		76	0.767677	1.991004	1.223328	103.2765	2.619609	0.003168	93.35139		1.210527	103.7622	2.579634	0.003297	92.93981	
		78	0.787879	1.991004	1.203126	104.0366	2.5571	0.003373	92.72385		1.190325	104.5003	2.519093	0.003503	92.29092	
		80	0.808081	1.991004	1.182924	104.7622	2.497666	0.003579	92.0625		1.170123	105.2047	2.461523	0.003711	91.60852	
		82	0.828283	1.991004	1.162722	105.4547	2.441146	0.003787	91.36782		1.149921	105.8772	2.40677	0.00392	90.89318	
		84	0.848485	1.991004	1.14252	106.1158	2.387388	0.003996	90.64046		1.129719	106.5191	2.354687	0.004129	90.14565	
		86	0.868687	1.991004	1.122318	106.7469	2.336247	0.004206	89.88119		1.109516	107.1318	2.305135	0.004339	89.36677	
		88	0.888889	1.991004	1.102116	107.3493	2.28759	0.004416	89.09091		1.089314	107.7167	2.257984	0.004548	88.5575	
		90	0.909091	1.991004	1.081914	107.9243	2.241287	0.004625	88.27062		1.069112	108.275	2.213111	0.004757	87.71893	
		92	0.929293	1.991004	1.061712	108.4731	2.197219	0.004833	87.42144		1.04891	108.8079	2.1704	0.004965	86.85224	
		94	0.949495	1.991004	1.041509	108.997	2.155273	0.00504	86.54459		1.028708	109.3166	2.129742	0.005171	85.95868	
		96	0.969697	1.991004	1.021307	109.4971	2.11534	0.005246	85.64135		1.008506	109.8022	2.091032	0.005375	85.0396	
		98	0.989899	1.991004	1.001105	109.9745	2.077319	0.005449	84.71309		0.988304	110.2657	2.054173	0.005577	84.0964	
		100	1.010101	1.991004	0.980903	110.4302	2.041115	0.00565	83.76123		0.968102	110.7081	2.019072	0.005776	83.13053	
		102	1.030303	1.991004	0.960701	110.8652	2.006636	0.005848	82.78724		0.9479	111.1305	1.985643	0.005972	82.14349	
		104	1.050505	1.991004	0.940499	111.2803	1.973798	0.006043	81.79263		0.927698	111.5336	1.953801	0.006165	81.13681	
		106	1.070707	1.991004	0.920297	111.6767	1.942518	0.006235	80.77894		0.907496	111.9184	1.923469	0.006355	80.11203	
		108	1.090909	1.991004	0.900095	112.055	1.912719	0.006424	79.74773		0.887294	112.2857	1.894571	0.006542	79.07073	
		110	1.111111	1.991004	0.879893	112.4161	1.88433	0.006609	78.70057		0.867092	112.6364	1.867039	0.006724	78.01446	
		112	1.131313	1.991004	0.859691	112.7608	1.857281	0.00679	77.63901		0.84689	112.9711	1.840805	0.006903	76.9448	
		114	1.151515	1.991004	0.839489	113.0898	1.831506	0.006967	76.56461		0.826688	113.2905	1.815805	0.007078	75.86328	
		2	0.02	2.184407	2.164407	113.3	1.815065	0.007083	75.48765	2.172943	2.152943	113.6	1.791628	0.007251	74.77106	2.10063
		4	0.04	2.184407	2.144407	113.6	1.791628	0.007251	74.40083		2.132943	113.9	1.768228	0.007423	73.66926	
		6	0.06	2.184407	2.124407	113.9	1.768228	0.007423	73.30449		2.112943	114.2	1.744864	0.007598	72.55823	
		8	0.08	2.184407	2.104407	114.1	1.752648	0.007539	72.20747		2.092943	114.4	1.729308	0.007717	71.44693	
		10	0.1	2.184407	2.084407	114.3	1.737084	0.007657	71.11005		2.072943	114.6	1.713768	0.007838	70.33565	
		12	0.12	2.184407	2.064407	114.5	1.721536	0.007777	70.0125		2.052943	114.7	1.706004	0.007899	69.2332	
		14	0.14	2.184407	2.044407	114.7	1.706004	0.007899	68.91511		2.032943	114.8	1.698244	0.007961	68.13964	
		16	0.16	2.184407	2.024407	114.9	1.690488	0.008023	67.81816		2.012943	114.8	1.698244	0.007961	67.06335	
		18	0.18	2.184407	2.004407	115.1	1.674988	0.008148	66.72196		1.992943	114.8	1.698244	0.007961	66.00406	

	Center + 2						Center + 3						Center + 4			
	Temp (°C)	Rate Const		Retention (%)	log term		Temp (°C)		Rate Const	Retention (%)	log term		Temp (°C)		Rate Const	Retention (%)
				100						100						100
.879591	70	5.627749	0.000156	99.96871	1.661813	1.641611	76.18623	5.025139	0.000286	99.94285	1.171434	1.151232	105.8345	2.410242	0.003906	99.22184
.859389	70	5.627749	0.000156	99.93744		1.621409	78.17763	4.83567	0.000345	99.87382		1.13103	106.4783	2.35799	0.004116	98.40849
.839187	70	5.627749	0.000156	99.90617		1.601207	80.07852	4.656807	0.000413	99.79134		1.110828	107.0929	2.308277	0.004325	97.56087
.818985	70	5.627749	0.000156	99.87491		1.581005	81.89301	4.487861	0.000489	99.69376		1.090626	107.6796	2.260974	0.004535	96.68002
.798783	70	5.627749	0.000156	99.84367		1.560803	83.62503	4.328197	0.000574	99.57942		1.070424	108.2396	2.215957	0.004744	95.76713
.778581	70	5.627749	0.000156	99.81243		1.540601	85.27833	4.177231	0.000667	99.4466		1.050222	108.7741	2.173109	0.004951	94.82347
.758379	70	5.627749	0.000156	99.7812		1.520399	86.85649	4.03442	0.00077	99.29362		1.03002	109.2843	2.132321	0.005157	93.8504
.738177	70	5.627749	0.000156	99.74999		1.500197	88.36291	3.899264	0.000881	99.11878		1.009818	109.7714	2.093487	0.005362	92.84939
.717975	70	5.627749	0.000156	99.71878		1.479995	89.80086	3.771298	0.001001	98.92045		0.989616	110.2363	2.056511	0.005564	91.82196
.697773	70.13763	5.614106	0.000159	99.68715		1.459792	91.17346	3.650091	0.001131	98.69704		0.969414	110.68	2.021299	0.005763	90.76969
.677571	72.40395	5.391012	0.000198	99.64763		1.43959	92.48367	3.535242	0.001268	98.44705		0.949212	111.1037	1.987764	0.00596	89.69422
.657369	74.56726	5.180772	0.000245	99.59889		1.419388	93.73433	3.426379	0.001414	98.16904		0.92901	111.508	1.955821	0.006153	88.5972
.637167	76.63225	4.982515	0.000298	99.53949		1.399186	94.92815	3.323154	0.001568	97.86172		0.908808	111.894	1.925393	0.006343	87.48034
.616965	78.60338	4.795441	0.00036	99.46793		1.378984	96.0677	3.225244	0.001729	97.52391		0.888606	112.2624	1.896405	0.00653	86.34533
.596763	80.48492	4.618817	0.000429	99.3826		1.358782	97.15546	3.132346	0.001897	97.15456		0.868404	112.6141	1.868786	0.006713	85.19389
.576561	82.28094	4.451965	0.000507	99.28187		1.33858	98.19378	3.044179	0.002072	96.75275		0.848202	112.9498	1.84247	0.006892	84.02771
.556359	83.99533	4.294263	0.000594	99.16407		1.318378	99.18491	2.960477	0.002253	96.31776		0.828	113.2702	1.817392	0.007067	82.8485
.536157	85.6318	4.145136	0.000689	99.0275		1.298176	100.131	2.880995	0.002439	95.84898		0.807798	113.5761	1.793492	0.007237	81.65791
.515955	87.19389	4.00405	0.000794	98.87046		1.277974	101.0341	2.805501	0.002631	95.346		0.787596	113.8681	1.770715	0.007404	80.4576
.495753	88.68498	3.870514	0.000907	98.6913		1.257772	101.8961	2.733777	0.002826	94.80856		0.767393	114.1468	1.749004	0.007567	79.24917
.475551	90.10829	3.744071	0.001029	98.48838		1.23757	102.7189	2.665621	0.003026	94.23657		0.747191	114.4128	1.72831	0.007725	78.03419
.455349	91.46692	3.624296	0.00116	98.26014		1.217368	103.5044	2.60084	0.003228	93.6301		0.726989	114.6668	1.708583	0.007879	76.8142
.435147	92.76379	3.510794	0.001299	98.0051		1.197166	104.2541	2.539255	0.003433	92.98939		0.706787	114.9092	1.689776	0.008028	75.59066
.414945	94.00172	3.4032	0.001447	97.72186		1.176964	104.9698	2.480697	0.00364	92.31483		0.686585	115.1406	1.671846	0.008174	74.365
.394743	95.18338	3.301172	0.001603	97.40916		1.156762	105.653	2.425007	0.003849	91.60697		0.666383	115.3614	1.654751	0.008315	73.1386
.374541	96.31133	3.204389	0.001765	97.06583		1.13656	106.3051	2.372036	0.004058	90.86647		0.646181	115.5723	1.638451	0.008451	71.91277
.354339	97.38802	3.112556	0.001935	96.69088		1.116358	106.9275	2.321641	0.004268	90.09415		0.625979	115.7735	1.622908	0.008584	70.68876
.334137	98.41577	3.025393	0.002111	96.28343		1.096156	107.5217	2.273691	0.004478	89.29096		0.605777	115.9656	1.608087	0.008712	69.46778
.313935	99.39681	2.94264	0.002294	95.84277		1.075954	108.0888	2.22806	0.004687	88.45793		0.585575	116.149	1.593953	0.008836	68.25097
.293733	100.3333	2.864055	0.002481	95.36835		1.055752	108.6302	2.18463	0.004895	87.59622		0.565373	116.324	1.580474	0.008956	67.03938
.273531	101.2271	2.789408	0.002673	94.8598		1.03555	109.147	2.143288	0.005101	86.70708		0.545171	116.4911	1.567619	0.009072	65.83405
.253329	102.0804	2.718486	0.00287	94.31688		1.015348	109.6403	2.103929	0.005306	85.79181		0.524969	116.6506	1.555359	0.009183	64.63592
.233126	102.8949	2.651088	0.00307	93.73955		0.995146	110.1111	2.066454	0.005509	84.85182		0.504767	116.8028	1.543665	0.009291	63.44589
.212924	103.6723	2.587025	0.003273	93.12792		0.974944	110.5606	2.030768	0.005709	83.88854		0.484565	116.9481	1.532511	0.009396	62.26478
.192722	104.4144	2.52612	0.003479	92.48225		0.954742	110.9896	1.996782	0.005906	82.90347		0.464363	117.0868	1.521872	0.009496	61.09339
.17252	105.1228	2.468206	0.003686	91.80297		0.93454	111.3992	1.964412	0.0061	81.89813		0.444161	117.2193	1.511723	0.009593	59.93242
.152318	105.799	2.413127	0.003895	91.09065		0.914338	111.7901	1.933577	0.006291	80.87408		0.423959	117.3456	1.502042	0.009686	58.78254
.132116	106.4445	2.360734	0.004104	90.34599		0.894136	112.1633	1.904202	0.006479	79.83288		0.403757	117.4663	1.492807	0.009776	57.64436
.111914	107.0606	2.310889	0.004314	89.56983		0.873934	112.5194	1.876215	0.006663	78.77611		0.383555	117.5814	1.483998	0.009863	56.51844
.091712	107.6487	2.263459	0.004524	88.76313		0.853732	112.8594	1.849548	0.006843	77.70534		0.363353	117.6914	1.475593	0.009946	55.40529
.07151	108.2101	2.218322	0.004732	87.92697		0.83353	113.184	1.824137	0.007019	76.62213		0.343151	117.7963	1.467575	0.010026	54.30536
.051308	108.746	2.17536	0.00494	87.06249		0.813328	113.4938	1.799921	0.007191	75.52802		0.322949	117.8965	1.459925	0.010103	53.21907
.031106	109.2575	2.134464	0.005146	86.17097		0.793126	113.7895	1.776842	0.007359	74.42455		0.302747	117.9921	1.452626	0.010177	52.1468
.010904	109.7457	2.095528	0.005351	85.25373		0.772924	114.0718	1.754845	0.007523	73.31319		0.282545	118.0833	1.445663	0.010248	51.08886
.990702	110.2118	2.058454	0.005553	84.31217		0.752722	114.3412	1.733877	0.007682	72.19541		0.262343	118.1705	1.439019	0.010316	50.04554
.9705	110.6567	2.02315	0.005752	83.34774		0.73252	114.5984	1.713889	0.007837	71.07262		0.242141	118.2536	1.43268	0.010382	49.01711
.950298	111.0814	1.989526	0.005949	82.36194		0.712318	114.8439	1.694835	0.007988	69.94621		0.221939	118.333	1.426632	0.010445	48.00375
.930096	111.4867	1.9575	0.006143	81.35628		0.692116	115.0783	1.67667	0.008134	68.81748		0.201737	118.4088	1.42086	0.010506	47.00567
.909894	111.8737	1.926992	0.006333	80.33233		0.671914	115.302	1.65935	0.008276	67.68774		0.181535	118.4811	1.415353	0.010564	46.02299
.889692	112.243	1.897929	0.00652	79.29164		0.651712	115.5155	1.642836	0.008414	66.55819		0.161333	118.5501	1.410099	0.010619	45.05585
.86949	112.5956	1.870238	0.006703	78.23578		0.63151	115.7194	1.62709	0.008548	65.43001		0.141131	118.616	1.405085	0.010673	44.10431
.849288	112.9321	1.843853	0.006882	77.16632		0.611308	115.9139	1.612075	0.008677	64.30432		0.120929	118.6789	1.4003	0.010724	43.16846
.829086	113.2534	1.81871	0.007057	76.08481		0.591106	116.0996	1.597756	0.008802	63.18218		0.100727	118.739	1.395734	0.010773	42.24831
.808884	113.56	1.794749	0.007228	74.99279		0.570904	116.2769	1.584101	0.008923	62.0646		0.080525	118.7963	1.391377	0.01082	41.34389
.788682	113.8527	1.771912	0.007395	73.89176		0.550702	116.4461	1.571078	0.00904	60.95253		0.060323	118.851	1.38722	0.010865	40.45519
.76848	114.1321	1.750146	0.007558	72.7832		0.5305	116.6077	1.558658	0.009153	59.84686		0.040121	118.9032	1.383252	0.010908	39.58216
.748278	114.3988	1.729398	0.007717	71.66856		0.510298	116.7618	1.546811	0.009262	58.74843		0.019919	118.9531	1.379466	0.01095	38.72477
.080637	114.7	1.706004	0.007899	70.54521	1.875393	1.855393	106.6792	2.341722	0.004183	58.25899	1.3965	1.3765	58.79581	6.776371	4.96E-05	38.72093
.060637	115	1.682736	0.008085	69.41365		1.835393	103.4531	2.605059	0.003215	57.88563		1.3565	57.72482	6.890241	4.43E-05	38.7175
.040637	115.2	1.667244	0.008211	68.283		1.815393	100.3722	2.860792	0.002489	57.59817		1.3365	56.70204	6.999676	3.97E-05	38.71443
.020637	115.4	1.651768	0.008339	67.15357		1.795393	97.43	3.108985	0.001942	57.37488		1.3165	55.72528	7.104822	3.57E-05	38.71167
.000637	115.5	1.644036	0.008404	66.03426		1.775393	94.62019	3.349717	0.001527	57.19997		1.2965	54.79249	7.20582	3.23E-05	38.70917
.980637	115.6	1.636308	0.008469	64.92515		1.755393	91.93684	3.583075	0.001209	57.06184		1.2765	53.90168	7.302811	2.93E-05	38.7069
.960637	115.6	1.636308	0.008469	63.83467		1.735393	89.37426	3.809156	0.000964	56.9519		1.2565	53.05097	7.395931	2.67E-05	38.70483
.940637	115.6	1.636308	0.008469	62.7625		1.715393	86.92702	4.028067	0.000775	56.86373		1.2365	52.23854	7.485315	2.44E-05	38.70294
.920637	115.6	1.636308	0.008469	61.70834		1.695393	84.58992	4.239922	0.000627	56.79249		1.2165	51.46268	7.571093	2.24E-05	38.70121

(Continued)

Table A.8.1.d Temperature and retention distribution histories for heating medium at 120°C (Continued)

		Center						Center + 1						
Time (t)	f_h/t	log term		Temp (°C)	Rate Const		Retention (%)	log term		Temp (°C)	Rate Const		Retention (%)	log ter
20	0.2	2.184407	1.984407	115.2	1.667244	0.008211	65.63515		1.972943	114.8	1.698244	0.007961	64.96151	
22	0.22	2.184407	1.964407	115.3	1.659504	0.008275	64.55781		1.952943	114.8	1.698244	0.007961	63.93542	
24	0.24	2.184407	1.944407	115.4	1.651768	0.008339	63.48999		1.932943	114.8	1.698244	0.007961	62.92554	
26	0.26	2.184407	1.924407	115.4	1.651768	0.008339	62.43983		1.912943	114.8	1.698244	0.007961	61.93161	
28	0.28	2.184407	1.904407	115.243	1.663916	0.008239	61.41942		1.892943	113.1524	1.826605	0.007002	61.0704	
30	0.3	2.184407	1.884407	111.6315	1.946082	0.006213	60.66091		1.872943	109.635	2.104349	0.005304	60.42602	
32	0.32	2.184407	1.864407	108.1825	2.220539	0.004722	60.09073		1.852943	106.2759	2.374402	0.004049	59.93872	
34	0.34	2.184407	1.844407	104.8887	2.487321	0.003616	59.6577		1.832943	103.0679	2.636804	0.003114	59.56656	
36	0.36	2.184407	1.824407	101.7432	2.746474	0.002791	59.32565		1.812943	100.0044	2.891609	0.002414	59.27971	
38	0.38	2.184407	1.804407	98.73927	2.998056	0.00217	59.06874		1.792943	97.0787	3.138884	0.001885	59.05665	
40	0.4	2.184407	1.784407	95.87053	3.242141	0.0017	58.86825		1.772943	94.28469	3.378707	0.001483	58.88175	
42	0.42	2.184407	1.764407	93.13091	3.47881	0.001342	58.71049		1.752943	91.61644	3.611168	0.001175	58.7435	
44	0.44	2.184407	1.744407	90.51459	3.708158	0.001067	58.58537		1.732943	89.06828	3.836364	0.000938	58.63336	
46	0.46	2.184407	1.724407	88.01602	3.930287	0.000854	58.48536		1.712943	86.63481	4.054404	0.000755	58.54494	
48	0.48	2.184407	1.704407	85.62991	4.145307	0.000689	58.40483		1.692943	84.31086	4.265403	0.000611	58.47344	
50	0.5	2.184407	1.684407	83.35118	4.353339	0.00056	58.3395		1.672943	82.09151	4.469484	0.000498	58.4152	
52	0.52	2.184407	1.664407	81.17502	4.554506	0.000458	58.28613		1.652943	79.97204	4.666775	0.000409	58.36744	
54	0.54	2.184407	1.644407	79.0968	4.74894	0.000377	58.24223		1.632943	77.94796	4.857412	0.000338	58.32799	
56	0.56	2.184407	1.624407	77.11212	4.936778	0.000312	58.20587		1.612943	76.01499	5.041532	0.000281	58.2952	
58	0.58	2.184407	1.604407	75.21676	5.118161	0.00026	58.17556		1.592943	74.16901	5.219279	0.000235	58.26776	
60	0.6	2.184407	1.584407	73.40671	5.293233	0.000219	58.15013		1.572943	72.40611	5.3908	0.000198	58.24466	
62	0.62	2.184407	1.564407	71.67812	5.462142	0.000185	58.12866		1.552943	70.72256	5.556244	0.000168	58.22509	
64	0.64	2.184407	1.544407	70.02734	5.625038	0.000157	58.11043		1.532943	69.11478	5.715762	0.000143	58.20841	
66	0.66	2.184407	1.524407	68.45085	5.782073	0.000134	58.09485		1.512943	67.57936	5.869506	0.000123	58.19411	
68	0.68	2.184407	1.504407	66.94531	5.933399	0.000115	58.08146		1.492943	66.11305	6.01763	0.000106	58.18178	
70	0.7	2.184407	1.484407	65.50753	6.079172	9.96E-05	58.06988		1.472943	64.71273	6.160288	9.19E-05	58.17109	

Center + 2							Center + 3					Center + 4			
	Temp (°C)	Rate Const		Retention (%)	log term		Temp (°C)		Rate Const	Retention (%)	log term	Temp (°C)		Rate Const	Retention (%)
00637	114.5495	1.717692	0.007807	60.75226		1.675393	82.358	4.444844	0.000511	56.73452	1.1965	50.72173	7.653395	2.06E-05	38.69961
80637	110.9692	1.998403	0.005896	60.04002		1.655393	80.22654	4.64296	0.000419	56.68701	1.1765	50.01414	7.732345	1.91E-05	38.69813
60637	107.55	2.271411	0.004488	59.50354		1.635393	78.19101	4.834404	0.000346	56.64781	1.1565	49.33839	7.808066	1.77E-05	38.69676
40637	104.2847	2.536751	0.003442	59.09534		1.615393	76.2471	5.019315	0.000287	56.61525	1.1365	48.69306	7.880675	1.64E-05	38.69549
20637	101.1664	2.794471	0.00266	58.7818		1.595393	74.39067	5.197836	0.00024	56.58803	1.1165	48.07677	7.950289	1.53E-05	38.69431
00637	98.18839	3.044635	0.002071	58.53881		1.575393	72.6178	5.370112	0.000202	56.56512	1.0965	47.48822	8.01702	1.43E-05	38.69319
80637	95.34445	3.287315	0.001625	58.34888		1.555393	70.92472	5.536292	0.000171	56.54573	1.0765	46.92616	8.080976	1.35E-05	38.69215
60637	92.6285	3.522597	0.001284	58.1992		1.535393	69.30784	5.696528	0.000146	56.52921	1.0565	46.38939	8.142264	1.27E-05	38.69117
40637	90.03479	3.750576	0.001022	58.08031		1.515393	67.76373	5.850971	0.000125	56.51507	1.0365	45.87678	8.200986	1.19E-05	38.69025
20637	87.55782	3.971356	0.00082	57.98515		1.495393	66.28912	5.999776	0.000108	56.50288	1.0165	45.38725	8.257242	1.13E-05	38.68938
00637	85.19232	4.18505	0.000662	57.90842		1.475393	64.88088	6.143095	9.34E-05	56.49232	0.9965	44.91974	8.311128	1.07E-05	38.68855
80637	82.9333	4.391778	0.000538	57.84609		1.455393	63.53602	6.281084	8.14E-05	56.48312	0.9765	44.47328	8.362736	1.02E-05	38.68776
60637	80.77594	4.591665	0.000441	57.7951		1.435393	62.25169	6.413896	7.13E-05	56.47507	0.9565	44.04691	8.412158	9.66E-06	38.68702
40637	78.71569	4.784846	0.000363	57.7531											
20637	76.74816	4.971456	0.000302	57.71828											
00637	74.86918	5.151638	0.000252	57.68921											
80637	73.07477	5.325537	0.000212	57.66479											
60637	71.36112	5.493301	0.000179	57.64416											
40637	69.7246	5.655081	0.000152	57.62661											
20637	68.16174	5.811028	0.00013	57.6116											

Table A.8.1.e Temperature and retention distribution histories for heating medium at 115°C

							Center						Center + 1		
	Time (t)	f_h/t		log term		Temp (°C)		Rate Const	Retention (%)	log term		Temp (°C)		Rate Const	Retention (%)
Input Values															
Medium Temp	115°C														
Initial Temp	70°C								100						100
j - value	1.959	2	0.020202	1.945247	1.925045	70	5.627749	0.000156	99.96871	1.932446	1.912244	70	5.627749	0.000156	99.96871
Heating rate	99	4	0.040404	1.945247	1.904843	70	5.627749	0.000156	99.93744		1.892042	70	5.627749	0.000156	99.93744
k - ref	0.0435/min	6	0.060606	1.945247	1.884641	70	5.627749	0.000156	99.90617		1.87184	70	5.627749	0.000156	99.90617
Act Energy	97 kJ/mole	8	0.080808	1.945247	1.864439	70	5.627749	0.000156	99.87491		1.851638	70	5.627749	0.000156	99.87491
R - value	8.314	10	0.10101	1.945247	1.844237	70	5.627749	0.000156	99.84367		1.831436	70	5.627749	0.000156	99.84367
Ref T	138°C	12	0.121212	1.945247	1.824035	70	5.627749	0.000156	99.81243		1.811234	70	5.627749	0.000156	99.81243
Time Increment	2 min	14	0.141414	1.945247	1.803833	70	5.627749	0.000156	99.7812		1.791032	70	5.627749	0.000156	99.7812
Cmed Temp	35°C	16	0.161616	1.945247	1.783631	70	5.627749	0.000156	99.74999		1.77083	70	5.627749	0.000156	99.74999
j2-value	1.9021	18	0.181818	1.945247	1.763429	70	5.627749	0.000156	99.71878		1.750628	70	5.627749	0.000156	99.71878
j3-value	1.5879	20	0.20202	1.945247	1.743227	70	5.627749	0.000156	99.68758		1.730426	70	5.627749	0.000156	99.68758
j4-value	0.918	22	0.222222	1.945247	1.723025	70	5.627749	0.000156	99.65639		1.710224	70	5.627749	0.000156	99.65639
j5-value	0.2968	24	0.242424	1.945247	1.702823	70	5.627749	0.000156	99.62521		1.690022	70	5.627749	0.000156	99.62521
Cooling rate - f	100	26	0.262626	1.945247	1.682621	70	5.627749	0.000156	99.59404		1.66982	70	5.627749	0.000156	99.59404
Cooling - j	1.958	28	0.282828	1.945247	1.662419	70	5.627749	0.000156	99.56288		1.649618	70.37096	5.591001	0.000162	99.56172
		30	0.30303	1.945247	1.642217	70	5.627749	0.000156	99.53174		1.629416	72.39942	5.391455	0.000198	99.52227
		32	0.323232	1.945247	1.622015	73.11923	5.321207	0.000213	99.48943		1.609214	74.33568	5.203153	0.000239	99.47467
		34	0.343434	1.945247	1.601813	75.02278	5.136836	0.000256	99.43858		1.589012	76.18393	5.025358	0.000286	99.41783
		36	0.363636	1.945247	1.581611	76.8398	4.962717	0.000304	99.37809		1.568809	77.94818	4.857391	0.000338	99.35064
		38	0.383838	1.945247	1.561409	78.57424	4.798191	0.000359	99.30683		1.548607	79.63225	4.698626	0.000396	99.27195
		40	0.40404	1.945247	1.541207	80.22985	4.64265	0.000419	99.22365		1.528405	81.23977	4.548485	0.00046	99.18059
		42	0.424242	1.945247	1.521005	81.81021	4.495533	0.000485	99.12737		1.508203	82.77422	4.406434	0.000531	99.07538
		44	0.444444	1.945247	1.500803	83.31874	4.35632	0.000558	99.01682		1.488001	84.23893	4.271978	0.000607	98.95518
		46	0.464646	1.945247	1.4806	84.7587	4.22453	0.000637	98.89086		1.467799	85.63707	4.144658	0.000689	98.81883
		48	0.484848	1.945247	1.460398	86.13321	4.099715	0.000721	98.74833		1.447597	86.97166	4.024047	0.000778	98.66523
		50	0.505051	1.945247	1.440196	87.44525	3.981462	0.000812	98.58817		1.427395	88.24559	3.909749	0.000872	98.49331
		52	0.525253	1.945247	1.419994	88.69766	3.869383	0.000908	98.40931		1.407193	89.46162	3.801396	0.000972	98.30207
		54	0.545455	1.945247	1.399792	89.89314	3.763121	0.00101	98.21079		1.386991	90.62238	3.698644	0.001077	98.09057
		56	0.565657	1.945247	1.37959	91.03429	3.662339	0.001117	97.99168		1.366789	91.73038	3.601172	0.001187	97.85794
		58	0.585859	1.945247	1.359388	92.12356	3.566726	0.001229	97.75115		1.346587	92.78802	3.508681	0.001302	97.6034
		60	0.606061	1.945247	1.339186	93.16333	3.475989	0.001346	97.48846		1.326385	93.79759	3.420892	0.001422	97.32627
		62	0.626263	1.945247	1.318984	94.15584	3.389855	0.001467	97.20293		1.306183	94.76127	3.337543	0.001545	97.02593
		64	0.646465	1.945247	1.298782	95.10324	3.30807	0.001592	96.89403		1.285981	95.68115	3.258388	0.001673	96.70191
		66	0.666667	1.945247	1.27858	96.00758	3.230394	0.00172	96.56127		1.265779	96.55922	3.183199	0.001803	96.35379
		68	0.686869	1.945247	1.258378	96.87082	3.156603	0.001852	96.20431		1.245577	97.39739	3.111759	0.001937	95.98129
		70	0.707071	1.945247	1.238176	97.69482	3.086486	0.001986	95.82288		1.225375	98.19745	3.043868	0.002073	95.58421
		72	0.727273	1.945247	1.217974	98.48136	3.019846	0.002123	95.41685		1.205173	98.96115	2.979335	0.002211	95.16248
		74	0.747475	1.945247	1.197772	99.23216	2.956498	0.002262	94.98615		1.184971	99.69014	2.917981	0.002351	94.71611
		76	0.767677	1.945247	1.17757	99.94883	2.896267	0.002402	94.53084		1.164769	100.386	2.85964	0.002492	94.2452
		78	0.787879	1.945247	1.157368	100.6329	2.838989	0.002544	94.05108		1.144567	101.0502	2.804153	0.002634	93.74997
		80	0.808081	1.945247	1.137166	101.2859	2.78451	0.002687	93.54709		1.124365	101.6843	2.751372	0.002777	93.23072
		82	0.828283	1.945247	1.116964	101.9093	2.732684	0.002829	93.01921		1.104163	102.2895	2.701155	0.00292	92.68782
		84	0.848485	1.945247	1.096762	102.5043	2.683375	0.002972	92.46786		1.083961	102.8672	2.653372	0.003063	92.12176
		86	0.868687	1.945247	1.07656	103.0722	2.636452	0.003115	91.89353		1.063759	103.4187	2.607898	0.003205	91.53306
		88	0.888889	1.945247	1.056358	103.6143	2.591793	0.003258	91.29679		1.043557	103.945	2.564615	0.003347	90.92233
		90	0.909091	1.945247	1.036156	104.1318	2.549284	0.003399	90.67826		1.023355	104.4475	2.523412	0.003488	90.29025
		92	0.929293	1.945247	1.015954	104.6258	2.508816	0.003539	90.03864		1.003153	104.9271	2.484183	0.003628	89.63754
		94	0.949495	1.945247	0.995752	105.0973	2.470286	0.003678	89.37868		0.982951	105.385	2.44683	0.003766	88.96498
		96	0.969697	1.945247	0.97555	105.5474	2.433597	0.003816	88.69916		0.962749	105.822	2.41126	0.003902	88.2734
		98	0.989899	1.945247	0.955348	105.9771	2.398657	0.003952	88.00093		0.942547	106.2391	2.377382	0.004037	87.56363
		100	1.010101	1.945247	0.935146	106.3872	2.365379	0.004085	87.28485		0.922345	106.6373	2.345114	0.004169	86.83658
		102	1.030303	1.945247	0.914944	106.7786	2.33368	0.004217	86.55182		0.902143	107.0174	2.314376	0.004299	86.09316
		104	1.050505	1.945247	0.894742	107.1523	2.303483	0.004346	85.80275		0.881941	107.3802	2.285092	0.004427	85.3343
		106	1.070707	1.945247	0.87454	107.509	2.274713	0.004473	85.03859		0.861739	107.7266	2.25719	0.004552	84.56094
		108	1.090909	1.945247	0.854338	107.8495	2.247302	0.004597	84.26029		0.841537	108.0572	2.230605	0.004675	83.77403
		110	1.111111	1.945247	0.834136	108.1745	2.221182	0.004719	83.46879		0.821335	108.3727	2.205271	0.004795	82.97455
		112	1.131313	1.945247	0.813934	108.4847	2.196291	0.004838	82.66507		0.801133	108.6739	2.181127	0.004912	82.16343
		114	1.151515	1.945247	0.793732	108.7808	2.172569	0.004954	81.85007		0.780931	108.9615	2.158116	0.005026	81.34165
		116	1.171717	1.945247	0.77353	109.0635	2.149959	0.005067	81.02475		0.760729	109.2359	2.136184	0.005138	80.51013
		118	1.191919	1.945247	0.753328	109.3333	2.128408	0.005178	80.19004		0.740527	109.4979	2.115277	0.005246	79.66982
		120	1.212121	1.945247	0.733126	109.5909	2.107865	0.005285	79.34687		0.720325	109.748	2.095348	0.005352	78.82163
		122	1.232323	1.945247	0.712924	109.8367	2.088282	0.00539	78.49617		0.700123	109.9867	2.076349	0.005454	77.96646
		124	1.252525	1.945247	0.692722	110.0714	2.069612	0.005491	77.6388		0.679921	110.2146	2.058235	0.005554	77.1052
		126	1.272727	1.945247	0.67252	110.2954	2.051812	0.00559	76.77566		0.659719	110.4321	2.040965	0.005651	76.23869
		128	1.292929	1.945247	0.652318	110.5093	2.034841	0.005686	75.90759		0.639517	110.6397	2.024497	0.005745	75.36778
		130	1.313131	1.945247	0.632116	110.7134	2.018658	0.005778	75.03541		0.619315	110.8379	2.008795	0.005836	74.49327
		132	1.333333	1.945247	0.611914	110.9082	2.003227	0.005868	74.15993		0.599113	111.0271	1.993822	0.005924	73.61595
		134	1.353535	1.945247	0.591712	111.0942	1.988512	0.005955	73.28191		0.578911	111.2076	1.979543	0.006009	72.73656
		136	1.373737	1.945247	0.57151	111.2717	1.97448	0.006039	72.40209		0.558708	111.38	1.965926	0.006091	71.85584
		138	1.393939	1.945247	0.551308	111.4412	1.961097	0.006121	71.5212		0.538506	111.5445	1.952939	0.006171	70.97448
		140	1.414141	1.945247	0.531106	111.6029	1.948333	0.006199	70.63992		0.518304	111.7016	1.940552	0.006248	70.09315
		142	1.434343	1.945247	0.510904	111.7573	1.936159	0.006275	69.75891		0.498102	111.8515	1.928738	0.006322	69.21248

log term		Temp (°C)		Rate Const	Retention (%)
Center + 2					
C + 2					100
1.854036	1.833834	70	5.627749	0.000156	99.96871
	1.813632	70	5.627749	0.000156	99.93744
	1.79343	70	5.627749	0.000156	99.90617
	1.773228	70	5.627749	0.000156	99.87491
	1.753026	70	5.627749	0.000156	99.84367
	1.732824	70	5.627749	0.000156	99.81243
	1.712622	70	5.627749	0.000156	99.7812
	1.69242	70	5.627749	0.000156	99.74999
	1.672217	70	5.627749	0.000156	99.71878
	1.652015	70.12386	5.61547	0.000158	99.68719
	1.631813	72.16355	5.414537	0.000194	99.6486
	1.611611	74.11054	5.22494	0.000234	99.60196
	1.591409	75.96903	5.045935	0.00028	99.54621
	1.571207	77.74304	4.876835	0.000332	99.48023
	1.551005	79.43643	4.717008	0.000389	99.40287
	1.530803	81.05285	4.565873	0.000452	99.31296
	1.510601	82.5958	4.422888	0.000522	99.20934
	1.490399	84.06862	4.287555	0.000598	99.09083
	1.470197	85.4745	4.159411	0.000679	98.95629
	1.449995	86.81648	4.038025	0.000767	98.80461
	1.429793	88.09746	3.922998	0.000861	98.63471
	1.409591	89.32022	3.813958	0.00096	98.44558
	1.389389	90.48741	3.710558	0.001064	98.23628
	1.369187	91.60155	3.612475	0.001174	98.00592
	1.348985	92.66504	3.519408	0.001288	97.75372
	1.328783	93.6802	3.431075	0.001407	97.47896
	1.308581	94.64922	3.347212	0.00153	97.18105
	1.288379	95.57419	3.267571	0.001657	96.85947
	1.268177	96.45712	3.191923	0.001788	96.5138
	1.247975	97.29993	3.120049	0.001921	96.14376
	1.227773	98.10442	3.051747	0.002057	95.74913
	1.207571	98.87235	2.986825	0.002194	95.32981
	1.187369	99.60538	2.925103	0.002334	94.88582
	1.167167	100.3051	2.866413	0.002475	94.41724
	1.146965	100.973	2.810595	0.002617	93.92429
	1.126763	101.6105	2.7575	0.00276	93.40724
	1.106561	102.2191	2.706986	0.002903	92.86647
	1.086359	102.8	2.658921	0.003046	92.30244
	1.066157	103.3545	2.613179	0.003189	91.71568
	1.045955	103.8838	2.569642	0.00333	91.1068
	1.025753	104.3891	2.528197	0.003471	90.47645
	1.005551	104.8714	2.488739	0.003611	89.82536
	0.985349	105.3317	2.451169	0.003749	89.1543
	0.965147	105.7712	2.415392	0.003886	88.46408
	0.944945	106.1906	2.381318	0.004021	87.75557
	0.924743	106.591	2.348863	0.004153	87.02964
	0.904541	106.9732	2.317947	0.004284	86.28721
	0.884339	107.3381	2.288494	0.004412	85.52921
	0.864137	107.6863	2.260432	0.004537	84.75658
	0.843935	108.0187	2.233694	0.00466	83.97028
	0.823733	108.336	2.208214	0.004781	83.17126
	0.803531	108.6389	2.183932	0.004898	82.36049
	0.783329	108.928	2.16079	0.005013	81.53892
	0.763127	109.204	2.138732	0.005124	80.7075
	0.742925	109.4675	2.117707	0.005233	79.86716
	0.722723	109.7189	2.097664	0.005339	79.01883
	0.702521	109.959	2.078557	0.005442	78.1634
	0.682318	110.1881	2.06034	0.005542	77.30177
	0.662116	110.4068	2.042972	0.005639	76.43479
	0.641914	110.6156	2.026411	0.005734	75.56329
	0.621712	110.8148	2.01062	0.005825	74.6881
	0.60151	111.0051	1.995563	0.005913	73.81001
	0.581308	111.1866	1.981203	0.005999	72.92976
	0.561106	111.36	1.967509	0.006082	72.04808
	0.540904	111.5254	1.954448	0.006161	71.16569
	0.520702	111.6833	1.941992	0.006239	70.28325
	0.5005	111.8341	1.930112	0.006313	69.40139
	0.480298	111.978	1.91878	0.006385	68.52075
	0.460096	112.1153	1.907971	0.006455	67.64189
	0.439894	112.2464	1.89766	0.006521	66.76537
	0.419692	112.3716	1.887825	0.006586	65.89171

log term		Temp (°C)		Rate Const	Retention (%)
Center + 3					
C + 3					100
1.616055	1.595853	75.5676	5.084437	0.000269	99.94614
	1.575651	77.35987	4.913214	0.00032	99.88226
	1.555449	79.07067	4.7514	0.000376	99.80721
	1.535247	80.70371	4.5984	0.000438	99.71983
	1.515045	82.26253	4.453667	0.000506	99.61893
	1.494843	83.7505	4.316692	0.00058	99.50335
	1.474641	85.17084	4.187004	0.000661	99.37192
	1.454439	86.52662	4.064167	0.000747	99.22353
	1.434237	87.82078	3.947774	0.000839	99.05709
	1.414035	89.05612	3.837447	0.000937	98.87156
	1.393833	90.23531	3.732835	0.001041	98.66597
	1.373631	91.3609	3.63361	0.001149	98.43944
	1.353429	92.43533	3.539464	0.001263	98.19114
	1.333227	93.46093	3.450113	0.001381	97.92036
	1.313025	94.43992	3.365288	0.001503	97.62645
	1.292823	95.3744	3.284739	0.001629	97.30888
	1.272621	96.26642	3.208232	0.001759	96.96722
	1.252419	97.11789	3.135546	0.001891	96.60114
	1.232217	97.93066	3.066475	0.002026	96.21042
	1.212015	98.70648	3.000825	0.002164	95.79493
	1.191813	99.44705	2.938414	0.002303	95.35466
	1.171611	100.154	2.879071	0.002444	94.88968
	1.151409	100.8287	2.822634	0.002586	94.40017
	1.131207	101.4728	2.768953	0.002729	93.8864
	1.111005	102.0877	2.717883	0.002872	93.34873
	1.090803	102.6746	2.66929	0.003015	92.7876
	1.070601	103.2348	2.623047	0.003157	92.20354
	1.050399	103.7695	2.579035	0.003299	91.59711
	1.030197	104.28	2.537139	0.003441	90.969
	1.009995	104.7672	2.497253	0.003581	90.31989
	0.989793	105.2323	2.459276	0.003719	89.65056
	0.969591	105.6763	2.423112	0.003856	88.96182
	0.949389	106.1	2.388671	0.003991	88.25452
	0.929187	106.5045	2.355867	0.004124	87.52954
	0.908984	106.8907	2.324619	0.004255	86.78779
	0.888782	107.2593	2.294851	0.004384	86.0302
	0.86858	107.6111	2.266489	0.00451	85.25772
	0.848378	107.9469	2.239465	0.004633	84.4713
	0.828176	108.2675	2.213714	0.004754	83.6719
	0.807974	108.5735	2.189174	0.004872	82.8605
	0.787772	108.8656	2.165786	0.004988	82.03804
	0.76757	109.1444	2.143494	0.0051	81.20548
	0.747368	109.4106	2.122246	0.00521	80.36377
	0.727166	109.6646	2.101991	0.005316	79.51382
	0.706964	109.9071	2.082682	0.00542	78.65656
	0.686762	110.1386	2.064273	0.005521	77.79288
	0.66656	110.3595	2.046721	0.005618	76.92363
	0.646358	110.5705	2.029987	0.005713	76.04968
	0.626156	110.7718	2.01403	0.005805	75.17184
	0.605954	110.964	1.998814	0.005894	74.2909
	0.585752	111.1474	1.984303	0.00598	73.40764
	0.56555	111.3225	1.970466	0.006064	72.5228
	0.545348	111.4897	1.957268	0.006144	71.63708
	0.525146	111.6492	1.944682	0.006222	70.75116
	0.504944	111.8015	1.932677	0.006297	69.8657
	0.484742	111.9469	1.921227	0.00637	68.98131
	0.46454	112.0857	1.910305	0.00644	68.09859
	0.444338	112.2181	1.899887	0.006507	67.2181
	0.424136	112.3446	1.889949	0.006572	66.34038
	0.403934	112.4653	1.880469	0.006635	65.46591
	0.383732	112.5805	1.871425	0.006695	64.59519
	0.36353	112.6904	1.862798	0.006753	63.72865
	0.343328	112.7954	1.854567	0.006809	62.86672
	0.323126	112.8956	1.846714	0.006862	62.00979
	0.302924	112.9913	1.839222	0.006914	61.15823
	0.282722	113.0826	1.832075	0.006964	60.31238
	0.26252	113.1697	1.825255	0.007011	59.47256
	0.242318	113.2529	1.818748	0.007057	58.63906
	0.222116	113.3323	1.812539	0.007101	57.81217
	0.201914	113.4081	1.806615	0.007143	56.99212
	0.181712	113.4805	1.800962	0.007184	56.17916

log term		Temp (°C)		Rate Const	Retention (%)
Center + 4					
C + 4					100
1.125676	1.105474	102.2511	2.70434	0.002911	99.41954
	1.085272	102.8305	2.656403	0.003054	98.81419
	1.06507	103.3836	2.610782	0.003196	98.18453
	1.044868	103.9116	2.56736	0.003338	97.53122
	1.024666	104.4156	2.526025	0.003479	96.85496
	1.004464	104.8967	2.486672	0.003619	96.15653
	0.984262	105.3559	2.4492	0.003757	95.43676
	0.96406	105.7942	2.413516	0.003893	94.69652
	0.943858	106.2126	2.379532	0.004028	93.93674
	0.923656	106.612	2.347161	0.00416	93.15837
	0.903454	106.9933	2.316326	0.004291	92.36237
	0.883252	107.3572	2.28695	0.004419	91.54975
	0.86305	107.7046	2.258961	0.004544	90.72152
	0.842848	108.0362	2.232292	0.004667	89.87871
	0.822646	108.3527	2.206878	0.004787	89.02233
	0.802444	108.6548	2.182659	0.004904	88.15343
	0.782242	108.9432	2.159577	0.005019	87.27301
	0.76204	109.2185	2.137576	0.00513	86.3821
	0.741838	109.4813	2.116604	0.005239	85.48169
	0.721636	109.7321	2.096613	0.005345	84.57277
	0.701434	109.9716	2.077555	0.005448	83.6563
	0.681232	110.2001	2.059385	0.005548	82.73324
	0.66103	110.4183	2.042061	0.005645	81.8045
	0.640828	110.6265	2.025543	0.005739	80.87098
	0.620626	110.8253	2.009792	0.00583	79.93354
	0.600424	111.015	1.994773	0.005918	78.99304
	0.580222	111.1962	1.98045	0.006003	78.05027
	0.56002	111.3691	1.96679	0.006086	77.10602
	0.539818	111.5341	1.953763	0.006166	76.16103
	0.519616	111.6916	1.941339	0.006243	75.21604
	0.499414	111.842	1.929488	0.006317	74.27171
	0.479212	111.9855	1.918185	0.006389	73.3287
	0.45901	112.1225	1.907404	0.006458	72.38764
	0.438808	112.2533	1.89712	0.006525	71.44912
	0.418606	112.3782	1.887309	0.006589	70.5137
	0.398404	112.4973	1.877951	0.006651	69.58189
	0.378202	112.6111	1.869023	0.006711	68.65421
	0.358	112.7197	1.860506	0.006768	67.73113
	0.337798	112.8233	1.85238	0.006824	66.81307
	0.317596	112.9222	1.844628	0.006877	65.90046
	0.297394	113.0167	1.837232	0.006928	64.99368
	0.277192	113.1068	1.830176	0.006977	64.09309
	0.25699	113.1929	1.823443	0.007024	63.19901
	0.236788	113.275	1.817019	0.007069	62.31177
	0.216586	113.3534	1.81089	0.007113	61.43165
	0.196383	113.4282	1.805041	0.007154	60.5589
	0.176181	113.4997	1.799461	0.007194	59.69377
	0.155979	113.5679	1.794136	0.007233	58.83648
	0.135777	113.633	1.789054	0.00727	57.98723
	0.115575	113.6951	1.784206	0.007305	57.1462
	0.095373	113.7544	1.779579	0.007339	56.31355
	0.075171	113.811	1.775164	0.007371	55.48943
	0.054969	113.8651	1.770951	0.007402	54.67397
	0.034767	113.9167	1.76693	0.007432	53.86727
	0.014565	113.9659	1.763093	0.007461	53.06945
	-0.00564	114.0129	1.759431	0.007488	52.28058
	-0.02584	114.0578	1.755937	0.007514	51.50074
	-0.04604	114.1006	1.752602	0.00754	50.72998
	-0.06624	114.1415	1.749419	0.007564	49.96836
	-0.08644	114.1805	1.746382	0.007587	49.21591
	-0.10665	114.2177	1.743483	0.007609	48.47265
	-0.12685	114.2533	1.740717	0.00763	47.7386
	-0.14705	114.2872	1.738077	0.00765	47.01378
	-0.16725	114.3196	1.735557	0.007669	46.29817
	-0.18745	114.3506	1.733152	0.007688	45.59177
	-0.20766	114.3801	1.730857	0.007705	44.89456
	-0.22786	114.4082	1.728667	0.007722	44.20652
	-0.24806	114.4351	1.726576	0.007738	43.52762
	-0.26826	114.4608	1.724581	0.007754	42.85782
	-0.28847	114.4853	1.722676	0.007769	42.19708
	-0.30867	114.5087	1.720859	0.007783	41.54535

(Continued)

Table A.8.1.e Temperature and retention distribution histories for heating medium at 115°C (Continued)

144	1.454545	1.945247	0.490701	111.9047	1.924548	0.006348	68.87878		0.4779	111.9946	1.91747	0.006394	68.33309
2	0.02	2.177766	2.157766	112.05	1.913111	0.006421	67.99983	2.165186	2.145186	112.15	1.905244	0.006472	67.45426
4	0.04	2.177766	2.137766	112.2	1.901312	0.006498	67.12186		2.125186	112.3	1.893451	0.006549	66.5765
6	0.06	2.177766	2.117766	112.35	1.889522	0.006575	66.24502		2.105186	112.4	1.885594	0.006601	65.70338
8	0.08	2.177766	2.097766	112.45	1.881667	0.006627	65.37285		2.085186	112.5	1.877741	0.006653	64.83497
10	0.1	2.177766	2.077766	112.55	1.873816	0.006679	64.50543		2.065186	112.55	1.873816	0.006679	63.97468
12	0.12		2.057766	112.65	1.865969	0.006731	63.64281		2.045186	112.6	1.869892	0.006705	63.12249
14	0.14		2.037766	112.7	1.862048	0.006758	62.78842		2.025186	112.6	1.869892	0.006705	62.28166
16	0.16		2.017766	112.75	1.858127	0.006784	61.9422		2.005186	112.6	1.869892	0.006705	61.45202
18	0.18		1.997766	112.8	1.854207	0.006811	61.10413		1.985186	112.6	1.869892	0.006705	60.63344
20	0.2		1.977766	112.8	1.854207	0.006811	60.2774		1.965186	112.6	1.869892	0.006705	59.82576
22	0.22		1.957766	112.8	1.854207	0.006811	59.46186		1.945186	112.6	1.869892	0.006705	59.02884
24	0.24		1.937766	112.8	1.854207	0.006811	58.65735		1.925186	112.6	1.869892	0.006705	58.24254
26	0.26		1.917766	112.8	1.854207	0.006811	57.86373		1.905186	112.6	1.869892	0.006705	57.4667
28	0.28		1.897766	112.8	1.854207	0.006811	57.08084		1.885186	112.6	1.869892	0.006705	56.70121
30	0.3		1.877766	110.4685	2.038076	0.005667	56.43752		1.865186	112.6	1.869892	0.006705	55.94591
32	0.32		1.857766	107.0718	2.30998	0.004318	55.95223		1.845186	105.0142	2.477071	0.003654	55.5386
34	0.34		1.837766	103.8281	2.574223	0.003315	55.58246		1.825186	101.8631	2.73652	0.002819	55.2264
36	0.36		1.817766	100.7303	2.830854	0.002565	55.29807		1.805186	98.85373	2.988396	0.002191	54.98493
38	0.38		1.797766	97.77195	3.079938	0.001999	55.07739		1.785186	95.97984	3.232771	0.001716	54.79654
40	0.4		1.777766	94.94675	3.321551	0.00157	54.9047		1.765186	93.23529	3.469728	0.001354	54.64836
42	0.42		1.757766	92.2487	3.555778	0.001242	54.76845		1.745186	90.61427	3.699359	0.001076	54.53086
44	0.44		1.737766	89.67208	3.782717	0.00099	54.6601		1.725186	88.11122	3.921767	0.000862	54.43698
46	0.46		1.717766	87.21143	4.002472	0.000795	54.57329		1.705186	85.72082	4.137062	0.000695	54.3614
48	0.48		1.697766	84.86153	4.215158	0.000642	54.50321		1.685186	83.43801	4.345363	0.000564	54.30011
50	0.5		1.677766	82.61739	4.420896	0.000523	54.44622		1.665186	81.25794	4.546795	0.000461	54.25005
52	0.52		1.657766	80.47426	4.619812	0.000429	54.39956		1.645186	79.17599	4.741489	0.00038	54.20888
54	0.54		1.637766	78.42758	4.812041	0.000354	54.36109		1.625186	77.18774	4.929582	0.000314	54.1748
56	0.56		1.617766	76.47302	4.997719	0.000294	54.32917		1.605186	75.28898	5.111213	0.000262	54.14639
58	0.58		1.597766	74.60642	5.17699	0.000246	54.30249		1.585186	73.47568	5.286529	0.00022	54.12256
60	0.6		1.577766	72.82384	5.349999	0.000207	54.28006		1.565186	71.74399	5.455675	0.000186	54.10245
62	0.62		1.557766	71.12149	5.516895	0.000175	54.26109		1.545186	70.09024	5.618802	0.000158	54.08537
64	0.64		1.537766	69.49575	5.677828	0.000149	54.24494		1.525186	68.51091	5.776063	0.000135	54.07078
66	0.66		1.517766	67.94319	5.83295	0.000127	54.23112		1.505186	67.00267	5.927609	0.000116	54.05825
68	0.68		1.497766	66.4605	5.982415	0.00011	54.21922		1.485186	65.56232	6.073595	0.0001	54.04742
70	0.7		1.477766	65.04454	6.126377	9.5E-05	54.20892		1.465186	64.18679	6.214175	8.7E-05	54.03801

	0.39949	112.4911	1.878443	0.006648	65.02141		0.16151	113.5495	1.795569	0.007222	55.3735		-0.32887	114.531	1.719124	0.007796	40.90259
2.090075	2.070075	112.62	1.868323	0.006716	64.15393	1.857986	1.837986	103.863	2.571349	0.003325	55.0065	1.373001	1.353001	57.54242	6.909707	4.34E-05	40.89903
	2.050075	112.72	1.860479	0.006769	63.29133		1.817986	100.7637	2.828064	0.002572	54.72427		1.333001	56.52785	7.018381	3.89E-05	40.89585
	2.030075	112.82	1.852639	0.006822	62.43367		1.797986	97.80385	3.077231	0.002005	54.50529		1.313001	55.55893	7.122791	3.51E-05	40.89298
	2.010075	112.87	1.848721	0.006849	61.58434		1.777986	94.97721	3.318926	0.001574	54.33394		1.293001	54.63363	7.223078	3.17E-05	40.89038
	1.990075	112.92	1.844804	0.006875	60.74329		1.757986	92.27779	3.553234	0.001245	54.19877		1.273001	53.74997	7.319382	2.88E-05	40.88803
	1.970075	112.92	1.844804	0.006875	59.91373		1.737986	89.69986	3.780253	0.000993	54.09129		1.253001	52.90608	7.411838	2.63E-05	40.88588
	1.950075	112.92	1.844804	0.006875	59.0955		1.717986	87.23797	4.000087	0.000797	54.00517		1.233001	52.10018	7.500582	2.4E-05	40.88391
	1.930075	112.92	1.844804	0.006875	58.28845		1.697986	84.88687	4.21285	0.000644	53.93566		1.213001	51.33054	7.585743	2.21E-05	40.88211
	1.910075	112.92	1.844804	0.006875	57.49242		1.677986	82.64159	4.418664	0.000524	53.87914		1.193001	50.59555	7.667449	2.03E-05	40.88044
	1.890075	112.6381	1.866905	0.006725	56.7243		1.657986	80.49737	4.617655	0.00043	53.83287		1.173001	49.89363	7.745826	1.88E-05	40.8789
	1.870075	109.1438	2.143544	0.0051	56.14867		1.637986	78.44965	4.809956	0.000354	53.79472		1.153001	49.22331	7.820993	1.75E-05	40.87748
	1.850075	105.8068	2.412497	0.003897	55.71272		1.617986	76.49409	4.995706	0.000294	53.76306		1.133001	48.58315	7.89307	1.62E-05	40.87615
	1.830075	102.6199	2.673806	0.003001	55.37933		1.597986	74.62655	5.175047	0.000246	53.73661		1.113001	47.97181	7.962171	1.52E-05	40.87491
	1.810075	99.57654	2.927527	0.002329	55.12203		1.577986	72.84306	5.348124	0.000207	53.71438		1.093001	47.38798	8.028409	1.42E-05	40.87375
	1.790075	96.67011	3.173728	0.00182	54.92171		1.557986	71.13984	5.515087	0.000175	53.69557		1.073001	46.83043	8.091891	1.33E-05	40.87266
	1.770075	93.8945	3.41249	0.001434	54.76445		1.537986	69.51328	5.676084	0.000149	53.67956		1.053001	46.29797	8.152722	1.25E-05	40.87164
	1.750075	91.24381	3.643903	0.001138	54.64		1.517986	67.95993	5.83127	0.000128	53.66586		1.033001	45.78948	8.211006	1.18E-05	40.87067
	1.730075	88.71242	3.868066	0.000909	54.54074		1.497986	66.47649	5.980797	0.00011	53.65406		1.013001	45.30388	8.26684	1.12E-05	40.86976
	1.710075	86.29497	4.08509	0.000732	54.46098		1.477986	65.05981	6.124819	9.52E-05	53.64385		0.993001	44.84012	8.320321	1.06E-05	40.86889
	1.690075	83.98631	4.295089	0.000593	54.39642		1.457986	63.7069	6.26349	8.28E-05	53.63496		0.973001	44.39725	8.37154	1.01E-05	40.86807
	1.670075	81.78157	4.498188	0.000484	54.34377		1.437986	62.41487	6.396964	7.25E-05	53.62719		0.953001	43.9743	8.420588	9.58E-06	40.86729
	1.650075	79.67605	4.694516	0.000398	54.30055		1.417986	61.181	6.525395	6.38E-05	53.62035		0.933001	43.57039	8.467551	9.14E-06	40.86654
	1.630075	77.66529	4.88421	0.000329	54.26483		1.397986	60.00266	6.648933	5.64E-05	53.61431		0.913001	43.18466	8.512511	8.74E-06	40.86583
	1.610075	75.74504	5.067407	0.000274	54.2351												
	1.590075	73.91121	5.244253	0.00023	54.2102												
	1.570075	72.15992	5.414893	0.000194	54.18922												
	1.550075	70.48745	5.579478	0.000164	54.17143												
	1.530075	68.89025	5.738158	0.00014	54.15625												
	1.510075	67.36494	5.891087	0.00012	54.14323												
	1.490075	65.90827	6.038418	0.000104	54.13199												
	1.470075	64.51717	6.180305	9E-05	54.12225												
	1.450075	63.18868	6.316902	7.85E-05	54.11374												
	1.430075	61.91998	6.448363	6.89E-05	54.10629												
	1.410075	60.70838	6.57484	6.07E-05	54.09973												
	1.390075	59.55131	6.696485	5.37E-05	54.09391												

Table A.8.2 Comparison of quality retention for PATP and traditional thermal process

		Pressure (MPa)	Time (min)	Temp (°C)		k (1/min)	N
		0	0	40			1000
		50	30	42.5	10.15118	9.76E-05	999.9024
		100	60	45			999.9024
Density	900 kg/m^3	150	90	47.5	7.772188	0.001053	998.8499
Specific Heat	3880 kJ/kg K	200	120	50			998.8499
Microbial		250	150	52.5	3.329017	0.089571	913.272
Initial population	1000	300	180	55			913.272
Rate constant	2.5/min	350	210	57.5	1.092885	0.83812	395.011
E-constant	400 kJ/mole	400	240	60			395.011
Temp	65°C	400	270	60	0	2.5	32.42448
Vitamin		400	300	60			32.42448
Rate constant	0.25/min	400	330	60	0	2.5	2.661563
E-constant	80 kJ/mole	400	360	60			2.661563
Temp	95°C	400	390	60	0	2.5	0.218474
j-value	2.048	400	420	60			0.218474
Heating rate	38.4 min	400	450	60	0	2.5	0.017933
Med Temp	65.5°C	400	480	60			0.017933
Cool Med	20°C	400	510	60	0	2.5	0.001472
cooling rate-f	62.7 min	400	540	60			0.001472
cooling j	1.815	400	570	60	0	2.5	0.000121
jm-value	0.551	400	600	60			0.000121
jm cool	0.813	400	630	60	0	2.5	9.92E-06
		400	660	60			9.92E-06
		400	690	60	0	2.5	8.14E-07
		400	720	60			8.14E-07
		300	750	55	2.202429	0.276336	8.08E-07
		200	780	50			8.08E-07
		100	810	45	6.815063	0.002743	8.06E-07
		0	840	40			8.06E-07

	k (1/min)	Retention (%)	Time (min)	Temp (°C)		k (1/min)			k (1/min)	Retention (%)
4.59463	0.002526	100	0	40			1000			100
4.35103	0.003223	99.67818	30	42.5	10.15118	9.76E-05	999.9024	4.35103	0.003223	99.67818
		99.67818	60	45			999.9024			99.67818
3.875232	0.005187	99.16245	90	47.5	7.772188	0.001053	998.8499	3.875232	0.005187	99.16245
		99.16245	120	50			998.8499			99.16245
3.414052	0.008227	98.35	150	52.5	3.329017	0.089571	913.272	3.414052	0.008227	98.35
		98.35	180	55			913.272			98.35
2.966825	0.012867	97.09267	210	57.5	1.092885	0.83812	395.011	2.966825	0.012867	97.09267
		97.09267	240	60			395.011			97.09267
2.748248	0.01601	95.5506	270	60	0	2.5	32.42448	2.748248	0.01601	94.03301
		95.5506	300	60			32.42448			94.03301
2.748248	0.01601	94.03301	330	60	0	2.5	2.661563	2.748248	0.01601	91.06977
		94.03301	360	60			2.661563			91.06977
2.748248	0.01601	92.53953	390	60	0	2.5	0.218474	2.748248	0.01601	88.19991
		92.53953	420	60			0.218474			88.19991
2.748248	0.01601	91.06977	450	60	0	2.5	0.017933	2.748248	0.01601	85.42049
		91.06977	480	60			0.017933			85.42049
2.748248	0.01601	89.62336	510	60	0	2.5	0.001472	2.748248	0.01601	82.72866
		89.62336	540	60			0.001472			82.72866
2.748248	0.01601	88.19991	570	60	0	2.5	0.000121	2.748248	0.01601	80.12165
		88.19991	600	60			0.000121			80.12165
2.748248	0.01601	86.79908	630	60	0	2.5	9.92E-06	2.748248	0.01601	77.59679
		86.79908	660	60			9.92E-06			77.59679
2.748248	0.01601	85.42049	690	60	0	2.5	8.14E-07	2.748248	0.01601	76.36436
		85.42049	720	60			8.14E-07			76.36436
3.188734	0.010306	84.54467	750	57.5	1.092885	0.83812	3.52E-07	2.966825	0.012867	75.38811
		84.54467	780	55			3.52E-07			75.38811
4.111261	0.004097	84.19902	810	52.5	3.329017	0.089571	3.22E-07	3.414052	0.008227	74.77044
		84.19902	840	50			3.22E-07			74.77044
			900	45			3.19E-07			74.38358
			930	42.5	8.013909	0.000827	3.19E-07	4.35103	0.003223	74.1442
			960				3.19E-07			

Bibliography

Ahmed, M. N., Connor, D. E., & Huffman, D. L. (1995). Heat-resistance of *Escherichia coli* O157:H7 in meat and poultry as affected by product composition. *Journal of Food Science*, *60*, 606–610.

Ahn, J., Balasubramaniam, V. M., & Yousef, A. E. (2007). Inactivation kinetics of selected aerobic and anaerobic bacterial surrogate spores by pressure-assisted thermal processing. *International Journal of Food Microbiology*, *113*, 321–329.

Amiali, M., Ngadi, M. O., Smith, J. P., & Raghaven, V. G. S. (2006). Inactivation of *Escherichia coli* O157:H7 and *Salmonella enteritidis* in liquid egg white using pulsed electric field. *Journal of Food Science*, *71*(3), M88–M94.

Ananth, V., Dickson, J. S., Olson, D. G., & Murano, E. A. (1998). Shelf-life extension, safety and quality of fresh pork loin treated with high hydrostatic pressure. *Journal of Food Protection*, *61*(12), 1649–1656.

Avila, I. M. L. B., & Silva, C. L. M. (1981). Photochemical degradation of betanine and selected anthocyanins. *Journal of Food Science*, *46*(6), 1934–1937.

Bendix, G. H., Herberlein, D. G., Ptak, L. R., & Clifcorn, L. E. (1951). Factors influencing the stability of thiamine during heat sterilization. *Food Research*, *46*(6), 494–503.

Bolton, D. J., McMahon, C. M., Doherty, A. M., Sheridan, J. J., McDowell, D. A., & Blair, J. S. (2000). Thermal inactivation of *Listeria monocytogenes* and *Yersinia enterocolitica* in minced beef under laboratory conditions and in sous-vide prepared minced and solid beef cooked in a commercial retort. *Journal of Applied Microbiology*, *79*, 626–632.

Calderon-Miranda, M. L., Barbosa-Canovas, G. V., & Swanson, B. G. (1999a). Inactivation of *Listeria innocua* in liquid whole egg by pulsed electric fields and nisin. *International Journal of Food Microbiology*, *51*, 7–17.

Calderon-Miranda, M. L., Barbosa-Canovas, G. V., & Swanson, B. G. (1999b). Inactivation of *Listeria innocua* in skimmed milk by pulsed electric fields and nisin. *International Journal of Food Microbiology*, *51*, 19–30.

Carlson, T. R., Marks, B. P., Booren, A. M., Ryser, E. T., & Orta-Ramirez, A. (2005). Effect of water activity on thermal inactivation of *Salmonella* in ground turkey. *Journal of Food Science*, *70*(7), M363–M366.

Cemeroglu, B., Velioglu, S., & Isik, S. (1994). Degradation kinetics of anthocyanins in sour cherry juice and concentrate. *Journal of Food Science*, *59*(6), 1216–1218.

Cho, H.-Y., Sastry, S. K., & Yousef, A. E. (1999). Kinetics of inactivation of *Bacillus subtilis* spores by continuous or intermittent ohmic and conventional heating. *Biotechnology and Bioengineering*, *62*(3), 368–372.

Dietrich, W. C., & Neumann, H. J. (1965). Blanching brussel sprouts. *Food Technology*, *19*(7), 1174–1177.

Dietrich, W. C., Olsen, R. L., Nutting, M. D., Neumann, H. J., & Boggs, M. M. (1959). Time-temperature tolerances of frozen foods. XVIII. Effect of blanching condition on color stability of frozen beans. *Food Technology*, *13*(5), 258–261.

Dunn, J. E., & Pearlman, J. S. (1987). *Methods and apparatus for extending the shelf-life of fluid food products*. Maxwell Laboratories Inc. U.S. Patent 4,695,472.

Edelson-Mammel, S. G., & Buchanan, R. L. (2004). Thermal inactivation of *Enterobacter sakazakii* in rehydrated infant formula. *Journal of Food Protection*, *67*(1), 60–63.

Enache, E., Chen, Y., Awuah, G., Economides, A., & Scott, V. (2006). Thermal resistance parameters for pathogens in white grape juice concentrate. *Journal of Food Protection*, *69*(3), 564–569.

Evan, S. R., Gregory, J. F., & Kirk, J. R. (1981). Thermal degradation kinetics of pyridoxine hydrochloride in dehydrated model food systems. *Journal of Food Science*, *46*(2), 555–558, 563.

Farber, J. M. (1989). Thermal resistance of *Listeria* monocytogenes in foods. *International Journal of Food Microbiology*, *8*, 285–291.

Feliciotti, E., & Esselen, W. B., Jr. (1957). Thermal destruction rates of thiamine in pureed meats and vegetables. *Food Technology*, *11*(2), 77–84.

Fernandez-Molina, J. J., Barkstrom, E., Torstenssen, P., Barbosa-Canovas, G. V., & Swanson, B. G. (1999). Shelf-life extension of raw skim milk by combining heat and pulsed electric fields. *Food Research International*, 32.

Gaze, J. E., Brown, G. D., Gaskell, D. E., & Banks, J. G. (1989). Heat resistance of *Listeria* monocytogenes in homogenates of chicken, beef steak and carrot. *Food Microbiology*, *6*, 251–259.

Gervilla, R., Felipe, X., Ferragut, V., & Guamis, G. (1997). Effect of high hydrostatic pressure on *Escherichia coli* and *Pseudomonas fluorescens* strains in bovine milk. *Journal of Dairy Science*, *80*, 2297–2303.

Grahl, T., Sitzman, W., & Marki, H. (1992). Killing of microorganisms in fluid media by high voltage pulses. *DECHEMA Biotechnology Conferences*, 675–679.

Gregory, J. F., & Hiner, M. E. (1983). Thermal stability of vitamin B_6 compounds in liquid model food systems. *Journal of Food Science*, *48*(4), 1323–1327, 1339.

Gupta, R. P., & Murray, W. (1988). Pulsed high electric field sterilization. *IEEE Pulsed Power Conference Record National Research Council*, 58–64.

Herrmann, J. (1970). Calculation of the chemical and sensory alterations in food during heating and storage processes. *Ernaehrungsforschung*, *15*(4), 279–299.

Herrmann, J., & Nour, S. (1977). Modelluntersuchungen zur Bildung von Carbonylverbindungen und Braunung durch die Mailard-Reaktion beim Backprozeb. *Die Nahrung*, *21*(4), 319–330.

Himelbloom, B. H., Rutledge, J. E., & Biede, S. L. (1983). Color changes in blue crabs (Callinectes sapidus) during cooking. *Journal of Food Science*, *48*(2), 652–653.

Hulsheger, H., & Nieman, E. G. (1980). Lethal effect of high voltage pulsed on *E. coli* K12. *Radiation and Environmental Biophysics*, *18*(4), 281–288.

Hulsheger, H., Pottel, J., & Niemann, E. G. (1983). Killing of bacteria with electric pulses of high field intensity. *Radiation and Environmental Biophysics*, *20*, 149–162.

Huang, L., & Juneja, V. K. (2003). Thermal inactivation of *Escherichia coli* O157:H7 in ground beef supplemented with sodium lactate. *Journal of Food Protection*, *66*(4), 664–667.

Ibarz, A., Pagan, J., & Garza, S. (1999). Kinetic models for color changes in pear puree during heating at relatively high temperatures. *Journal of Food Engineering*, *39*(4), 415–422.

ICMSF. (1996). *Microorganisms in foods. Microbiological specification of food pathogens*. London: Blackie Academic & Professional Publishers (Book 5).

Juneja, V. K. (2003). Predictive model for the combined effect of temperature, sodium lactate, and sodium diacetate on the heat resistance of *Listeria monocytogenes* in beef. *Journal of Food Protection*, *66*(5), 804–811.

Juneja, V. K., Synder, O. P., & Manner, B. S. (1997). Thermal destruction of *Escherichia coli* O157:H7 in beef and chicken: determination of D- and z-values. *International Journal of Food Microbiology*, *35*, 231–237.

Kotrola, J. S., & Connor, D. F. (1997). Heat inactivation *Escherichia coli* O157:H7 in turkey meat as affected by sodium chloride, sodium lactate, polyphosphate, and fat content. *Journal of Food Protection*, *60*, 898–902.

Lathrop, P. J., & Leung, H. K. (1980). Rates of ascorbic acid degradation during thermal processing of canned peas. *Journal of Food Science*, *45*(1), 152–153.

Lau, M. H., Tang, J., & Swanson, B. G. (2000). Kinetics of texture and color changes in green asparagus during thermal treatments. *Journal of Food Engineering*, *45*(4), 23–236.

Lenz, M. K., & Lund, D. B. (1977). The lethality-Fourier number method: Experimental verification of a model for calculating average quality factor retention in conduction heating canned foods. *Journal of Food Science*, *42*(4), 997–1001.

Line, J. E., Fain, A. R., Moran, A. B., Martin, L. M., Lechowich, R. V., & Carosella, J. M. (1991). Lethality of heat to *Escherichia coli* O157:H7: D-value and z-value determinations in ground beef. *Journal of Food Protection*, *54*, 762–786.

Mackey, B. M., Forestier, K., & Issacs, N. (1995). Factors affecting the resistance of *Listeria monocytogenes* to high hydrostatic pressure. *Food Biotechnology*, *9*, 1–11.

Marquez, V. O., Mittal, G. S., & Griffiths, M. W. (1997). Destruction and inhibition of bacterial spores by high voltage pulsed electric field. *Journal of Food Science*, *62*(2), 399–401, 409.

Martin-Belloso, O., Vega-Mercado, H., Qin, B. L., Chang, E. J., Barbosa-Canovas, G. V., & Swanson, B. G. (1997a). Inactivation of *Escherichia coli* suspended in liquid egg using pulsed electric fields. *Journal of Food Processing and Preservation*, *21*(3), 193–208.

Martin-Belloso, O., Qin, B. L., Chang, E. J., Barbosa-Canovas, G. V., & Swanson, B. G. (1997b). Inactivation of *Escherichia coli* in skim milk by high intensity pulsed electric fields. *Journal of Food Process Engineering*, *20*, 317–336.

Mattick, K. L., Legan, J. D., Humphrey, T. J., & Peleg, M. (2001). Calculating Salmonella inactivation in nonisothermal heat treatments from isothermal nonlinear survival curves. *Journal of Food Protection*, *64*(5), 606–613.

Metrick, C., Hoover, D. G., & Farkas, D. F. (1989). Effects of high hydrostatic pressure on heat-resistant and heat-sensitive strains of Salmonella. *Journal of Food Science*, *54*, 1547–1564.

Miki, N., & Akatsu, K. (1970). Effect of heating sterilization on color of tomato juice. *Nippon Shokuhn Kogyo Gakkai-Shi [Japan]*, *17*(5), 175–181.

Mishkin, M., & Saguy, I. (1982). Thermal stability of pomegranate juice. *Zeitschrift fur Lebensmittel-Untersuchung und – Forschung*, *175*(6), 410–412.

Mnkeni, A. P., & Beveridge, T. (1982). Thermal destruction of pteroy glutamic acid in buffer and model food system. *Journal of Food Science*, *47*(6), 2038–2063.

Murano, E. A., Murano, P. S., Brennan, R. E., Shenoy, K., & Moreira, A. G. (1999). Applications of High Hydrostatic Pressure to Eliminate *Listeria monocytogenes* from Fresh Pork Sausage. *Journal of Food Protection*, *62*(5), 480–483.

Mulley, E. A., Stumbo, C. R., & Hunting, W. M. (1975). Kinetics of thiamine destruction by heat. Effect of pH and form of vitamin on its rate of destruction. *Journal of Food Science*, *40*(5), 989–992.

Murphy, R. Y., Duncan, L. K., Johnson, E. R., Davis, M. D., & Smith, J. N. (2002). Thermal inactivation D- and z-values of Salmonella serotypes and *Listeria innocua* in chicken patties, chicken tenders, franks, beef patties, and blended beef and turkey patties. *Journal of Food Protection*, *65*(1), 53–60.

Murphy, R. Y., Duncan, L. K., Beard, B. L., & Driscoll, K. H. (2003). D and z values of salmonella, *Listeria innocua* and *Listeria monocytogenes* in fully cooked poultry products. *Journal of Food Science*, *68*(4), 1443–1447.

Murphy, R. Y., Beard, R. L., Martin, E. M., Duncan, L. K., & Marcy, J. A. (2004). Comparative study of thermal inactivation of *Escherichia coli* O157:H7, salmonella, and *Listeria monocytogenes* in ground pork. *Journal of Food Science*, *69*(4), FSM97–FSM101.

Navankasattusas, S., & Lund, D. B. (1982). Thermal destruction of vitamin B_6 viamers in buffer solution and cauliflower puree. *Journal of Food Science*, *47*(5), 1512–1518.

Nguyen, M. T., Indrawati, & Hendrickx, M. (2003). Model studies on the stability of folic acid and 5-methyltetrahydrofolic acid degradation during thermal treatment in combination with high hydrostatic pressure. *Journal of Agricultural and Food Chemistry*, *5*(11), 3352–3357.

Noma, S., Shimoda, M., & Hayakawa, I. (2002). Inactivation of vegetative bacteria by rapid decompression treatment. *Journal of Food Science*, *67*(90), 3408–3411.

O'Bryan, C. A., Crandall, P. G., Martin, E. M., Griffis, C. L., & Johnson, M. G. (2006). Heat resistance of salmonella spp., *Listeria monocytogenes, Escherichia coli* O157:H7, and *Listeria innocua* M1, A potential surrogate for *Listeria monocytogenes*, in meat poultry: A review. *Journal of Food Science*, *71*(3), R23–R30.

Osaili, T. M., Griffis, C. L., Martin, E. M., Beard, B. L., Keener, A., & Marcy, J. A. (2006). Thermal inactivation studies of *Escherichia coli* O157:H7, Salmonella, and *Listeria monocytogenes* in ready-to-eat chicken-fried beef patties. *Journal of Food Protection*, *69*(5), 1080–1086.

Osaili, T. M., Griffis, C. L., Martin, E. M., Beard, B. L., Keener, A. E., & Marcy, J. A. (2007). Thermal inactivation of *Escherichia coli* O157:H7, Salmonella, and *Listeria monocytogenes* in breaded pork patties. *Journal of Food Science*, *72*(2), M56–M61.

Patterson, M. F., Quinn, M., Simpson, R., & Gilmore, A. (1995). Sensitivity of vegetative pathogens to high hydrostatic pressure treatment in phosphate buffered saline and foods. *Journal of Food Protection*, *58*(5), 524–529.

Patterson, M. F., & Kirpatrick, D. J. (1998). The combined effect of high hydrostatic pressure and mild heat on inactivation of pathogens in milk and poultry. *Journal of Food Protection*, *61*(4), 432–436.

Pina Perez, M. C., Aliaga, D. R., Reyes, D. S., & Martinez Lopez, A. (2007). Pressure inactivation of *Enterobacter sakazakii* in infant formula milk. *Journal of Food Protection*, *70*(10), 2281–2289.

Ponce, E., Pia, R., Mor-Mur, M., Gervilla, R., & Guamis, B. (1998). Inactivation of *Listeria innocua* inoculated into liquid whole egg by high hydrostatic pressure. *Journal of Food Protection*, *61*(1), 119–122.

Qin, B. L., Pothakamury, U. R., Vega, H., Barbosa-Canovas, G. V., & Swanson, B. G. (1995). Food pasteurization using high intensity pulsed electric fields. *Journal of Food Technology*, *49*(12), 55–60.

Ramaswamy, H. S., Riahi, E., & Idziak, E. (2003). High-pressure destruction kinetics of *E. coli* (29055) in apple juice. *Journal of Food Science*, *68*(5), 1750–1756.

Reddy, N. R., Solomon, H. M., Fingerhut, G. A., Rhodehamel, E. J., Balasubramaniam, V. M., & Palaniappan, S. (1999). Inactivation of *Clostridium botulium* Type E spores by high pressure processing. *Journal of Food Safety*, *19*, 277–288.

Reina, L. D., Jin, Z. T., Yousef, A. E., & Zhang, Q. H. (1998). Inactivation of *Listeria monocytogenes* in milk by pulsed electric field. *Journal of Food Protection*, *61*(9), 1203–1206.

Rodrigo, D., Barbosa-Canovas, G. V., Martinez, A., & Rodrigo, M. (2003). Weibull distribution function based on an empirical mathematical model for inactivation of *Escherichia coli* by pulsed electric fields. *Journal of Food Protection*, *66*(60), 1007–1012.

Rovier, P., Carpi, G., Dall'Aglio, G., Gola, S., Maggi, A., Miglioli, I., & Scaramuzza, N. (1996). High pressure heat treatments: Evaluation of the sterilizing effect and of thermal damage. *Industrial Conservation*, *71*, 473–483.

Saguy, I. (1979). Thermostability of red beet pigments (betanine and vulgaxanthin-1): influence of pH and temperature. *Journal of Food Science*, *44*(5), 1554–1555.

Saguy, I., Kopelman, I. J., & Mizarhi, S. (1978a). Extent of nonenzymatic browning in grapefruit juice during thermal and concentration processes: kinetics and prediction. *Journal of Food Processing and Preservation*, *2*(3), 175–184.

Saguy, I., Kopelman, I. J., & Mizrahi, S. (1978b). Simulation of ascorbic acid stability during heat processing and concentration of grapefruit juice. *Journal of Food Process Engineering*, *2*(3), 213–225.

Sanchez, A. H., Rejano, L., & Montano, A. (1991). Kinetics of destruction by heat of color and texture pickled green olives. *Journal of the Science of Food and Agriculture*, *54*(3), 379–385.

Sastry, L. V. L., & Tischer, R. G. (1952). Behavior of anthocyanin pigments in Concord grapes during heat processing and storage. *Food Technology*, *6*(3), 82–86.

Schwartz, S. J., & von Elbe, J. H. (1983). Kinetics of chlorophyll degradation to pyropheophytin in vegetables. *Journal of Food Science*, *48*(4), 1303–1306.

Sela, S., Pinto, R., Merin, U., & Rosen, B. (2003). Thermal inactivation of *Escherichia coli* in camel milk. *Journal of Food Protection*, *66*(9), 1708–1711.

Sensoy, L., Zhang, Q. H., & Sastry, S. K. (1997). Inactivation kinetics of *Salmonella dublin* by pulsed electric fields. *Journal of Food Process Engineering*, *20*, 367–381.

Shi, J., LeMaguer, M., Bryan, M., & Kakuda, Y. (2003). Kinetics of lycopene degradation in tomato puree by heat and light irradiation. *Journal of Food Process Engineering*, *25*(6), 485–498.

Skjoldebrand, C., Anas, A., Oste, R., & Sjodin, P. (1983). Prediction of thiamine content in convective heated meat products. *Journal of Food Technology*, *18*(1), 61–83.

Smith, S. E., Maurer, J. L., Orta-Ramirez, A., Ryser, E. T., & Smith, S. M. (2001). Thermal inactivation of *Salmonella spp. Salmonella typhimurium* DT104 and *Escherichia coli* O157:H7 in ground beef. *Journal of Food Science*, *66*, 1164–1168.

Steenstrup, L. L., & Floros, J. D. (2002). Statistical modeling of D- and z-value of *E. coli* O157:H7 and pH in apple cider containing preservatives. *Journal of Food Science*, *67*(2), 793–796.

Steet, J. A., & Tong, C. H. (1996a). Degradation kinetic of green color and chlorophylls in peas by colorimetry and HPLC. *Journal of Food Science*, *61*(5), 924–927, 931.

Steet, J. A., & Tong, C. H. (1996b). Quantification of color change resulting from pheophytinization and non-enzymatic browning reactions in thermally processed green peas. *Journal of Agriculture and Food Chemistry*, *44*(6), 1531–1537.

Stefanovich, A. F., & Karel, M. (1982). Kinetics of beta-carotene degradation at temperatures typical of air drying of foods. *Journal of Food Processing and Preservation*, *6*(4), 227–242.

Stopforth, J. D., Surslim, R., Kottapalli, B., Hill, W. E., & Samadpour, M. (2008). Thermal inactivation D- and z-values of multidrug-resistant and non-multidrug-resistant Salmonella serotypes and survival in ground beef exposed to consumer-style cooking. *Journal of Food Protection*, *71*(3), 509–515.

Stumbo, C. R. (1965). *Thermobacteriology in Food Processing*. New York: Academic Press.

Tanchev, S. (1972). Kinetics of the thermal degradation of anthocyanins of the raspberry. *Zeitschrift fur Lebensmittel-Untersuchung und – Forschung*, *150*(1), 28–30.

Tanchev, S. (1983). Kinetics of thermal degradation of anthocyanins. Research in food science and nutrition. Vol. 2. Basic Studies in Food Science. *Proceedings of 6th International Congress of Food Science and Technology*, 96.

van den Broeck, L., Weemaes, L. C., van Loey, A., & Hendrickx, M. (1998). *Journal of Agriculture and Food Chemistry*, *46*(5), 2001–2006.

Vega-Mercado, H., Martin-Belloso, O., Chang, E.-J., Barbosa-Canovas, G. V., & Swanson, B. G. (1996). Inactivation of *Escherichia coli* and *Bacillus subtilis* suspended in pea soup using pulsed electric fields. *Journal of Food Processing and Preservation*, *20*(6), 501–510.

Viberg, U., Jagerstad, M., Oste, R., & Sjoholn, I. (1997). Thermal processing of 5-methyltetrahydrofolic acid in the UHT region in the presence of oxygen. *Food Chemistry*, *59*(3), 381–386.

von Elbe, J. H., Maing, I. Y., & Amundson, C. H. (1974). Color stability of betanine. *Journal of Food Science*, *39*(2), 334–337.

Watanabe, F., Abe, K., Fujita, T., Goto, M., Hiemori, M., & Nakano, Y. (1998). Effect of microwave heating on the loss of vitamin B_{12} in foods. *Journal of Agriculture and Food Chemistry*, *46*(1), 206–210.

Weemans, C. A., Ooms, V., Indrawati, I. L., van den Broeck, A., van Loey, A. M., & Hendrickx, M. E. (1999a). Pressure-temperature degradation of green color in broccoli juice. *Journal of Food Science*, *64*(3), 504–508.

Weemans, C. A., Ooms, V., van Loey, A. M., & Hendrickx, M. E. (1999b). Kinetics of chlorophyll degradation and color loss in broccoli juice. *Journal of Agriculture and Food Chemistry*, *47*(6), 2404–2409.

Wilkinson, S. A., Earle, M. D., & Cleland, A. C. (1982). Effects of food composition, pH, and copper on the degradation of vitamin A in beef liver puree during heat processing. *Journal of Food Science*, *47*(3), 844–848.

Zhang, Q. H., Monsalve-Gonzales, A., Barbosa-Canovas, G. V., & Swanson, B. G. (1994). Inactivation of *E. coli* and *S. cerevisiae* by pulsed electric fields under controlled temperature conditions. *Transactions ASAE*, *37*(2), 581–587.

Food Science and Technology International Series

Altschul, A. M. (Ed.). *New protein foods*: Volume 1, *Technology, Part A*—1974. Volume 2, *Technology, Part B*—1976. Volume 3, *Animal protein supplies, Part A*—1978. Volume 4, *Animal protein supplies, Part B*—1981. Volume 5, *Seed storage proteins*—1985.

Amerine, M. A., Pangborn, R. M., & Roessler, E. B. (1965). *Principles of sensory evaluation of food.*

Arvanitoyannis, I. S. (2008). *Waste management for the food industries.*

Arendt, E. K., & Bello, F. D. (Eds.). (2008). *Gluten-free cereal products and beverages.*

Bagchi, D. (Eds.). (2008). *Nutraceutical and functional food regulations in the United States and around the World.*

Barbosa-Cánovas, G. V., Góngora-Nieto, M. M., Pothakamury, U. R., & Swanson, B. G. (1999). *Preservation of foods with pulsed electric fields.*

Barnes, P. J. (Ed.). (1983). *Lipids in cereal technology.*

Bauernfeind, J. C. (Ed.). (1981). *Carotenoids as colorants and vitamin A precursors: technological and nutritional applications.*

Bechtel, P. J. (1986). *Muscle as food.*

Bemiller, J., & Whistler, R. (Eds.) (2009). *Starch: Chemistry and Technology* (3rd ed.).

Bender, A. E. (1978). *Food processing and nutrition.*

Berk, Z. (2009). *Food process engineering and technology.*

Blankenship, L. C. (1991). *Colonization control of human bacterial enteropathogens in poultry.*

Bourne, M. C. (2002). *Food texture and viscosity: concept and measurement* (2nd ed.).

Caballero, B., & Popkin, B. M. (Eds.), (2002). *The nutrition transition: diet and disease in the developing world.*

Chan, H. W.-S. (1986). *Autoxidation of unsaturated lipids.*

Cliver, D. O., & Riemann, H. P. (Eds.), (2002). *Foodborne diseases* (2nd ed.).

Clydesdale, F. M., & Wiemer, K. L. (Eds.), (1985). *Iron fortification of foods.*

Cunningham, F. E., & Cox, N. A. (Eds.), (1987). *Microbiology of poultry meat products.*

David, K., & Thompson, P. (Eds.). (2008). *What can nanotechnology learn from biotechnology?*

Decareau, R. V. (1985). *Microwaves in the food processing industry.*

Dennis, C. (Ed.). (1983). *Post-harvest pathology of fruits and vegetables.*

Duckworth, R. B. (Ed.). (1975). *Water relations of food.*

Florkowski, W. J., Prussia, S. E., Shewfelt, R. L., & Brueckner, B. (Eds.) (2009). *Postharvest handling* (2nd ed.).

Gacula, M., Jr., & Singh, J. Bi, J., & Altan, S. (2009). *Statistical methods in food and consumer research* (2nd ed.).

Glicksman, M. (1970). *Gum technology in the food industry.*

Goldblith, S. A., Rey, L., & Rothmayr, W. W. (1975). *Freeze drying and advanced food technology.*

Hafs, H. D., & Zimbelman, R. G. (1994). *Low-fat meats.*

Han, J. H. (2005). *Innovations in food packaging.*

Herschdoerfer, S. M. (Ed.). *Quality control in the food industry* (2nd ed.). Volume 1—1985. Volume 2—1985. Volume 3—1986. Volume 4—1987.

Hoover, D. G., & Steenson, L. R. (1993). *Bacteriocins.*

Iglesias, H. A., & Chirife, J. (1982). *Handbook of food isotherms: water sorption parameters for food and food components.*

Jackson, R. S. (2008). *Wine science: principles and applications* (3rd ed.).

Jackson, R. S. (2009). *Wine tasting: a professional handbook* (2nd ed.).

Japtiani, J., Chan, H. T., Jr., & Sakai, W. S. (1987). *Tropical fruit processing.*

Jensen, R. G. (1995). *Handbook of milk composition.*

Joslyn, M. A. (1970). *Methods in food analysis* (2nd ed.).

Kohlmeier, M. (2003). *Nutrient metabolism.*

Loncin, M., & Merson, R. L. (1979). *Food engineering: Principles and selected applications.*

Macrae, R. (1988). *HPLC in food analysis* (2nd ed.).

McCorkle, C. O., Jr. (1987). *Economics of food processing in the United States.*

Markakis, P. (Ed.). (1982). *Anthocyanins as food colors.*

Nagodawithana, T., & Reed, G. (1993). *Enzymes in food processing* (3rd ed.).

Osborne, D. R., & Voogt, P. (1978). *The analysis of nutrients in foods.*

Pearson, A. M., & Young, R. B. (1989). *Muscle and meat biochemistry.*

Penfield, M. P., & Campbell, A. M. (1990). *Experimental food science* (3rd ed.).

Phillips, L. G., Whitehead, D. M., & Kinsella, J. (1994). *Structure-function properties of food proteins.*

Pimentel, D., & Hall, C. W. (Eds.), (1984). *Food and energy resources.*

Pollock, J. R. A. (Ed.). *Brewing science.* Volume 1—1979. Volume 2—1980. Volume 3—1987.

Pomeranz, Y. (1991). *Functional properties of food components* (2nd ed.).

Regenstein, J. M., & Regenstein, C. E. (1984). *Food protein chemistry: an introduction for food scientists.*

Riemann, H., & Cliver, D. (Eds.), (2006). *Foodborne infections and intoxications* (3rd ed.).

Roos, Y. H. (1995). *Phase transitions in foods.*

Sapers, G. M., Solomon, E. B., & Mathews, K. R. (2009). *The produce contamination problem: Causes and solutions.*

Shewfelt, R. L., & Prussia, S. E. (1993). *Postharvest handling: a systems approach.*

Shibamoto, T., & Bjeldanes, L. (2009). *Introduction to food toxicology* (2nd ed.).

Singh, R. P., & Heldman, D. R. (2008). *Introduction to food engineering* (4th ed.).

Solms, J., Booth, D. A., Dangborn, R. M., & Raunhardt, O. (1987). *Food acceptance and nutrition.*

Stewart, G. F., & Amerine, M. A. (Eds.), (1982). *Introduction to food science and technology* (2nd ed.).

Stone, H., & Sidel, J. L. (1993). *Sensory evaluation practices* (2nd ed.).

Stone, H., & Sidel, J. L. (2004). *Sensory evaluation practices* (3rd ed.).

Stumbo, C. R. (1973). *Thermobacteriology in food processing* (2nd ed.).

Sun, D.-W. (Ed.). (2005). *Emerging technologies for food processing.*

Sun, D.-W. (Ed.). (2008). *Computer vision technology for food quality evaluation.*

Thompson, A., Boland, M., & Singh, H. (Eds.). (2009). *Milk proteins: from expression to food.*

Troller, J. A. (1993). *Sanitation in food processing* (2nd ed.).

Troller, J. A., & Christian, J. H. B. (1978). *Water activity and food.*

Urbain, W. M. (1986). *Food irradiation.*

Vaughan, J. G. (Ed.). (1979). *Food microscopy.*

Walter, R. H. (1991). *The chemistry and technology of pectin.*

Walter, R. H. (1997). *Polysaccharide dispersions.*

Index

Acidity, preservation process, 3–4
Activation energy constant
  chlorophyll retention, 32, 33
  kinetic food systems
    agent intensity models, 31t, 33
    Thermal Resistance Coefficient, 41
    uniform parameters, 45
  microbial inactivation rate, 45
  preservation process optimization, 260
  quality retention kinetics, 90, 105–106
  survivor curve measurements, vegetative populations, 42
Activation Volume Coefficient, microbial inactivation kinetics, pressure magnitude, 69, 77
Activation Volume Constant
  kinetic food systems, 34–36
    uniform parameters, 45–46
  microbial inactivation rate, 45–46
  preservation technology, 34
Additives, food preservation, 4
Agent intensity models
  kinetic food systems, 30–36
  preservation process validation, 12
Anthocyanins, quality retention kinetics, 94
Appert, Nicholas, 6, 7f
Arrhenius equation, kinetic food systems, agent intensity models, 30
Ascorbic acid
  first-order model, 25–26
  quality retention kinetics, 92
    thermal process and, 98
    ultra-high pressure process, 102–104
Aseptic processing systems, ohmic heating process, 193–197
Atmospheric composition, food preservation, 4

Ball, C. Olin, 6–11, 7f
Ball's Formula method, lethal rate curves, 162–163
Bessel functions
  process design-preservation integration, 174, 175t
  thermal process design, 174, 175t
Biosensors, preservation process validation, time-temperature indicators (TTIs), 231–233

Biot number
physical transport models
containerized heating and cooling, 126
heating rate constant and, 132, 133f
inverse Biot number, 131
mass average lag constant, 133, 134f
spherical temperature-time chart, 129f, 130
thermal process design, 174

Carotenoids, quality retention kinetics, 94
Channel length parameters, pulsed-electric field process design, 203–209
Characteristic dimension, physical transport models
heat transfer resistance, 126
infinite cylinder temperature-time chart, 130, 131f
infinite slab temperature-time chart, 130, 130f
Chemical reactions, quality retention kinetics, 89
Chlorophyll
quality retention kinetics, 93–94, 98
thermal process, 97–98
validation and evaluation, 234–235
retention rate constants, 32–33
thermal preservation process validation, 234–236
Clostridium spp., survivor curve kinetics, 54–56, 56t, 57
Code of Federal Regulations, 8, 8f
Color pigment intensity, quality retention kinetics, 88–89
Color retention, quality retention kinetics, 93–95, 94t
Combined process design, basic parameters, 209–212
Commercial-scale process
evolving preservation technologies and, 274
validation of, 223–229
Concentration-time relationship, second-order model, primary reactant, 28
Conduction-heating foods
pressure-assisted thermal process, 209–212
thermal process design, 165–166
optimization, 250–251, 251f
Container geometry, preservation process optimization, 259
Containerized heating and cooling
physical transport models, 125–138
thermal process design, 181
Container size and shape, process validation and evaluation, 218
Convective heat transfer coefficients, physical transport models, containerized heating and cooling, 126, 126t, 129, 135–138
Cooling in containers, physical transport models, 125–138, 126t
Cylindrical containers, thermal process design, integrated impact parameters, 172, 172f, 173

Decimal Reduction Time
kinetic systems models
microbial survivor curves, 44, 50–51
thermal process, 37, 41–43, 42f
uniform parameters, 43
microbial survivor curves, 53–54
survivor curve measurements, vegetative populations, 42, 43f

Dehydration, food preservation, 3, 3f
Density, physical transport models, 112–114, 113t
Deterioration reaction, primary reactant, 22–24
Dielectric properties, physical transport models, 123–125, 124t

Economic impact assessment, process design models, 169
Electrical conductivity
  ohmic heating process design, 193
  physical transport models, 121–123, 122t
    electrolytes and, 123
    microwave heating of food products, 140
  potato example, 123
Electric field strength, microwave heating of food products, physical transport models, 139, 140–141
Electrolytes, physical transport models, electrical conductivity and, 123
Endpoint targeting
  microbial survivor curves, laboratory evaluation, 219–221
  process design models, 166–171
Escherichia coli spp.
  pulsed-electric field processing, 73
  survivor curve kinetics, 54–56, 55t
Evolving technologies, preservation processes, 273–276
Eyring relationship, preservation technology, 33–34

Finite geometries
  physical transport models, transient heat transfer, temperature prediction, 134
  as process design parameter, time-step calculation, 151
First-order model, kinetic food systems, 25–27
First-order rate constants
  quality retention kinetics, 90
  time-step calculation, 150
Folates, quality retention kinetics, 92
Folic acid, quality retention kinetics, 92
  thermal process design, microwave heating, 181–192
Food, Drug and Cosmetic Act, history of, 8
Food and Drug Administration (FDA), food preservation technology review, 15
Food-borne illness, successful prevention of, 14
Food components/attributes, preservation process validation, 12
Food poisoning, history of outbreaks, 8
Food preservation process
  goals of, 14–15
  microbial inactivation kinetics, 59–75
Food quality attributes
  future kinetic parameters, 270
  preservation process design validation, 234–236
    chlorophyll content example, 234–236
  preservation process optimization, 252–260
  reduction of preservation impact on, 16
  retention kinetics
    applications, 95
    basic principles, 88–90
    first-order reactions and rate constants, 87–88

Food quality attributes (*Continued*)
heat-sensitive vitamins, 91–93, 91t
high-temperature preservation and, 17f
overview, 87
preservation products, impact of, 105–106
process design models, 170, 171
product color retention, 93–95, 94t
thermal preservation processes and, 170, 171
microwave heating, 181–192
Food safety, history of, 8
Forced convection, physical transport models, containerized heating and cooling, 125–126
Formula Method, process design models, 162
Fourier number, physical transport models, spherical temperature-time chart, 129f, 130
Frequency, microwave heating of food products, physical transport models, 139
F-value. *See* Thermal death times

General Method, process design models, 157–162

Halvorson/Zeigler (H-Z) expression
commercial-scale process validation, 223–225
microwave process validation, 238–239
pilot-scale preservation process verification, 221
process validation, 219
Heating curve, canned food, 9–10, 10f
Heating curve, physical transport models, 10f
Heating curve equation, process design models, Formula Method, 162
Heating in containers, physical transport models, 125–138
Heating rate constant, physical transport models, transient heat transfer, temperature prediction, 132, 133f
Heat preservation process, 1–2, 2f
food quality and, 16
See also Thermal process design
Heat-sensitive products
quality retention kinetics
thermal process design, 175–180
vitamins, 91–93, 91t
thermal preservation process design, microwave heating, 181–192
Heat transfer
physical transport models
containerized heating and cooling, internal resistance, 126–127
transient heat transfer, temperature prediction, 131–138
process design-preservation integration, 174–180
High-temperature short-time (HTST) preservation process
liquid products, 247–249
nonliquid products, 249–263
optimization, 246–247, 246f

IFT/FDA Task Force
evolving technologies opportunities list, 273–274
food preservation technology review, 15

Inactivation rate constants, kinetic food systems models, 35–36
Index parameters, kinetic food system models, 45
Infinite cylinder model
  process design-preservation integration, 172, 172f
  temperature-time chart, 130, 131f
  thermal process design, integrated impact parameters, 173–174
Infinite slab model
  process design-preservation integration, 173–174
  temperature-time chart, 130, 130f
  thermal process design, integrated impact parameters, 173–174
Initial concentration, kinetic food system models, 21
Integrated impact parameters
  process design models, future trends, 272–273, 273f
  thermal process design, 171–181

Kinetic food systems models
  agent intensity models, 30–36
  evolving preservation technologies, 275
  first-order model, 25–27
  list of symbols, 46–47
  multiple-order model, 27–29
  overview, 19–20
  rate equations and constants, 20–24
  thermal process models, 37–43
  uniform parameters, 43–46
Kinetic parameters
  assembly, 268–270, 268t
  food quality attributes, 16
  high-temperature short-time preservation optimization, 246–247, 246f
  microbial populations, 54–59
  process design and, 9
  process validation and evaluation, 218
  pulsed electric fields (PEF) process design, 207–208

Laboratory evaluation, process validation, 219–221
Lag factor
  microbial survival curves, 51–53, 52f
  physical transport models, transient heat transfer, temperature prediction, 131–132, 133f
Lethal rate curve
  Ball's Formula method, 162–163
  basic properties, 11f
  process design models, Formula Method, 162–163
  thermal process, 158–162, 158f
    General Method example, 159–162, 161f
Lethal rate parameter, process design models, General Method, 158, 158f
Liquid products, high-temperature short-time preservation optimization, 247–249
Listeria spp.
  pulsed-electric field preservation process and, 71–72
  survivor curve kinetics, 54–57, 55t
  ultra-high pressure process for reduction of, 70–71
Location, thermal process design, integrated impact parameters, 173–174
Log-linear probability, microbial survival curves, 50–51, 51, 52f
  process time prediction, 62–64
Loss tangent, dielectric properties, physical transport models, 123, 124

Maillard browning, quality retention kinetics, 88–89
color retention, 94–95
Mass average lag constant, physical transport models, 133, 134f
Mathematical methods, process design models, 162–166
Maximum reaction rate, kinetic food systems, multiple-order models, 29
Measurement precision, preservation process validation, 12–13
5-Methyltetrahydrofolate, quality retention kinetics, 92
pasteurization and, 100–102
Michaelis-Menten equation, kinetic food systems, multiple-order models, 29, 30f
Microbial populations
food product influence, future kinetic parameters, 269–270
inactivation technology
defined, 76
food preservation and, 15
process time calculations, 61
process validation, 219–229
shelf-stable product, 65–66
ultra-high pressure process design, 198–202
kinetic parameters, 54–59
alternative preservation technologies, 77
applications, 59–75
survivor curves, 55t, 56–57, 56t
log-linear survivor curve model, 62–63
overview, 49
pilot-scale preservation process verification, 221–223
population variability, 57–58
preservation process times, survivor curve model, 64
process design models, 147, 148
reduction, 152
target identification, 166–171
ultra-high pressure process, 197–202
process technology, 58–59
spore survivor curve, 10f, 58
survival curves, 50–54, 60–61, 60f, 64
laboratory evaluation, 219–221
thermal preservation process, microwave heating, 181–192
time-step calculation, 150
time-step calculation and reduction of, 152
variability, 57–58
vegetative *vs.* spore populations, 58
Microwave heating of food products
dielectric properties, 123–125
physical transport models, 139–141
dielectric properties, 124
electric field strength, 139, 140–141
process design parameters, 181–192
process validation and evaluation, 237–239
thermal preservation process design and, 181–192
Mixed-culture population, microbial survival curve, 53, 53f
Multiple-order models
kinetic food systems, 27–29
quality retention kinetics, 105–106

National Canners Association (NCA), 8
Newtonian fluids, physical transport models, viscosity, 114–116, 115t
Nonenzymatic browning, quality retention kinetics, 88–89
  color retention, 94–95
Nonliquid products, high-temperature short-time preservation process optimization, 249–263
Non-log-linear microbial survivor curves
  microbial spores example, 51–53, 52f
  thermal process models, kinetic food systems, 38, 38f
  time *vs.* microbial survivors, 39, 40f
Nonthermal preservation processes, process design parameters, time-step calculation, 151
Normal probability plot, commercial-scale process validation, pathogen survivors, 227–228, 228f
Nutrients, quality retention kinetics, thermal process, 97

Ohmic heating
  aseptic practices, 193–197
  food preservation, 4–5
  physical transport models
    applications, 138–139
    electrical conductivity, 121–123
  process design parameters, 192–197
  validation and verification, 239–240
Operator process time
  preservation process optimization, 251
  process design models, Formula Method, 162–163
  as process design parameter, 149
Optimization of preservation process
  high-temperature short-time processes, 246–247, 246f
    shelf stability example, 247–249
  liquid food products, 247–249
  nonliquid food applications, 249–263
    thermal process example, 252–261
  overview, 245
  viscous food products, 261–263

PA 3679 surrogate microorganism, preservation process validation, 230
  microwave process, 238–239
Pasteur, Louis, 6
Pasteurization, quality retention kinetics, 100–102
Pathogenic microorganisms, preservation process validation, 13–14
Physical properties of products
  future transport models, 271–272
  preservation process optimization, 247
  process validation and evaluation, 218
Pilot-scale preservation processes, verification, 221–223
  microwave process validation, 238–239
  ohmic heating, 239–240
Prediction models
  food preservation, 8–9
  quality retention kinetics, 99

Preservation process design
acidity, 3–4
additives, 4
alternatives, 4–5
atmospheric composition, 4
current trends in, 5, 6
dehydration, 3, 3f
emerging technologies, 15–16
experimental validation, 12–14
heat, 1–2, 2f
history of, 1, 6–9
models, 10–11
process design integration, 171–181
quality retention kinetics, 95, 105–106
radiation, 4
refrigeration, 2, 3f
smoking, 4
success rates, 14–15
ultra-high pressurization, 4–5, 5f
water activity, 4
Pressure
physical transport models
density and, 113–114
specific heat and, 118
thermal conductivity and, 121
viscosity, 116
preservation technology and, 33
Pressure-assisted thermal process (PATP)
basic parameters, 209, 212
conduction-heating food product, 209–212
Pressure magnitude, microbial inactivation kinetics, 69
Pressure z-value, microbial inactivation technology, 69
Primary reactant
deterioration reaction, 22–24
kinetic food systems, multiple-order models, 28
second-order model, 28–29
Probability of survivors
commercial-scale process validation, 223–225, 226f
normal probability plot, 227–228, 228f
future kinetic parameters, 269–270
microbial inactivation kinetics, 60–61, 60f
as process design parameter, endpoint targeting, 167
thermal preservation process, endpoint prediction, 167
Process design models
basic parameters, 148–150
combined processes design, 209–212
future trends, 270
future trends, 267, 272–273, 273f
evolving technologies, 273–276
kinetic parameter assembly, 268–270, 268t
General Method, 157–162
integrated preservation process impacts, 171–181
mathematical methods, 162–166
microwave process design, 181–192
Ohmic heating process, 192–197
overview, 9–11
pulsed-electric-field processes, 202–209
targets, 166–171
time-step calculation, 150–157
transport models
future trends, 271–272
new transport models, 272
physical properties of foods, 271–272
ultra-high pressure processes, 197–202

validation and evaluation, chlorophyll content preservation, 234–235
Process time prediction
log-linear survivor model, 62–63
microbial inactivation technology, 61
pulsed electric field experiment, 71–73
shelf-stable product, 65–66
UHP process, 67–68, 70–71
process design models
endpoint targeting, 168–169
Formula Method, 162–163
thermal process design, mathematical method, 165–166
Product color retention, quality retention kinetics, 93–95, 94t
Product composition
density estimation and, 113–114
quality retention kinetics, heat-sensitive vitamins, 91–93, 91t
thermal conductivity predictions, 120–121
Product temperature profile
thermal process design, 181
ultra-high pressure preservation process, 69
Pulsed electric fields (PEF) process
design parameters, 202–209
kinetic parameters, 147
temperature, 205–207
velocity determination, 203
food preservation, 4–5
future kinetic parameters, 269
microbial inactivation kinetics, 58–59
commercial applications, 72, 73
kinetic parameters, 77
process time calculations, 71–73
survivor curves, 71
process design and, 202–209
transport models, future trends, 272
validation and evaluation, 242–243
Pulsed electric field technology, food preservation, 15

Quality attribute retention kinetics
applications, 95
basic principles, 88–90
first-order reactions and rate constants, 87–88
future parameters, 270
heat-sensitive vitamins, 91–93, 91t
high-temperature preservation and, 17f
overview, 87
pasteurization and, 100–102
preservation process design validation, 233–234
preservation process optimization, thermal process, 252–260
preservation products, impact of, 105–106
pressure-assisted thermal process, 209–212
process design models, 170, 171
process validation and evaluation, 218
product color retention, 93–95, 94t
thermal preservation process, microwave heating, 181–192
Quality control, food products, 16
Quantitative approach, 9–16

Radiation, consumer concerns about, 5
Radiation, food preservation, 4, 15

Rate constants
commercial-scale process validation, pathogen survivors, 227–228, 228f
kinetic food system models, 20–24
first-order model, 25–27
multiple-order models, 29
uniform parameters, 44
microbial survival curves, kinetic parameter measurements, 64–66
as process design parameter, time-step calculation, 151, 152
time-step calculation, 152
zero-order relationship, 24, 24f
Rate equations, kinetic food system models, 20–24
Rate of reaction
concentration *vs.*, 23f
kinetic food system models, 20–24
Reaction kinetics, 9, 10f
preservation process validation, time-temperature indicators (TTIs), 231
quality attributes in food, 88
Reaction order, kinetic food system models, 22–24, 22f
Reference temperature, thermal death times and, 162–163, 164f
Refrigeration, food preservation, 2, 3f
Riboflavin (vitamin B_2), quality retention kinetics, 92
Risk assessment, process design models, 169–170
microbial inactivation kinetics, 62–63

Safety record, food preservation, 14–15
Safety risk assessment, thermal process design, 169–170
Salmonella spp.
safety risk assessment, 62–63
survivor curve kinetics, 54–56, 55t
Scale-up, preservation process, 14
Schmidt approach, commercial-scale process validation, 223–225, 227, 228f
Second-order reaction, kinetic food systems, 27, 27f
Sensors, preservation process validation, time-temperature indicators (TTIs), 232
Shear stress and rate, physical transport models, viscosity, 116
Shelf-stable product
commercial-scale process validation, 223–225
microbial inactivation technology, 65–66
ultra-high pressure experiments, 74–75
"Shoulders" concept, microbial survival curves, 51
Smoking, food preservation, 4
Specific heat, physical transport models, 116, 117t
prediction in ground beef, 117–118
pressure and, 118
Sphere
physical transport model, temperature-time chart, 129f, 130
thermal process design, integrated impact parameters, 173–174
Spoilage microorganism population

thermal preservation process, time-temperature calculations, 168–169
thermal process design, 180–181
ultra-high pressure process validation, 240–242
Spore populations
survival curves, 51
laboratory evaluation and validation, 220–221
thermal processes, 74
survivor curves, 58
Sterilization in Food Technology, 6–7
Stumbo, C. R., 6–11
Sugar content, microbial survival curves, process time prediction, 64
Surrogate microorganisms, preservation process validation, 229–230
PA3679 example, 230
Survivor curve data
characteristics, 50–54
future trends in evaluation of, 269
microbial inactivation technology
characteristics, 50–54
laboratory evaluation, 219–221
process time prediction, 64
survivor probability, 60–61, 60f
microbial populations, 9, 10f, 39–40, 40f
pilot-scale preservation process verification, 222
as process design parameter, 148–150
endpoint targeting, 166–171
time-step calculation, 150–151
ultra-high pressure process, 197–202
thermal preservation process, endpoint prediction, 167
time-step calculation, 150–151
vegetative populations
Activation Energy Constant, 42
vs. spores, 58

Temperature
microbial survival curves, process times and, 66
physical transport models
density and, 112–114, 113t
microwave heating of food products, 140
specific heat, 116, 117t
thermal conductivity and, 120, 120t
transient heat transfer, temperature prediction, 131–138
ultra-high pressure process, 142, 143
preservation process optimization, 251
thermal process, 252–259
process design parameters, Ohmic heating and, 193
pulsed electric fields (PEF) process design, 205–207
thermal death times, microbial survival curves, 50, 51f
thermal process design
ohmic heating systems, 193
product temperature profile, 181
Temperature distribution history
as process design parameter
preservation integration, 174
time-step calculation, 151
thermal process design, 174
transport models, future trends, 272

Temperature-time charts
physical transport models
infinite cylinder temperature-time chart, 130, 131f
infinite slab model, 130, 130f
spheres, 129, 129f
thermal process design, 149, 149f
spoilage microorganism population, 168–169
time-step computations, 153–157, 156t
Texture attributes, thermal preservation processes, 170–171
Thermal conductivity, physical transport models, 120, 120t
pressure and, 121
product composition and, 120–121
Thermal death times
lethal rate curve, 162–163
microbial survival curves, 50, 51f
process design models
Formula Method, 162–163, 164f
General Method, 157
Thermal death times, microbial spore curve, 13f
Thermal energy balance
physical transport models, ohmic heating, 138–139
process design parameters, ohmic heating and, 193
Thermal Inactivation Coefficient, microbial inactivation kinetics, 77
Thermal preservation process design
combined processes design, 209–212
conduction-heating food, 165–166
ground beef product example, 61
historical evolution of, 6–7
kinetic food systems models, 37–43
lethal rate curve, 158–162, 158f
General Method evaluation, 159–162, 160t, 161f
microbial inactivation kinetics, 58–59
microwave process, 181–192
optimization
food quality retention, 252–258
nonliquid foods, 249–263
physical transport models
convective heat-transfer, 129
transient heat transfer, temperature prediction, 131–138
preservation integration, 173
quality retention kinetics, 94–95
ascorbic acid loss, 98
chlorophyll retention, 97–98
maximum retention regions, 181
microbial population, 102
nutrients, 97
pasteurization, 100–102
texture attributes, 170–171
thiamine loss, 95–97
temperature distribution history, 151
time parameters, 149
time-step calculation, 150–157
time-temperature indicators, 233
uniform integration, 171–181
validation, 234–235
Thermal Resistance Coefficient, thermal process models, kinetic food systems, 41
Thermobacteriology in Food Processing, 6–7
Thiamin (vitamin B_1), quality retention kinetics, 92
Thiamine concentration, quality retention kinetics, 88, 89f

optimization, 250–251, 251f
thermal process, 95–97
Time intervals, process design parameters, time-step calculation, 151
Time parameters
kinetic food system models, 21
physical transport models, transient heat transfer, temperature prediction, 131–132
as process design parameter, 148–150
time-step calculation, 150
Time-step calculation, thermal process design, 150–157
temperature time charts, 153–157, 156t
Time-temperature indicators (TTIs), preservation process validation, 231–232
biosensors, 231–233
Transient heat transfer, physical transport models
container heating and cooling, 125–138
temperature prediction, 131–138
Transport models, 9–10
Biot numbers
heating/cooling lag constant, 132, 133f
heating rate constant, 132, 133f
mass average lag constant, 133, 134f
unsteady-state heat transfer, 131
containerized heating and cooling, 125–138, 126t
convective heat transfer coefficients, 126t
density, 112–114, 113t
dielectric properties, 123–125, 124t
electrical conductivity of foods, 121–123, 122t
future trends in, 271–272
evolving preservation technologies, 273–276
microwave heating, 139–141
Ohmic heating, 138–139
overview, 111
specific heat, 116, 117t
temperature prediction, transient heat transfer, 131–138
thermal conductivity, 120, 120t
ultra-high pressure applications, 141–143
unsteady-state heat transfer
infinite cylinder, 130, 131f
infinite slab, 130, 130f
sphere, 129f, 130
viscosity, 114–116, 115t

Ultra-high pressure (UHP) process
alternative preservation technologies, 77
design parameters, 197–202
microbial population reduction, 198–202
food preservation, 4–5, 5f
future kinetic parameters, 269
microbial inactivation kinetics, 58–59
design criteria, 66–75
injured cell survival, 76
shelf-stable product, 74–75
time determination, 67–68
physical transport models, 111
applications, 141–143
density effects, 113–114
future transport models, 271–272
specific heat, 118
process design model integration with, 172, 197–202
quality retention kinetics, 104
ascorbic acid content, 102–104
color retention, 95
heat-sensitive vitamins, 93

Ultra-high pressure (UHP) process (*Continued*)
survivor curve data analysis, 76
thermal process design, integration uniformity, 172
verification and evaluation, 240
Ultra-high pressure (UHP) process, food preservation, 4–5, 5f, 15
Uniform parameters, kinetic food systems models, 43–46
Unsteady-state transfer, physical transport models
container heating and cooling, 125–138
infinite cylinder temperature-time chart, 130, 131f
infinite slab temperature-time chart, 130, 130f
spherical temperature-time chart, 129f, 130
thermal energy balance, 126–127

Validation and evaluation, preservation process
alternative approaches, 229–236
commercial-scale verification, 223–229
kinetic constants acceptability, 218
laboratory evaluation, 219–221
microbial inactivation, 219–229
microwave processes, 237–239
Ohmic heating processes, 239–240
overview, 12–14
PA 3679 evaluation, 229–230
Halvorson-Ziegler expression, 230
microwave processes, 238–239
package and container shape and size, 218
pilot-scale verification, 221–223, 239
product physical properties, 218
pulsed-electric-field processes, 242–243
quality attributes, 234–236
quality controls, 218
research background, 217
safety issues, 217
surrogate microorganism identification, 229
thermal process evaluation, 236
time-temperature-integrators, 231–233
biosensor applications, 232–233
ultra-high pressure processes, 240
Vegetative populations, survivor curves, 58
Velocity determination, pulsed electric fields (PEF) process design, 203–205
Viscosity
physical transport models, 114–116, 115t
preservation process optimization, 259–260
quality attribute retention, 261–263
Vitamin A, quality retention kinetics, 93
Vitamin B_6, quality retention kinetics, 92–93
Vitamin B_{12}, quality retention kinetics, 92–93
Volume fraction, physical transport models, thermal conductivity and, 119

Water activity, preservation process, 4

Zero-order model, kinetic food system models, 24, 24f